Signals and Transforms in Linear Systems Analysis

Wasyl Wasylkiwskyj

Signals and Transforms in Linear Systems Analysis

 Springer

Wasyl Wasylkiwskyj
Professor of Engineering and Applied Science
The George Washington University
Washington, DC, USA

ISBN 978-1-4899-8710-5   ISBN 978-1-4614-3287-6 (eBook)
DOI 10.1007/978-1-4614-3287-6
Springer New York Heidelberg Dordrecht London

Printed on acid-free paper

Springer is part of Springer Science+Business Media (www.springer.com)

Preface

This book deals with aspects of mathematical techniques and models that constitute an important part of the foundation for the analysis of linear systems. The subject is classical and forms a significant component of linear systems theory. These include Fourier, Z-transforms, Laplace, and related transforms both in their continuous and discrete versions. The subject is an integral part of electrical engineering curricula and is covered in many excellent textbooks. In light of this, an additional book dealing with the same topics would appear superfluous. What distinguishes this book is that the same topics are viewed from a distinctly different perspective. Rather than dealing with different transforms essentially in isolation, a methodology is developed that unifies the classical portion of the subject and permits the inclusion of topics that usually are not considered part of the linear systems theory. The unifying principle here is the least mean square approximation, the normal equations, and their extensions to the continuum. This approach gives equal status to expansions in terms of special functions (that need not be orthogonal), Fourier series, Fourier integrals, and discrete transforms. As a by-product one also gains new insights. For example, the Gibbs phenomenon is a general property of LMS convergence at step discontinuities and is not limited to Fourier series.

This book is suitable for a first year graduate course that provides a transition from the level the subject is presented in an undergraduate course in signals and systems to a level more appropriate as a prerequisite for graduate work in specialized fields. The material presented here is based in part on the notes used for a similar course taught by the author in the School of Electrical and Computer Engineering at The George Washington University. The six chapters can be covered in one semester with sufficient flexibility in the choice of topics within each chapter. The exception is Chap. 1 which, in the spirit of the intended unity, sets the stage for the remainder of the book. It includes the mathematical foundation and the methodology applied in the chapters to follow.

The prerequisites for the course are an undergraduate course in signals and systems, elements of linear algebra, and the theory of functions of a complex variable. Recognizing that frequently the preparation, if any, in the latter is sketchy, the necessary material is presented in the Appendix.

Wasyl Wasylkiwskyj

Contents

The subsections marked with are supplements and not parts of the main text

Introduction

Although the book's primary purpose is to serve as a text, the basic nature of the subject and the selection of topics should make it also of interest to a wider audience. The general idea behind the text is two fold. One is to close the gap that usually exists between the level of student's preparation in transform calculus acquired in undergraduate studies and the level needed as preparation for graduate work. The other is to broaden the student's intellectual horizon.

The approach adopted herein is to exclude many of the topics that are usually covered in undergraduate linear systems texts, select those that in the opinion of the author serve as the common denominator for virtually all electrical engineering disciplines and present them within the unifying framework of the generalized normal equations. The selected topics include Fourier analysis, both in its discrete and continuous formats, its ramifications to time- frequency analysis, frequency dispersion and its ties to linear systems theory, wherein equal statues is accorded the LTI and time-varying systems . The Laplace and Z-transforms are presented with special emphasis on their connection with Fourier analysis.

The book begins within a rather abstract mathematical framework that could be discouraging for a beginner. Nevertheless, to pave the path to the material in the following chapters, I could not find a simpler approach. The introductory mathematics is largely contained in the first chapter. The following is the synopsis.

Starting on familiar ground, a signal is defined as a piecewise differentiable functions of time and the system input/output relation as a mapping by an operator from its domain unto its range. Along with the restriction to linear operators the representation of a signal as a sum of canonical expansion functions is introduced. The brief discussion of error criteria with focus on the LMS (Least Mean Squared approximation) is followed by an examination of the basic linear algebra concepts: norm, inner products linear independence and orthogonality. To emphasize the conceptual unity of the subject, analogue signals and their discrete counterparts are given equal status.

The LMS problem is viewed from the standpoint of the normal equations. The regularization of ill conditioned matrices is studied using the SVD (singular value decomposition). Its noise suppression attributes are examined via numerical examples. The TLS (Total Least Square) solution technique is discussed briefly as is the Tikhonov regularization.

The normal equations present us with three algebraically distinct but conceptually identical representations. The first is the discrete form which solves the problem of minimizing the MS error in the solution of an overdetermined system:

$$\boldsymbol{\varphi}_k^H \mathbf{f} = \sum_{n=1}^{N} \boldsymbol{\varphi}_k^H \boldsymbol{\varphi}_n \hat{f}_n, \quad k = 1 \ldots N$$

$$\boldsymbol{\varphi}_k = [\varphi_{1k} \varphi_{2k} \cdots \varphi_{mk}]^T, \quad M > N \tag{a}$$

A representation wherein the expansion function set is discrete but the independent variable extends over a continuum is

$$\int_{T_1}^{T_2} \varphi_k^*(t) f(t) dt = \sum_{n=1}^{N} \left[\int_{T_1}^{T_2} \varphi_k^*(t) \varphi_n(t) dt \right] \hat{f}_n \quad k = 1 \ldots N \tag{b}$$

A further generalization follows by letting both the summation index and the independent variable form a continuum. In that case one obtains

$$\int_{T_1}^{T_2} \varphi^*(\omega, t) f(t) dt = \int_{-\Omega_1}^{\Omega_2} \left[\int_{T_1}^{T_2} \varphi^*(\omega, t) \varphi(\eta, t) dt \right] \hat{f}(\eta) d\eta \quad -\Omega_1 < \omega < \Omega_2 \tag{c}$$

The formal identity of the first and the second forms follows when one interprets the independent variable in the second equation as a row index. In the last version the dependent variable is continuous as well so that the sum is replaced by an integral.

In the formulation adopted herein the normal equations embody the essence of linear transform theory: the second form above leads to a representation as a sum of bases functions whereas the last equation encompasses integral transforms such as the Fourier integral.

Representations of signals by finite sums of orthogonal bases functions are illustrated using trigonometric functions and Legendre, Laguerre and Hermite polynomials. The mathematical rigor needed to examine the limiting forms as the number of expansion functions approaches infinity is outside the scope of this text. Instead, as in most engineering oriented works, we use the following semi-heuristic approach. The starting point is to use the analogy with linear algebra and argue that the LMS error is in general not equal to zero unless the given signal is in the subspace spanned by the basis functions. As the number of basis functions is allowed to approach infinity (assuming the series converges) the LMS error need not approach zero. If it does, the series will converge pointwise and the (infinite) set of basis functions is said to be complete. The first step in testing a given set of basis functions for completeness is to sum a truncated series of basis functions, multiply the resulting kernel by the specified function and integrate the product. If the set is complete the limit of this integral, as the set of basis functions approaches infinity, will converge to the given function on a dense set of points. This property defines a delta function, which is used liberally throughout the text.

Chapter 1

Signals and Their Representations

1.1 Signal Spaces and the Approximation Problem

We shall use the term signal to designate a function of time. Unless specified otherwise, the time shall be allowed to assume all real values. In order not to dwell on mathematical generalities that are of peripheral interest in engineering problems the mathematical functions which we shall employ to represent these signals will be restricted to those that are piecewise differentiable. We shall also be interested in treating a collection of such signals as a single entity, in which case we shall define these signals as components of a vector. For example, with $s_n(t)$, $n = 1, 2, 3, ..N$ the given signal set, the corresponding vector shall be denoted by $\mathbf{s}(t)$ with components denoted by

$$\mathbf{s}(t) = \begin{bmatrix} s_1(t) & s_2(t) & s_3(t) & \ldots & s_N(t) \end{bmatrix}^T, \tag{1.1}$$

where T is the transpose. We shall be interested in developing a representational methodology that will facilitate the study of the transformation of such signals by systems, more specifically by an important class of systems known as linear systems. In the most general sense a system may be defined by the operation it performs on the signal set (1.1), taken as the input to the system, which operation gives rise to another signal set,

$$\mathbf{y}(t) = \begin{bmatrix} y_1(t) & y_2(t) & y_3(t) & \ldots & y_M(t) \end{bmatrix}^T, \tag{1.2}$$

which we designate as the output signal set. Note that the dimension of the output set may differ from the dimension of the input set. This "black box" description can be represented schematically as in Fig. 1.1.

We could also formally indicate this transformation by writing

$$\mathbf{y}(t) = \mathfrak{T}\{\mathbf{s}(t)\}, \tag{1.3}$$

W. Wasylkiwskyj, *Signals and Transforms in Linear Systems Analysis*,
DOI 10.1007/978-1-4614-3287-6_1, © Springer Science+Business Media, LLC 2013

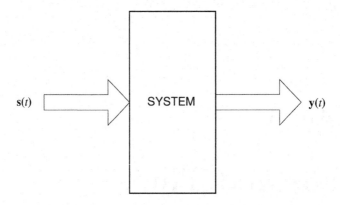

Figure 1.1: Input/output schematization of a system

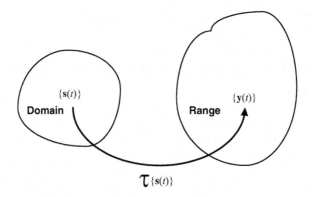

Figure 1.2: A system interpreted as a mapping

where the symbol $\mathfrak{T}$ stands for the an operator which defines the mathematical operation performed on the input. Another way of schematizing such transformations is to think of (1.3) as a mapping of a set inputs (domain of the operator) into a set of outputs (range of the operator). This mapping notion can be conveyed pictorially as in Fig. 1.2. The mapping of interest to us shall be exclusively linear by which we mean

$$T\{s_1(t) + s_2(t) + \ldots\} = T\{s_1(t)\} + T\{s_2(t)\} + \ldots$$

We shall return to this operator description and discuss it in detail in Chap. 3. Presently we shall focus entirely on alternative mathematical representations of signals.

It frequently happens that we would like to represent a given signal (time function) by a sum of other signals (functions). Such an approach may be desirable for several reasons. For example, we may be dealing with a system transformation which is especially simple in terms of a certain class of functions (as is exemplified by the use of exponentials for linear time-invariant systems) or a particular function class may render the transformation of the signal by the given system less susceptible to noise interference (which is routinely done through special choices of waveforms in communications and radar). Still other reasons may be related to certain desirable scaling properties possessed by the chosen function class (as, e.g., in the use of wavelet transforms in the interpretation of complex seismic signals). In any case, the general problem of signal representation may be phrased as follows. Given a function $f(t)$ in the interval $a \le t \le b$, and a set of functions $\phi_1(t), \phi_2(t), \ldots \phi_N(t)$ defined in the same interval, we should like to use a linear combination of these to approximate $f(t)$, i.e.,

$$f(t) \sim \sum_{n=1}^{N} \hat{f}_n \phi_n(t), \tag{1.4}$$

where the $\hat{f}_n$ are expansion coefficients. In what sense the sum on the right of (1.1) is to approximate the given function remains to be specified. Whatever our criterion, there will in general be a residual error depending on the chosen functions and, of course, on N. Denoting this error by $e_N(t)$, we can rewrite (1.1) using the equality sign as follows:

$$f(t) = \sum_{n=1}^{N} \hat{f}_n \phi_n(t) + e_N(t), \tag{1.5}$$

We should like to choose the coefficients $\hat{f}_n$ such that the error $e_N(t)$ is small within the interval $a \le t \le b$ in a sense that remains to be specified. For example, we could stipulate that the coefficients be chosen such that

$$\sup_{a \le t \le b} |e_N(t)| = \alpha, \tag{1.6}$$

where α is a specified positive constant which is to be minimized by a suitable choice of the coefficients. In this form the approximation problem can be solved explicitly, provided one constrains the expansion functions to polynomials. The solution can be phrased most succinctly in terms of Tschebytscheff polynomials, which finds extensive application in the design of frequency selective filters. For other classes of expansion functions, the approximation problem in this form can only be attempted numerically.

Another way in which the approximation problem can be phrased is to require that the error vanishes identically on a specific set of points, i.e.,

$$e_N(t_k) = 0; \quad k = 1, 2, 3, \ldots N. \tag{1.7}$$

The solution for the unknown coefficients is then formally reduced to the solution of the set of algebraic equations, viz.,

$$f(t_k) = \sum_{n=1}^{N} \hat{f}_n \phi_n(t_k) \tag{1.8}$$

for some specified set of $\phi_n(t)$. In this form the problem is usually referred to as the interpolation problem and the chosen functions $\phi_n(t)$ are referred to as interpolation functions. We shall touch on the interpolation problem only very briefly later. The approximation we shall focus on almost exclusively is the so-called least mean square (LMS) approximation. More than any other approximation technique it has become an indispensible tool in the most diverse applications and provides a unifying methodology with a truly amazingly large scope. The basic idea had its relatively humble origins as the method of least squares introduced by Karl Friedrich Gauss almost two centuries ago. Since then it has been substantially expanded in scope and plays a significant role not only in modern functional analysis but also in a variety of applied disciplines. Indeed, it appears difficult today to name an area of mathematical physics, statistics, estimation theory, or data analysis where it does not play at least some role. Our use of the principle of least squares in this work will incline toward its humbler aspects. Thus, we shall relinquish any pretense to mathematical generality. For example, we shall avoid the use of the theory of Lebesgue integration but instead stick to the pedestrian notion of the Riemann integral. Additionally, the class of functions we shall be concerned with shall be no larger than the class of piecewise differentiable functions. Significant as these restrictions may appear from a purely mathematical standpoint, they do not materially restrict the range of engineering problems to which our results are applicable.

In our terminology, the least squares approximation problem shall be phrased as follows. Given N functions $\phi_n(t)$ and a function $f(t)$ in the interval $a \leq t \leq b$, we seek N coefficients $\hat{f}_n$ in (1.5) such that the mean squared error ε_N,

$$\varepsilon_N \equiv \int_a^b |e_N(t)|^2 \, dt \tag{1.9}$$

is minimized, wherein $f(t)$ and the expansion functions and therefore the coefficients may in general be complex. The formal solution to this problem is straightforward and can be obtained, e.g., by substituting (1.5) into (1.9), differentiating with respect to the coefficients, and setting the result to zero. Before actually carrying this out it will be convenient to digress from the main theme and introduce a notation which will not only simplify the bookkeeping but will also provide interesting and useful geometrical interpretations of the results.

1.2 Inner Product, Norm and Representations by Finite Sums of Elementary Functions

1.2.1 Inner Product and Norm

Given two complex functions $f(t)$ and $g(t)$ we define their inner product (f, g) by

$$(f, g) \equiv \int_a^b f^*(t) g(t) \, dt, \tag{1.10}$$

where $*$ denotes the complex conjugate. When $f(t) = g(t)$, we introduce the special notation

$$(f, f) \equiv \|f\|^2, \tag{1.11}$$

where $\|f\|$ is defined as the norm of $f(t)$. Explicitly

$$\|f\| = \sqrt{\int_a^b |f(t)|^2\, dt} \qquad (1.12)$$

which may be taken as the generalization of the length of a vector. This terminology may be justified by recalling that for a real M-dimensional (X real) vector $\mathbf{x}$,

$$\mathbf{x}^T = \begin{bmatrix} x_1 & x_2 & x_3 & \ldots & x_M \end{bmatrix}, \qquad (1.13)$$

where the superscript T defines the transpose, the length of $\mathbf{x}$ is defined by $\sqrt{\sum_{n=1}^M x_n^2}$, which is nothing more than the Pythagorean theorem in M dimensions. Alternatively, in matrix notation this can be written as $\sqrt{\mathbf{x}^T \mathbf{x}}$. For vectors with complex components the last two definitions generalize to $\sqrt{\sum_{n=1}^M |x_n|^2}$ and $\sqrt{\mathbf{x}^H \mathbf{x}}$, respectively, where the superscript H denotes the complex conjugate transpose. Evidently if one interprets the integral in (1.12) as a limit of a sum of infinitessimal (Riemann sum), definition (1.12) appears to be a natural extension of the concept of length of a vector to an infinite-dimensional function space. Actually such analogies can be dispensed with if one accepts an axiomatic definition of the norm. Thus the norm of a function or a finite or even infinite-dimensional vector is defined by the following three postulates:

$$\|f\| > 0 \ and \ \|f\| = 0 \ \text{if and only if} \ f = 0, \qquad (1.14a)$$
$$\|f\| + \|g\| \geq \|f + g\| \ \text{(triangle inequality)}, \qquad (1.14b)$$
$$\|\lambda f\| = |\lambda|\,\|f\| \ , \ \text{for any complex constant } \lambda. \qquad (1.14c)$$

Clearly (1.12) satisfies (1.14a) and (1.14c). To prove that it also satisfies (1.14b) consider two functions f and g, each not identically zero, and define

$$w = \alpha f + \beta g, \qquad (1.15)$$

where α and β are constants. We form the inner product $(f, w) = \alpha(f, f) + \beta(f, g)$ and since $(f, f) \neq 0$ we can always find an α such that $(f, w) = 0$, to wit,

$$\alpha = -\beta \frac{(f, g)}{(f, f)}. \qquad (1.16)$$

Since $(w, w) \geq 0$ we have

$$(\alpha f + \beta g, \alpha f + \beta g) = |\alpha|^2 (f, f) + \beta^* \alpha (g, f) + \alpha^* \beta (f, g) + |\beta|^2 (g, g) \geq 0.$$

Substituting for α from (1.16) we note that the first and second terms on the right of the equality sign mutually cancel so that after multiplication by (f, f), the entire inequality may be replaced by $-|\beta|^2 |(f, g)|^2 + |\beta|^2 (f, f)(g, g) \geq 0$. After dividing both sides by $|\beta|^2$, moving the first term to the right of the inequality sign and using the definition of the norm, we obtain

$$|(f, g)|^2 \leq \|f\|^2 \|g\|^2 , \qquad (1.17)$$

where the equality sign applies if and only if $w = \alpha f + \beta g = 0$, or, equivalently, $f = \lambda g$, with λ a (complex) constant.

The inequality (1.17) is called the Schwarz (sometimes the Cauchy–Schwarz) inequality and is of considerable importance in its own right.[1] It implies the triangle inequality (1.14b), as can be seen in the following development. The distributive nature of the inner product permits us to write

$$\|f + g\|^2 = (f, f + g) + (g, f + g) = \|f\|^2 + \|g\|^2 + (f, g) + (g, f).$$

Since the last two terms are complex conjugates, we get the bound

$$\|f + g\|^2 \leq \|f\|^2 + \|g\|^2 + 2\,|(f, g)|. \tag{1.18}$$

By the Schwarz inequality (1.17) $|(f, g)| \leq \|f\|\,\|g\|$ so that the right side is bounded by the square of the sum of the norms, i.e.,

$$\|f + g\|^2 \leq (\|f\| + \|g\|)^2. \tag{1.19}$$

Upon taking the square root of both sides (1.14b) follows.

It is by no means true that (1.12) is the only definition that satisfies the norm postulates. Thus the following definition of the norm:

$$\|f\| = \left[\int_a^b |f(t)|^p \, dt \right]^{1/p}, \tag{1.20}$$

where $p \geq 1$ can also be shown to satisfy (1.14) (see problem (1.10)). Equation (1.20) is said to define the p-norm. In accordance with this terminology (1.12) can be said to correspond to the 2-norm. Our interest in norms for p other than two will be only peripheral. Unless specified otherwise we shall be dealing exclusively with 2-norms.

The concepts of inner product and norm apply equally well to finite-dimensional or denumerably infinitely dimensional vectors with the integrals replaced by sums. In the course of the following chapters we shall be dealing with functions defined over a continuum as well as with their discrete counterparts defined over discrete times only. The latter shall be described by vectors with a finite or a denumerably infinite number of components. Although the formal properties of norm and the inner product are identical in all these cases, so that, in principle, a uniform notation could be adopted we shall not do so. Instead, for the inner product of finite-dimensional or denumerably infinite-dimensional vectors matrix notation shall be employed. Thus for two M-dimensional vectors

---

[1] There are many other proofs of the Schwarz inequality. A particularly simple one is the following. For any two functions we have the identity

$$\|f\|^2 \|g\|^2 - |(f, g)|^2 = 1/2 \iint |f(x)g(y) - f(y)g(x)|^2 \, dx dy.$$

Since the right side is nonnegative (1.17) follows (from Leon Cohen, "Time-Frequency Analysis," Prentice Hall PTR, Englewood Cliffs, New Jersey (1995) p. 47).

x and **y** the inner product, defined by the sum $\sum_{n=1}^{M} x_n^* y_n$, shall be written in general as $\mathbf{x}^H \mathbf{y}$ or as $\mathbf{x}^T \mathbf{y}$ for real vectors and not $(\mathbf{x}, \mathbf{y})$; the latter symbol shall be reserved exclusively for functions defined on a continuous time interval. On the other hand, a common notation shall be used for the norm, e.g., $\|\mathbf{x}\|$ and $\|f\|$. For example, the Schwarz inequality for the vectors **x** and **y** reads

$$\left| \mathbf{x}^H \mathbf{y} \right|^2 \le \|\mathbf{x}\|^2 \|\mathbf{y}\|^2 . \tag{1.21}$$

The familiar interpretation of the inner product in three-dimensional Euclidean space as the product of the magnitude of two vectors and the cosine of the angle between the vectors can be extended without difficulty to spaces with higher dimensions and to function spaces. Thus if the vectors **x** and **y** are real, we have

$$\mathbf{x}^T \mathbf{y} = \|\mathbf{x}\| \, \|\mathbf{y}\| \cos \theta. \tag{1.22}$$

Since the vectors are assumed real, the factor $\cos \theta$ is necessarily real. Now the Schwarz inequality (1.21) guarantees that the magnitude of $\cos \theta$ is less than unity, i.e., that the angle θ is real. Also if f and g are real functions defined on a continuum of values $a \le t \le b$, (1.22) reads

$$(f, g) = \|f\| \, \|g\| \cos \theta. \tag{1.23}$$

Again the Schwarz inequality in (1.17) guarantees a real θ, even though the visualization of an angle between two functions may present a bit of a problem. With these geometrical analogies, we can extend the notion of orthogonality of two vectors from its familiar geometrical setting in three-dimensional Euclidean space to any number of dimensions or for that matter also to function space. For example, we can define two functions as orthogonal if $\theta = \pi/2$ or, equivalently, if the inner product vanishes, i.e.,

$$(f, g) = 0. \tag{1.24}$$

When the functions or (and) the vectors are complex, (1.22) and (1.23) still apply except that the factor $\cos \theta$ becomes a complex number, with magnitude less than unity, but which, unfortunately, no longer admits the interpretation as the cosine of an angle. Nevertheless, orthogonality of two complex-valued functions is still defined by the condition that their inner product vanishes, as stipulated by (1.24).

1.2.2 Orthogonality and Linear Independence

The concept of orthogonality defined by (1.24) for two functions can be generalized to any finite or infinite set. Thus the set of functions $\phi_n(t)$ with $n = 1, 2, \ldots N$ will be called orthogonal in $a \le t \le b$ if

$$(\phi_n, \phi_m) = \begin{cases} 0; & n \ne m, \\ Q_n; & n = m, \end{cases} \tag{1.25}$$

where Q_n is a normalization constant. It is sometimes convenient to normalize the functions such that $Q_n = 1$. In that case the functions shall be referred to as orthonormal. As a notational convenience the orthogonality condition shall be represented by

$$(\phi_n, \phi_m) = Q_n \delta_{nm},\tag{1.26}$$

where the δ_{nm} is the so called Kronecker symbol defined by

$$\delta_{nm} = \left\{\begin{array}{l} 0; \; n \neq m, \\ 1; \; n = m, \end{array}\right.\tag{1.27}$$

which is recognized as nothing more than an element of a unity matrix in N-dimensional space.

A more general concept than that of orthogonality is that of linear independence. This concept is suggestive of an independent coordinate set, or, to venture into the realm of communications theory, to the number of degrees of freedom of a signal. We say that the functions $\phi_n(t)$, $n = 1, 2, \ldots N$ are linearly independent in $a \leq t \leq b$ if the relationship

$$\sum_{n=1}^{N} \alpha_n \phi_n(t) = 0\tag{1.28}$$

implies $\alpha_n = 0$ for all n. Conversely, if (1.28) is satisfied with at least one of the $\alpha_n \neq 0$, the functions are said to be linearly dependent. A necessary and sufficient condition for N functions $\phi_n(t)$ to be linearly dependent is that

$$\det\left[(\phi_m, \phi_n)\right] = 0.\tag{1.29}$$

To prove this suppose the functions are linearly dependent so that (1.28) holds with at least one α_n not identically zero. Forming the inner product with $\phi_m(t)$ gives

$$\sum_{n=1}^{N} \alpha_n (\phi_m, \phi_n) = 0; \; m = 1, 2, \ldots N.\tag{1.30}$$

This system of equations can yield a set of nonzero α_n only if the determinant of the system vanishes, i.e., when (1.29) holds. On the other hand, suppose (1.29) holds. Then the system (1.30) will yield at least one α_n different from zero. Now multiply (1.30) by α_m^* and sum over m to obtain

$$\sum_{m=1}^{N} \sum_{n=1}^{N} \alpha_m^* \alpha_n (\phi_m, \phi_n) = 0.\tag{1.31}$$

Clearly this is equivalent to

$$\left\|\sum_{n=1}^{N} \alpha_n \phi_n(t)\right\|^2 = 0,\tag{1.32}$$

which implies (1.28). By analogy with finite-dimensional vector spaces we shall call a finite set of linearly independent functions a basis and the functions basis functions.

The determinant $\Gamma \equiv \det \left[(\phi_m, \phi_n) \right]$ in (1.29) is called the Gram determinant and the corresponding matrix

$$
\mathbf{G} \equiv
\begin{bmatrix}
(\phi_1, \phi_1) & (\phi_1, \phi_2) & (\phi_1, \phi_3) & \cdot & (\phi_1, \phi_N) \\
(\phi_2, \phi_1) & (\phi_2, \phi_2) & (\phi_2, \phi_3) & \cdot & (\phi_2, \phi_N) \\
(\phi_3, \phi_1) & (\phi_3, \phi_2) & (\phi_3, \phi_3) & \cdot & (\phi_3, \phi_N) \\
& & & \cdot & \\
(\phi_N, \phi_1) & (\phi_N, \phi_2) & (\phi_N, \phi_3) & \cdot & (\phi_N, \phi_N)
\end{bmatrix}
\tag{1.33}
$$

is known as the Gram matrix. From the definition of the inner product we see that $\mathbf{G} = \mathbf{G}^H$, i.e., the Gram matrix is Hermitian. As we have just shown $\Gamma \neq 0$ for a linearly independent set of functions. In fact since

$$
\sum_{m=1}^{N} \sum_{n=1}^{N} \alpha_m^* \alpha_n (\phi_m, \phi_n) = \left\| \sum_{n=1}^{N} \alpha_n \phi_n (t) \right\|^2 \geq 0,
\tag{1.34}
$$

the Gram matrix is non-negative definite and necessarily positive definite if the functions are linearly independent. Assuming linear independence, let us write (1.34) in "block matrix" form

$$
\boldsymbol{\alpha}^H \mathbf{G} \boldsymbol{\alpha} > 0
$$

with

$$
\boldsymbol{\alpha} = \begin{bmatrix} \alpha_1 & \alpha_2 & \cdots & \alpha_N \end{bmatrix}.
$$

Since $\mathbf{G}$ is Hermitian, it can be diagonalized by a unitary transformation[2] $\mathbf{U}$, so that $\mathbf{G} = \mathbf{U} \boldsymbol{\Lambda} \mathbf{U}^H$ with $\boldsymbol{\Lambda}$ a diagonal matrix comprised of Λ_n, $n = 1, 2, ..N$, the N eigenvalues of $\mathbf{G}$. Hence $\boldsymbol{\alpha}^H \mathbf{G} \boldsymbol{\alpha} = \boldsymbol{\alpha}^H \mathbf{U} \boldsymbol{\Lambda} \mathbf{U}^H \boldsymbol{\alpha} = \boldsymbol{\beta}^H \boldsymbol{\Lambda} \boldsymbol{\beta} > 0$ with $\boldsymbol{\beta} = \mathbf{U}^H \boldsymbol{\alpha}$. Because $\mathbf{G}$ is positive definite all elements of $\boldsymbol{\Lambda}$ must be positive so that

$$
\Gamma = \det (\mathbf{G}) = \det \left(\mathbf{U} \boldsymbol{\Lambda} \mathbf{U}^H \right) = \prod_{n=1}^{N} \Lambda_n > 0.
$$

Thus Γ is always positive for any linearly independent set of functions. We note in passing that $\Gamma \geq 0$ for $N > 2$ may be taken as the generalization of the Schwarz inequality (1.17).

When the expansion functions are differentiable, the necessary and sufficient conditions for linear independence can also be phrased in terms of the derivatives. This alternative formulation is of fundamental importance in the theory

---

[2] $\mathbf{U}^H \mathbf{U} = \mathbf{I}_{NN}$

of ordinary linear differential equations. To derive it consider again (1.28) under the assumption that the functions are linearly dependent. Assume that the functions posses $N - 1$ derivatives and denote the k-th derivative of $\phi_n(t)$ by $\phi_n^{(k)}(t)$. A k-fold differentiation then gives

$$\sum_{n=1}^{N} \alpha_n \phi_n^{(k)}(t) = 0, \quad k = 0, 1, 2 \ldots N - 1; \; a \leq t \leq b.$$

This system can have nontrivial solutions for α_n if and only if

$$W[\phi_1, \phi_2, \ldots \phi_N] \equiv \det \begin{bmatrix} \phi_1 & \phi_2 & \phi_3 & \cdots & \phi_N \\ \phi_1^{(1)} & \phi_2^{(1)} & \phi_3^{(1)} & \cdots & \phi_N^{(1)} \\ \phi_1^{(2)} & \phi_2^{(2)} & \phi_3^{(2)} & \cdots & \phi_N^{(2)} \\ \cdot & \cdot & \cdot & \cdot & \\ \phi_1^{(N-1)} & \phi_2^{(N-1)} & \phi_3^{(N-1)} & \cdots & \phi_N^{(N-1)} \end{bmatrix} = 0.$$

(1.35)

This determinant, known as the Wronskian, will be employed in Chap. 3 when we discuss systems described by linear differential equations. We note in passing that unlike the Gram determinant Γ, which depends on the values assumed by the functions throughout the entire interval, the Wronskian is a pointwise measure of linear independence. Consequently, it may vanish at isolated points within the interval even though $\Gamma \neq 0$. Consider, for example, the functions t and t^2 in the interval $(-1, 1)$. The corresponding Gram determinant is $4/15$ while the Wronskian is t^2, which vanishes at zero.

Even though mathematically the given function set is either linearly dependent or linearly independent, from a practical numerical standpoint one can speak of the degree of linear dependence. Thus based on the intuitive geometrical reasoning we would consider vectors that are nearly parallel as nearly linearly dependent. In fact our ability to decide numerically between linear dependence and independence will be vitiated in cases of nearly parallel vectors by the presence of noise in the form of round off errors. On the other hand, we would expect greater immunity to noise interference in cases of nearly orthogonal vectors. It is therefore useful to have a numerical measure of the degree of linear independence. A possible measure of the degree of this linear independence, or in effect, the degree of the singularity of the inverse of $\mathbf{G}$, is the numerical value of Γ. Other measures, referred to generally as matrix condition numbers, are sometimes more appropriate and will be discussed later.

From the foregoing it should be intuitively obvious that orthogonality of a function set implies linear independence. A formal proof is not difficult to construct. Thus suppose the N functions are linearly dependent and yet orthogonal. Since linear dependence means that (1.28) holds, we can form the inner product of its left side successively for $m = 1, 2, \ldots N$ with $\phi_m(t)$, which then yields $\alpha_m = 0; m = 1, 2 \ldots N$, thus contradicting the assumption of linear dependence.

The use of the language of three-dimensional Euclidean geometry in describing function spaces suggests that we also borrow the corresponding pictorial

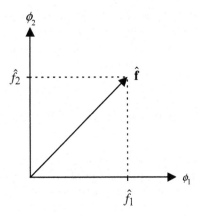

Figure 1.3: Projection vector

representation. For example, suppose the function $f(t)$ can be represented exactly by two orthonormal functions $\phi_1(t)$ and $\phi_2(t)$. We then have

$$f(t) = \hat{f}_1 \phi_1(t) + \hat{f}_2 \phi_2(t).\tag{1.36}$$

Since the two expansion functions are orthonormal, we can at once solve for the two expansion coefficients by multiplying (1.36) by $\phi_1^*(t)$ and $\phi_2^*(t)$ and integrating the result to obtain

$$\hat{f}_1 = (\phi_1, f),\tag{1.37a}$$
$$\hat{f}_2 = (\phi_2, f).\tag{1.37b}$$

Just as if we were dealing with finite-dimensional Euclidean vector space instead of function space, we can interpret the coefficients $\hat{f}_1$ and $\hat{f}_2$ in (1.37) as projections on a pair of unit basis vectors. Of course, formally these are projections of an infinite-dimensional vector $f(t)$ along the directions of the two "orthogonal unit vectors" ϕ_1 and ϕ_2 which are themselves infinite dimensional. Nevertheless, if we use these projections to define the two-dimensional vector $\hat{\mathbf{f}}$,

$$\hat{\mathbf{f}} = \begin{bmatrix} \hat{f}_1 \\ \hat{f}_2 \end{bmatrix}\tag{1.38}$$

we can interpret its relationship to the orthonormal basis functions ϕ_1 and ϕ_2 geometrically as shown in Fig. 1.3 where, to facilitate the graphical representation, we assume that $\hat{f}_1$ and $\hat{f}_2$ are real numbers.

We note that $\hat{\mathbf{f}}$ and $f(t)$ have the same norm. Depending on the notational preferences this result can be phrased in the following alternative ways:

$$\begin{aligned}
E &= \int_a^b |f(t)|^2\, dt = \|f(t)\|^2 = \int_a^b \left| \hat{f}_1 \phi_1(t) + \hat{f}_2 \phi_2(t) \right|^2 dt \\
&= \left| \hat{f}_1 \right|^2 + \left| \hat{f}_2 \right|^2 = \hat{\mathbf{f}}^H \hat{\mathbf{f}} = \left\| \hat{\mathbf{f}} \right\|^2,
\end{aligned}\tag{1.39}$$

where E is termed the signal energy.[3] The significance of (1.39) is that it shows that the signal energy can be computed indirectly by summing the squared magnitudes of the expansion coefficients. This is a special case of a more general result known as Parseval's theorem which holds generally for orthogonal expansions.

1.2.3 Representations by Sums of Orthogonal Functions

The preceding geometrical interpretation of the projections of the signal in function space can be employed for any number of orthonormal expansion functions even though the actual graphical construction is obviously limited to $N < 4$. Thus for a signal $f(t)$ that can be represented by N orthonormal functions (1.36) generalizes to

$$f(t) = \sum_{n=1}^{N} \hat{f}_n \phi_n(t).$$
(1.40)

As in (1.36) we take advantage of the orthogonality of the $\phi_n(t)$ so that the expansion coefficients $\hat{f}_n$ (or, speaking geometrically, the projections of the signal on hypothetical orthogonal axes) are given by

$$\hat{f}_n = (\phi_n, f).$$
(1.41)

Collecting these into the N-dimensional vector $\hat{\mathbf{f}}$, we have

$$\hat{\mathbf{f}}^T = \left[\hat{f}_1 \ \hat{f}_2 \ \dots \hat{f}_N\right],$$

This vector furnishes a representation that is fully equivalent to the direct display of the signal as a function of time because the time domain representation can always be reconstructed by appending the known expansion functions. Evidently the square of the norm of this vector is again the signal energy E as in (1.39) so that Parseval's theorem now generalizes to

$$
\begin{aligned}
E &= \int_a^b |f(t)|^2 \, dt = \|f(t)\|^2 = \int_a^b \left|\sum_{n=1}^{N} \hat{f}_n \phi_n(t)\right|^2 \, dt \\
&= \sum_{n=1}^{N} \left|\hat{f}_n\right|^2 = \hat{\mathbf{f}}^H \hat{\mathbf{f}} = \left\|\hat{\mathbf{f}}\right\|^2,
\end{aligned}
$$
(1.42)

---

[3]The term energy as used here is to be understood as synonymous with the square of the signal norm which need not (and generally does not) have units of energy.

which is Parseval's theorem for general orthogonal functions. Similarly the inner product of two functions $f(t)$ and $g(t)$ can be written as the inner product of vectors with expansion coefficients for their components:

$$\int_a^b f(t)^*g(t)dt \;=\; \int_a^b \sum_{n=1}^{N} \hat{f}_n^*(t)\,\phi_n^*(t) \sum_{n=1}^{N} \hat{g}_n\phi_n(t)\,dt$$

$$=\; \sum_{n=1}^{N} \hat{f}_n^*\hat{g}_n = \hat{\mathbf{f}}^H\hat{\mathbf{g}}. \qquad (1.43)$$

In case two or more signals are represented in terms of the same orthonormal set of functions their addition and subtraction can be reduced to operations with vectors defined by their expansion coefficients. For example, given signals $f_1(t)$ and $f_2(t)$ their direct addition $f_1(t) + f_2(t) = f_3(t)$ corresponds to the vector addition $\hat{\mathbf{f}}_1 + \hat{\mathbf{f}}_2 = \hat{\mathbf{f}}_3$.

In applications of signal analysis it is frequently important to have a measure of the difference between two signals. A measure frequently employed in detection theory is the "distance" between two signals, defined as follows:

$$d_{12} = \sqrt{\int_a^b |f_1(t) - f_2(t)|^2\,dt} = \sqrt{\|f_1\|^2 + \|f_2\|^2 - 2\Re\,(f_1, f_2)}. \qquad (1.44)$$

If each of the functions is represented by an orthonormal expansion with coefficients $\hat{f}_{1n}$, $\hat{f}_{2n}$, the preceding definition is easily shown to be equivalent to

$$d_{12} = \sqrt{\sum_{n=1}^{N} \left|\hat{f}_{1n} - \hat{f}_{2n}\right|^2} = \left\|\hat{\mathbf{f}}_1 - \hat{\mathbf{f}}_2\right\| = \sqrt{\left(\hat{\mathbf{f}}_1 - \hat{\mathbf{f}}_2\right)^H \left(\hat{\mathbf{f}}_1 - \hat{\mathbf{f}}_2\right)} \qquad (1.45)$$

Geometrically we can interpret this quantity as the (Euclidean) distance in N-dimensional space between points located by the position vectors, vectors $\hat{\mathbf{f}}_1$, and $\hat{\mathbf{f}}_2$ and think of it as a generalization of the 2-D construction as shown in Fig. 1.4.

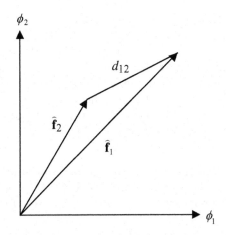

Figure 1.4: Vector representation of the distance between two signals

Clearly we can always represent the coefficients of signals that can be expressed as sums of N orthonormal functions as components of N-dimensional vectors, but graphical representations such as that in Fig. 1.4 are limited to real functions in at most three dimensions.

1.2.4 Nonorthogonal Expansion Functions and Their Duals

When the expansion functions are not orthogonal each coefficient $\hat{f}_n$ in (1.40) depends not just on ϕ_n but on the entire set $\phi_1, \phi_2, \phi_3 \ldots \phi_N$. We can see this directly from (1.40). After multiplying it from the left by $\phi_m^*(t)$ and integrating we get

$$(\phi_m, f) = \sum_{n=1}^{N} (\phi_m, \phi_n) \hat{f}_n \qquad (1.46)$$

so that the computation of the expansion coefficients requires the inversion of the Gram matrix which depends on the inner product of all the expansion functions. When the expansion functions are not independent, the Gram matrix is singular and the inverse does not exist. In case of orthonormal expansion function $(\varphi_m, \phi_n) = \delta_{nm}$ i.e., the Gram matrix reduces to a unit matrix, in which case the expansion coefficients are again given by (1.41).

A similar formula can be obtained by introducing a new set of expansion functions, referred to as dual or reciprocal basis functions $\psi_n(t), n = 1, 2, \ldots N$ constructed so as to be orthogonal to the $\varphi_n(t)$. This construction starts with the expansion

$$\psi_n(t) = \sum_{l=1}^{N} \alpha_{nl} \phi_l(t), n = 1, 2, \ldots N. \qquad (1.47)$$

Multiplying from the left by $\phi_k^*(t)$ and integrating we obtain

$$(\phi_k, \psi_n) = \sum_{l=1}^{N} \alpha_{nl} (\phi_k, \phi_l), \qquad (1.48)$$

where $k = 1, 2, \ldots N$. The sum over the index l is seen to be a product of two matrices: the transpose of the Gram matrix $\mathbf{G}^T$, with elements (ϕ_k, ϕ_l), and a matrix with elements α_{nl}. At this point the α_{nl} are at our disposal and as long as they represent the elements of a nonsingular matrix we can construct the corresponding set of reciprocal basis functions $\psi_n(t)$. If we choose the α_{nl} as elements of the inverse of $\mathbf{G}^T$, then the left side of (1.48) is necessarily the unit matrix, i.e.,

$$(\phi_k, \psi_n) = \delta_{kn}. \qquad (1.49)$$

Thus the direct basis functions ϕ_k are orthogonal to the reciprocal basis functions ψ_k. They will also have unit norm if the norm of ϕ_k is unity. Collectively the ψ_k and ϕ_k are also referred to as a biorthogonal set. In the special case of orthonormal ϕ_k the Gram matrix degenerates into a unit matrix and in accordance with (1.47) $\psi_k = \phi_k$.

In view of (1.49) we can express the expansion coefficients in (1.40) as inner products of $f(t)$ and $\psi_n(t)$. We obtain this by multiplying (1.40) from the left by $\psi_n^*(t)$, integrating and taking account of (1.49). This yields

$$\hat{f}_n = (\psi_n, f). \tag{1.50}$$

If a signal can be represented by a finite set of linearly independent functions, one can always construct the corresponding reciprocal basis and use it to represent the signal. Thus any given signal has two alternative representations, one in terms of the direct and other in terms of the reciprocal basis functions. For a function represented in terms of the direct basis functions[4]

$$f(t) = \sum_{n=1}^{N} \hat{f}_n \phi_n(t) \tag{1.51}$$

the expansion in terms of the reciprocal basis functions will be represented by

$$f(t) = \sum_{n=1}^{N} \tilde{f}_n \psi_n(t). \tag{1.52}$$

The two sets of coefficients are related by the Gram matrix. This follows from equating (1.51) to (1.52) and using (1.50). We get

$$\tilde{f}_m = \sum_{n=1}^{M} (\phi_m, \phi_n) \hat{f}_n. \tag{1.53}$$

Using the dual representations (1.51) and (1.52) the energy of the signal is

$$E = \int_a^b |f(t)|^2 \, dt = (f, f) = \sum_{n=1}^{N} \sum_{n=1}^{N} \hat{f}_n^* \tilde{f}_m (\phi_n, \psi_m) = \sum_{n=1}^{N} \hat{f}_n^* \tilde{f}_n, \tag{1.54}$$

which may be taken as an extension of Parseval's theorem (1.42) to biorthogonal functions. More generally, the inner product of two functions, one represented in the direct and the other in the reciprocal basis, reads

$$\int_a^b f(t)^* g(t) dt = (f, g) = \int_a^b \sum_{n=1}^{N} \hat{f}_n^*(t) \phi_n^*(t) \sum_{m=1}^{N} \tilde{g}_m \psi_m(t) \, dt$$

$$= \sum_{n=1}^{N} \sum_{m=1}^{N} \hat{f}_n^* \tilde{g}_m (\phi_n, \psi_m) = \sum_{n=1}^{N} \hat{f}_n^* \tilde{g}_n. \tag{1.55}$$

---

[4]Mathematically it is immaterial which set is designated as the direct and which as the reciprocal basis.

The representation of the inner product of two functions as an inner product of their expansion coefficients is an important attribute of orthogonal expansions. According to (1.55) this attribute is shared by biorthogonal expansions. Their drawback is the requirement of two sets of functions. Only one set of functions would be needed if the ϕ_n could be transformed into a symmetric orthogonal set. In the following we describe two techniques that provide such a transforming.

1.2.5 Orthogonalization Techniques

Orthogonalization via the Gram Matrix

A linear transformation of a nonorthogonal to an orthogonal basis can be constructed from eigenvectors of the Gram matrix. To this end we first solve the eigenvalue problem

$$\mathbf{G}\mathbf{v}_n = \lambda_n \mathbf{v}_n, \ n = 1, 2, 3, \ldots N \tag{1.56}$$

and take note of the fact that the Gram matrix is Hermitian so that the eigenvalues are real and the eigenvectors orthogonal. We assign them unit norm so that

$$\mathbf{v}_n^H \mathbf{v}_m = \delta_{nm} \tag{1.57}$$

with

$$\mathbf{v}_n^H = [v_1^* \ v_{n2}^* \ v_{n3}^* \ \ldots v_{nN}^*] \tag{1.58}$$

The orthonormal basis functions $w_n(t)$ are then given by

$$w_n(t) = \frac{1}{\sqrt{\lambda_n}} \sum_{k=1}^{N} v_{nk} \ \phi_k(t) \tag{1.59}$$

as can be demonstrated by computing the inner product

$$
\begin{aligned}
(w_n, w_m) &= \frac{1}{\sqrt{\lambda_n \lambda_m}} \left(\sum_{k=1}^{N} v_{nk} \ \phi_k, \sum_{l=1}^{N} v_{ml} \ \phi_l \right) \\
&= \frac{1}{\sqrt{\lambda_n \lambda_m}} \sum_{k=1}^{N} \sum_{l=1}^{N} v_{nk}^* v_{ml} \ (\phi_k, \phi_l) \\
&= \frac{1}{\sqrt{\lambda_n \lambda_m}} \mathbf{v}_n^H \mathbf{G} \mathbf{v}_m.
\end{aligned}
$$

With the aid of (1.56) and (1.57) we obtain the final result

$$\frac{1}{\sqrt{\lambda_n \lambda_m}} \mathbf{v}_n^H \mathbf{G} \mathbf{v}_m = \frac{\lambda_m}{\sqrt{\lambda_n \lambda_m}} \mathbf{v}_n^H \mathbf{v}_m = \delta_{nm}. \tag{1.60}$$

Since the key to this orthogonalization technique is the Hermitian structure of the Gram matrix it applies equally well to the discrete case. Here we start with a general set of linearly independent M-dimensional column vectors $\mathbf{a}_n, n = 1, 2, 3, \ldots N$ and for the $M \ X \ N$ matrix

$$\mathbf{A} = [\mathbf{a}_1 \ \mathbf{a}_2 \ \mathbf{a}_3 \ldots \mathbf{a}_N]. \tag{1.61}$$

The corresponding Gram matrix is

$$\mathbf{G} = \mathbf{A}^H \mathbf{A}$$

and (1.59) now reads

$$\mathbf{w}_n = \frac{1}{\sqrt{\lambda_n}} \sum_{k=1}^{N} \mathbf{a}_k v_{nk} \ .\tag{1.62}$$

The only difference in its derivation is the replacement of the inner products by $\mathbf{a}_n^H \mathbf{a}_m$ and $\mathbf{w}_n^H \mathbf{w}_m$ so that the orthogonality statement assumes the form

$$\mathbf{w}_n^H \mathbf{w}_m = \delta_{nm}.\tag{1.63}$$

The Gram–Schmidt Orthogonalization

One undesirable feature of the approach just presented is that every time the set of basis functions is enlarged the construction of the new orthogonal set requires a new solution of the eigenvalue problem. An alternative and less computationally demanding technique is the Gram–Schmidt approach where orthogonality is obtained by a successive subtraction of the projections of a vector (function) on a set of orthogonal subspaces. The geometrical interpretation of this process is best illustrated by starting with a construction in two dimensions. Thus for the 2 real vectors $\mathbf{a}_1$ and $\mathbf{a}_1$ as shown in Fig. 1.5 we denote their respective magnitudes by a_1 and a_2 and the included angle by θ. In this notation $\mathbf{a}_1/a_1$ is a vector of unit magnitude that points in the direction of $\mathbf{a}_1$ and $a_2 \cos \theta$ is the projection of $\mathbf{a}_2$ along $\mathbf{a}_1$. Thus

$$(\mathbf{a}_1/a_1)\, a_2 \cos \theta \tag{1.64}$$

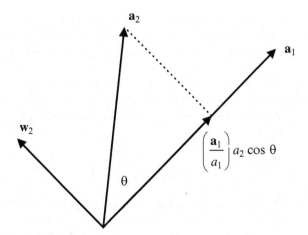

Figure 1.5: Orthogonalization in 2 dimensions

represents a component of the vector $\mathbf{a}_2$ in the direction of $\mathbf{a}_1$. If we define the "dot" product in the usual way by $\mathbf{a}_2 . \mathbf{a}_1 = a_1 a_2 \cos\theta$, solve for $\cos\theta$ and substitute into (1.64), we obtain

$$\frac{\mathbf{a}_2 . \mathbf{a}_1}{a_1^2}. \tag{1.65}$$

If we now form a new vector $\mathbf{w}_2$ by subtracting the projection (1.65) from $\mathbf{a}_2$

$$\mathbf{w}_2 = \mathbf{a}_2 - \frac{\mathbf{a}_2 . \mathbf{a}_1}{a_1^2}\mathbf{a}_1. \tag{1.66}$$

Forming the dot product of $\mathbf{a}_1$ with $\mathbf{w}_2$ we get orthogonality with $\mathbf{w}_2$. We also see directly from geometrical construction in Fig. 1.5 that $\mathbf{w}_2 . \mathbf{a}_1 = 0$. If we now define $\mathbf{w}_1 \equiv \mathbf{a}_1$, we can claim that we have succeeded in transforming a general (not necessarily orthogonal) basis $\mathbf{a}_1, \mathbf{a}_2$ into the basis $\mathbf{w}_1, \mathbf{w}_2$, which is orthogonal. This procedure, referred to as the Gram–Schmidt orthogonalization, works in any number of dimensions. For example, suppose we have three noncoplanar vectors in three dimensions: $\mathbf{a}_1, \mathbf{a}_2$, and $\mathbf{a}_3$. We can choose any one of them as the first member of the orthogonal set. For example, $\mathbf{w}_1 = \mathbf{a}_1$. We construct the second vector $\mathbf{w}_2$ precisely as in (1.66). With the replacement of $\mathbf{a}_1$ by $\mathbf{w}_1$ this reads

$$\mathbf{w}_2 = \mathbf{a}_2 - \frac{\mathbf{a}_2 . \mathbf{w}_1}{w_1^2}\mathbf{w}_1, \tag{1.67}$$

where w_1 is the magnitude of $\mathbf{w}_1$. In the next step we subtract from $\mathbf{a}_3$ its projection on the plane formed by the vectors $\mathbf{w}_2$ and $\mathbf{w}_1$. Thus

$$\mathbf{w}_3 = \mathbf{a}_3 - \frac{\mathbf{a}_3 . \mathbf{w}_2}{w_2^2}\mathbf{w}_2 - \frac{\mathbf{a}_3 . \mathbf{w}_1}{w_1^2}\mathbf{w}_1. \tag{1.68}$$

Evidently the procedure can be generalized to any number of dimensions. In the M-dimensional case it is customary to replace the dot product by the matrix inner product, i.e., write $\mathbf{a}_2^T \mathbf{a}_1$ instead of $\mathbf{a}_2 . \mathbf{a}_1$. Also, the restriction to real vectors is not necessary as long as the Hermitian inner product $\mathbf{a}_k^H \mathbf{a}_{k-1}$ is used. In fact the procedure is applicable as when the space is infinite dimensional. We merely have to employ the appropriate inner product. Thus given a set of linearly independent functions $\phi_1, \phi_2, \phi_3, \ldots \phi_N$ the corresponding orthogonal set $w_1, w_2, w_3, \ldots w_N$ is

$$w_k = \phi_k - \frac{(\phi_k, w_{k-1})}{(w_{k-1}, w_{k-1})}w_{k-1} - \frac{(\phi_k, w_{k-2},)}{(w_{k-2}, w_{k-2})}w_{k-2} \quad - \frac{(\phi_k, w_1)}{(w_1, w_1)}w_1 \tag{1.69}$$

with $k = 1, 2, \ldots, N$ and where quantities with subscripts less than unity are to be set to zero. Note that unlike the orthogonalization using the Gram matrix the Gram–Schmidt construction provides an orthogonal but not a orthonormal set. This is easily remedied dividing each w_k or $\mathbf{a}_k$ by its norm.

Clearly the Gram–Schmidt orthogonalization procedure can also be applied to functions that have been multiplied by a real weighting function, say $\Psi(t)$. This is equivalent to generalizing the definition of the inner product to

$$(f, g) \equiv \int_a^b \psi^2(t) f^*(t) g(t) \, dt, \qquad (1.70)$$

We shall discuss this generalized inner product definition in 1.3.3 in connection with the LMS error minimization. Using suitable weighing functions the Gram–Schmidt procedure can be employed to obtain special orthogonal functions using the polynomial set $1, t, t^2, \ldots, t^{N-1}$. Examples are the Legendre Polynomials [1] ($a = -1, b = 1, \psi = 1$), the Laguerre Polynomials [1] ($a = 0, b = \infty, \psi = e^{-t/2}$), and the Hermit Polynomials [1] ($a = -\infty, b = \infty, \psi = e^{-t^2/2}$) discussed in 1.5.3. For example, for Legendre Polynomials the first four terms are

$$w_1 = 1, w_2 = t - \frac{(1,t)}{(1,1)} 1 = t, w_2 = t - \frac{(1,t)}{(1,1)} 1 = t, \qquad (1.71a)$$

$$w_3 = t^2 - \frac{(t,t^2)}{(t,t)} t - \frac{(1,t^2)}{(1,1)} 1 = t^2 - \frac{1}{3}, \qquad (1.71b)$$

$$w_4 = t^3 - \frac{3}{5} t \qquad (1.71c)$$

Comparing these with (1.225) we note the following correspondence: $w_1 = P_0, w_2 = P_1, w_3 = \frac{2}{3} P_2, w_4 = \frac{2}{5} P_3$. Evidently, the normalization in Gram–Schmidt procedure differs from the standard normalization of the Legendre Polynomials.[5]

1.3   The LMS Approximation and the Normal Equations

1.3.1   The Projection Theorem

We now return to the minimization of the mean squared error (1.9), which we presently rewrite using the notation developed in the preceding section as

$$\varepsilon_N = (e_N, e_N). \qquad (1.72)$$

Solving (1.5) for $e_N(t)$ we have

$$e_N(t) = f(t) - f^N(t), \qquad (1.73)$$

where $f^N(t)$ is the partial sum

$$f^N(t) = \sum_{n=1}^N \hat{f}_n \phi_n(t), \qquad (1.74)$$

---

[5] Another set of orthogonal polynomials can be constructed using the eigenvectors of the Gram matix of t^k. The orthogonal polynomials are then given by (1.62).

$$\varepsilon_N = \left(f - \sum_{n=1}^{N} \hat{f}_n \phi_n, f - \sum_{n=1}^{N} \hat{f}_n \phi_n \right). \qquad (1.75)$$

We now seek the N complex coefficients $\hat{f}_n$ or, equivalently, $2N$ real coefficients that minimize (1.75). Alternatively, we can carry out the minimization with respect to the $2N$ complex coefficients $\hat{f}_n$ and $\hat{f}_n^*$. Using the latter approach we differentiate (1.75) with respect to $\hat{f}_m$ and $\hat{f}_m^*$ and set the derivatives to zero to obtain

$$\frac{\partial \varepsilon_N}{\partial \hat{f}_m} = -(e_N, \phi_m) = 0; \quad m = 1, 2, \dots N, \qquad (1.76a)$$

$$\frac{\partial \varepsilon_N}{\partial \hat{f}_m^*} = -(\phi_m, e_N) = 0; \quad m = 1, 2, \dots N. \qquad (1.76b)$$

We note that (1.76a) and (1.76b) are redundant since they are merely complex conjugates of each other. They both state that the minimization process leads to a residual error function which is orthogonal to all members of the chosen set of expansion functions. This result is usually referred to as the Projection Theorem. Interpreted geometrically it states that the best an LMS approximation can accomplish is to force the residual error function into a subspace orthogonal to that spanned by the given set of expansion functions. It is important to note that the expansion functions themselves need not be orthogonal.

The essence of the Projection Theorem is best captured with the aid of a geometrical construction. Assuming $N = 2$, we start by again representing $f(t)$ in terms of two expansion functions as in (1.36) but assume this time that they do not suffice for an exact representation of $f(t)$ and our objective is to minimize the error by adjusting the projection coefficients along the ϕ_1 and ϕ_2 "axes." To simplify matters suppose that the inclusion of one additional expansion function, say $\phi_3(t)$, orthogonal to both $\phi_1(t)$ and $\phi_2(t)$, would render the representation of $f(t)$ exact. In that case, according to the Projection Theorem, the residual error function $e_2(t) \equiv e_{2\min}(t)$ that corresponds to the optimum choice of the coefficients must be parallel to $\phi_3(t)$, as illustrated by the construction as shown in Fig. 1.6. Note that the vector $\hat{\mathbf{f}}$ is comprised of three projections of $f(t)$ that are nonvanishing on all three basis functions. The projections of $f(t)$ in the subspace spanned by the two basis functions ϕ_1 and ϕ_2 are represented by the vector $\hat{\mathbf{f}}^{(2)}$ which represents $\hat{\mathbf{f}}$ to within the error vector $\hat{\mathbf{e}}_2$. As may be seen from the figure, the norm of this error vector will be minimized whenever the projections in the subspace spanned by ϕ_1 and ϕ_2 are adjusted such that the resulting error vector approaches $\hat{\mathbf{e}}_{2\min}$ which lies entirely in the subspace orthogonal to ϕ_1 and ϕ_2. The corresponding projections (coefficients of the expansion of $f(t)$) yielding the best LMS approximation are then represented by the vector $\hat{\mathbf{f}}_{\min}$.

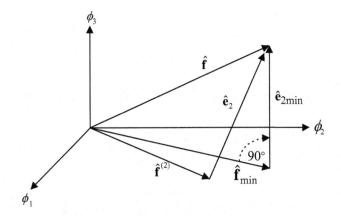

Figure 1.6: LMS error minimization and the orthogonality principle

1.3.2   The Normal Equations

Upon substituting (1.73) together with (1.74) into (1.76b) we obtain the system of linear algebraic equations that must be solved to find the projection coefficients $\hat{f}_n$ that afford the LMS approximation:

$$(\phi_m, f) = \sum_{n=1}^{N} (\phi_m, \phi_n)\, \hat{f}_n \; ; \; m = 1, 2, 3, \ldots N. \tag{1.77}$$

These equations are known as the normal equations for the unconstrained LMS minimization problem. We note that their formal solution requires the inversion of the Gram matrix. Clearly for a unique solution of (1.77) the Gram determinant must be nonvanishing, or, equivalently, the N expansion function must be linearly independent. The corresponding LMS error is found with the aid of (1.72) and (1.75) taking into account of (1.76b). We then obtain

$$\varepsilon_{N\,\min} = \left(f - \sum_{n=1}^{N} \hat{f}_n \phi_n, e_N\right) = (f, e_N) = (f, f) - \sum_{n=1}^{N} \hat{f}_n\, (f, \phi_n). \tag{1.78}$$

Since $\varepsilon_{N\,\min}$ is nonnegative, we have the inequality

$$(f, f) - \sum_{n=1}^{N} \hat{f}_n\, (f, \phi_n) \geq 0. \tag{1.79}$$

As will be discussed in 1.7.1 under certain circumstances an infinite set of expansion functions can be chosen that reduces the LMS to zero.

The normal equations are more commonly used with expansion functions (vectors) in discrete form. For an $M \times N$ matrix $\mathbf{A}$ we can identify these vectors as columns and collect the N unknown expansion coefficients into an N-dimensional vector $\mathbf{x}$, and attempt to solve the system $\mathbf{A}\mathbf{x} = \mathbf{y}$ for $\mathbf{x}$. As long as $N < M$ (which is the usual case), such a solution will in general not exist. We can, however, always obtain an $\mathbf{x}$ that approximates $\mathbf{y}$ by $\mathbf{A}\mathbf{x}$ in the LMS sense. It is not hard to see from (1.77) that the vector $\mathbf{x}$ in this case satisfies the normal equations in the form

$$\mathbf{A}^H \mathbf{y} = \mathbf{A}^H \mathbf{A} \mathbf{x}, \tag{1.80}$$

where

$$
\mathbf{A}^H \mathbf{A} =
\begin{bmatrix}
\mathbf{a}_1^H \\
\mathbf{a}_2^H \\
\mathbf{a}_3^H \\
\cdot \\
\mathbf{a}_N^H
\end{bmatrix}
\begin{bmatrix}
\mathbf{a}_1 & \mathbf{a}_2 & \mathbf{a}_3 & \cdot & \mathbf{a}_N
\end{bmatrix}
$$

$$
=
\begin{bmatrix}
\mathbf{a}_1^H \mathbf{a}_1 & \mathbf{a}_1^H \mathbf{a}_2 & \mathbf{a}_1^H \mathbf{a}_3 & \cdot & \mathbf{a}_1^H \mathbf{a}_N \\
\mathbf{a}_2^H \mathbf{a}_1 & \mathbf{a}_2^H \mathbf{a}_2 & \mathbf{a}_2^H \mathbf{a}_3 & \cdot & \mathbf{a}_2^H \mathbf{a}_N \\
\mathbf{a}_3^H \mathbf{a}_1 & \mathbf{a}_3^H \mathbf{a}_2 & \mathbf{a}_3^H \mathbf{a}_3 & \cdot & \mathbf{a}_3^H \mathbf{a}_N \\
\cdot & \cdot & \cdot & \cdot & \cdot \\
\mathbf{a}_N^H \mathbf{a}_1 & \mathbf{a}_N^H \mathbf{a}_2 & \mathbf{a}_N^H \mathbf{a}_3 & \cdot & \mathbf{a}_N^H \mathbf{a}_N
\end{bmatrix}. \tag{1.81}
$$

where $\mathbf{A}^H \mathbf{A}$ is the discrete form of the Gram matrix. Note that from a strictly algebraic viewpoint we could always append a finite number $(M - N)$ of linearly independent vectors to the columns of $\mathbf{A}$ transforming it into a nonsingular $M \times M$ matrix and thus ensuring an exact (rather than an LMS) solution of $\mathbf{A}\mathbf{x} = \mathbf{y}$ for $\mathbf{x}$.

1.3.3   Generalizations of the Normal Equations

In certain situations it is more meaningful to use as a measure of the goodness of fit to a given function a weighted MS error in the form

$$\varepsilon_{\psi,N} \equiv \int_a^b |\psi(t)|^2 \left| f(t) - f^N(t) \right|^2 dt = (\psi e_N, \psi e_N), \tag{1.82}$$

where ψ is a specified weighting function. Its choice could depend on achieving a desired amount of emphasis (or de-emphasis) of the contribution to the error from different regions of the time interval, or, as discussed in 1.2.5, it could serve as a means of achieving orthogonality among a selected set of expansion functions. Other reasons for the use of special weighting functions will be discussed later in connection with techniques known as windowing. Note that (1.82) implies a representation of the function $f(t)$ in the following form:

$$\psi(t) f(t) = \sum_{n=1}^N \hat{f}_n \psi(t) \phi_n(t) + e_{\psi,N}(t). \tag{1.83}$$

The corresponding normal equations for the expansion coefficients follow from (1.77) through the replacements $f \to \psi f$ and $\phi_n \to \psi \phi_n$:

$$(\psi \phi_m, \psi f) = \sum_{n=1}^N (\psi \phi_m, \psi \phi_n) \hat{f}_n \; ; \; m = 1, 2, 3, \ldots N. \tag{1.84}$$

In the discrete case, (1.80), the weighting factors appear as elements of an M-element diagonal matrix $\mathbf{D}$ so that the MS error becomes

$$\varepsilon_{\mathbf{D},N} = \|\mathbf{D}\,(\mathbf{y} - \mathbf{Ax})\|^2 \tag{1.85}$$

while the corresponding normal equations are

$$\mathbf{A}^H \mathbf{D}^H \mathbf{D} \mathbf{y} = \mathbf{A}^H \mathbf{D}^H \mathbf{D} \mathbf{Ax}. \tag{1.86}$$

We can also generalize the LMS approximation problem to expansion function sets defined over a "continuous summation index," i.e., we replace the sum in (1.73) by an integral. Thus

$$f(t) = \int_{\omega \in S_\Omega} \hat{f}(\omega)\,\phi(\omega,t)\,d\omega + e_\Omega(t), \tag{1.87}$$

where the summation variable ω ranges over an interval whose extent is denoted symbolically by S_Ω, wherein Ω is proportional to the total length of the interval (e.g., $S_\Omega = (-\Omega,\Omega)$), the $\phi(\omega,t)$ are the specified expansion functions, and the $\hat{f}(\omega)$ the unknown coefficients. The latter are to be chosen to minimize

$$\varepsilon_\Omega \equiv \int_a^b |e_\Omega(t)|^2\,dt = (e_\Omega, e_\Omega). \tag{1.88}$$

Since the summation index in (1.87) ranges over a continuum, the approach we used previously in (1.76) is not directly applicable. In such situations it is natural to use a variational approach. Before applying it to (1.88) we shall illustrate it by re-deriving (1.77) from (1.75).

To start with, we perturb each of the coefficients $\hat{f}_n$ in (1.75) by the small amount $\delta\hat{f}_n$. The corresponding change in the error $\delta\varepsilon_N$ follows by taking differentials

$$\delta\varepsilon_N = \left(f - \sum_{n=1}^N \hat{f}_n \phi_n, -\sum_{n=1}^N \delta\hat{f}_n \phi_n \right) + \left(-\sum_{n=1}^N \delta\hat{f}_n \phi_n, f - \sum_{n=1}^N \hat{f}_n \phi_n \right). \tag{1.89}$$

We now assume that the $\hat{f}_n$ are precisely those that minimize ε_N. It is then necessary for the left side in (1.89) to vanish. Upon expanding the two inner products on the right of (1.89) and relabeling the summation indices we obtain

$$0 = \sum_{m=1}^N \alpha_m \delta\hat{f}_m + \sum_{m=1}^N \beta_m \delta\hat{f}_m^*, \tag{1.90}$$

where

$$\alpha_m = -(f,\phi_m) + \sum_{n=1}^N (\phi_n,\phi_m)\,\hat{f}_n^*, \text{ and} \tag{1.91a}$$

$$\beta_m = -(\phi_m,f) + \sum_{n=1}^N (\phi_m,\phi_n)\,\hat{f}_n. \tag{1.91b}$$

Now the $\delta \hat{f}_m$ and $\delta \hat{f}_m^*$ are completely arbitrary so that (1.90) can be satisfied only by $\alpha_m = \beta_m = 0$. Thus (1.90) again yields (1.76) and hence the normal equations (1.77).

We now use the same approach to minimize (1.88) which we first rewrite as follows:

$$\varepsilon_\Omega = \left(f - f^{(\Omega)}, f - f^{(\Omega)} \right), \tag{1.92}$$

where $f^{(\Omega)}(t)$ is the partial "sum"

$$f^{(\Omega)}(t) = \int\limits_{\omega \in S_\Omega} \hat{f}(\omega) \, \phi(\omega, t) \, d\omega. \tag{1.93}$$

Proceeding as in (1.89) we obtain

$$\delta \varepsilon_\Omega = \left(f - f^{(\Omega)}, -\delta f^{(\Omega)} \right) + \left(-\delta f^{(\Omega)}, f - f^{(\Omega)} \right). \tag{1.94}$$

Upon setting $\delta \varepsilon_\Omega = 0$ and expanding the inner products results in a form analogous to (1.90) wherein the sums are replaced by integrals:

$$0 = \left[\int\limits_{\omega \in S_\Omega} \alpha(\omega) \, d\omega \right] \delta \hat{f}(\omega) + \left[\int\limits_{\omega \in S_\Omega} \beta(\omega) \, d\omega \right] \delta \hat{f}^*(\omega), \tag{1.95}$$

where

$$\alpha(\omega) = -(f, \phi(\omega)) + \int\limits_{\omega' \in S_\Omega} (\phi(\omega'), \phi(\omega)) \, \hat{f}^*(\omega') \, d\omega', \tag{1.96a}$$

$$\beta(\omega) = -(\phi(\omega), f) + \int\limits_{\omega' \in S_\Omega} (\phi(\omega), \phi(\omega')) \, \hat{f}(\omega') \, d\omega'. \tag{1.96b}$$

As in (1.90) we argue that $\delta \hat{f}(\omega)$ and $\delta \hat{f}^*(\omega)$ are arbitrary deviations. In fact, we could even choose $\delta \hat{f}(\omega) = \alpha^*(\omega)$ so that in view of $\alpha(\omega) = \beta^*(\omega)$ (1.95) becomes

$$0 = \int\limits_{\omega \in S_\Omega} |\alpha(\omega)|^2 \, d\omega, \tag{1.97}$$

which clearly requires $\alpha(\omega) = 0$. From (1.96b) we then have our final result

$$(\phi(\omega), f) = \int\limits_{\omega' \in S_\Omega} (\phi(\omega), \phi(\omega')) \, \hat{f}(\omega') \, d\omega'; \; \omega \in S_\Omega \, ; \; a \leq t \leq b, \tag{1.98}$$

as the desired generalization of the normal equations. In contrast to (1.77) which is simply a set of algebraic equations, (1.98) represents an infinite set of integral equations (one for each choice of the variable ω) for the unknown functions $\hat{f}(\omega)$. Writing out the inner products in (1.98) explicitly and setting

$$(\phi(\omega), \phi(\omega')) = \int_a^b \phi^*(\omega, t) \, \phi(\omega', t) \, dt \equiv G(\omega, \omega'), \tag{1.99}$$

(1.98) assumes the form

$$\int_a^b f(t) \phi^* (\omega, t) \, dt = \int_{\omega' \in S_\Omega} G(\omega, \omega') \hat{f}(\omega') \, d\omega' \; ; \; \omega \in S_\Omega. \tag{1.100}$$

Equation (1.100) will form the basis in our investigation of integral transforms of signals while (1.77) will provide the basis in the study of series representations. As a final generalization of the normal equations we consider expansions in functions that depend on two discrete indices, i.e., of the form $\phi_{nm}(t)$, so that

$$f(t) = \sum_{n=1}^{N} \sum_{m=1}^{M} \hat{f}_{nm} \phi_{nm}(t) + e_{N,M}(t) \tag{1.101}$$

The dependence of two indices is merely a notational detail and does not affect the general form of the Normal Equations. In fact these can be written down immediately from (1.77) by simply taking account of the extra index. Thus

$$(\phi_{pq}, f) = \sum_{n=1}^{N} \sum_{m=1}^{M} (\phi_{pq}, \phi_{nm}) \hat{f}_{nm} \; ; \; p = 1, 2, \ldots N; q = 1, 2, \ldots M. \tag{1.102}$$

1.3.4 LMS Approximation and Stochastic Processes*

The LMS approximation plays an important role in probability theory and the theory of stochastic processes and is treated in detail in standard works on the subject. In the following we confine ourselves to showing the connection between the LMS formulation just described and its extension to random variables.

The approximation problem may then be phrased as follows. Given a random variable $\underline{y}$ we are asked to approximate it by a linear combination of random variables $\underline{a}_n$, $n = 1, 2, ..N$. The error $\underline{e}_N$ is necessarily a random variable[6]

$$\underline{e}_N = \underline{y} - \sum_{n=1}^{N} x_n \underline{a}_n, \tag{1.103}$$

which may be taken as the analogue of the deterministic statement (1.73). The quantity to be minimized is the ensemble average (expectation) of the squared magnitude of the error $\underline{e}_N$

$$\varepsilon_N = < \underline{e}_N^* \underline{e}_N > = < |\underline{e}_N|^2 >, \tag{1.104}$$

---

[6]Here we distinguish a random variable by an underline and denote its expectation (or ensemble average) by enclosing it in $< >$

which may be interpreted as a generalization of the squared norm in the deterministic case (1.72). Substituting from (1.103) and expanding we get

$$
\varepsilon_N \;=\; < \left| \underline{y} - \sum_{n=1}^{N} x_n \underline{a}_n \right|^2 >
$$

$$
=\; < \underline{y}^* \underline{y} > - \sum_{n=1}^{N} < \underline{a}_n^* \underline{y} > x_n^* - \sum_{n=1}^{N} < \underline{a}_n \underline{y}^* > x_n
$$

$$
+ \sum_{n=1}^{N} \sum_{m=1}^{N} < \underline{a}_n \underline{a}_m^* > x_n x_m^*. \tag{1.105}
$$

Differentiating with respect to x_n and x_n^* and setting the derivatives to zero results in

$$
< \underline{a}_m^* \underline{e}_N >= 0 \;,\; m = 1,2,\ldots N. \tag{1.106}
$$

or, in expanded form,

$$
< \underline{a}_m^* \underline{y} >= \sum_{n=1}^{N} < \underline{a}_m^* \underline{a}_n > x_n \;,\; m = 1,2,\ldots N \tag{1.107}
$$

Equation (1.106) may be taken as an extension of the projection theorem (1.76a) and (1.107) of the normal equations (1.77). The fact that the algebraic forms of the deterministic and stochastic versions of the normal equations are identical is actually not unexpected if one recalls that ensemble averages are merely inner products with a weighting factor equal to the probability density function [c.f. (1.82)].

Typical applications of LMS estimation arise in statistical time series analyses. For example, suppose we would like to predict the value of the n-th sample of a random signal $\underline{f}(n\Delta t)$ at time $t = n\Delta t$ from its past values at $t = (n-\ell)\Delta t$, $\ell = 1,2,\ldots N$. We can formulate this as an LMS estimation problem and seek N linear prediction filter coefficients x_ℓ , $\ell = 1,2,\ldots N$ that minimize

$$
< \left| \underline{f}(n\Delta t) - \sum_{\ell=1}^{N} x_\ell \underline{f}[(n-\ell)\Delta t] \right|^2 > . \tag{1.108}
$$

If we assume samples from a stationary process with correlation coefficients

$$
< \underline{f}^*(n\Delta t) \underline{f}(m\Delta t) >= R[(n-m)\Delta t], \tag{1.109}
$$

the normal equations (1.107) become

$$
\begin{bmatrix} R(\Delta t) \\ R(2\Delta t) \\ . \\ R(N\Delta t) \end{bmatrix} = \begin{bmatrix} R(0) & R(-\Delta t) & . & R((1-N)\Delta t) \\ R(\Delta t) & R(0) & . & R((2-N)\Delta t) \\ . & . & . & \\ R((N-1)\Delta t) & R((N-2)\Delta t) & : & R(0) \end{bmatrix} \begin{bmatrix} x_1 \\ x_2 \\ \\ x_N \end{bmatrix}
$$

$$
\tag{1.110}
$$

A more general form of the normal equations arises in connection with multidimensional filtering. Here instead of a scalar random variable we approximate an M-dimensional vector $\underline{\mathbf{y}}$ (comprised of M random variables) by a linear combination of M-dimensional vector random variables $\underline{\mathbf{a}}_n$, $n = 1, 2, \ldots N$. Then

$$\underline{\mathbf{e}}_N = \underline{\mathbf{y}} - \sum_{n=1}^{N} x_n \underline{\mathbf{a}}_n \qquad (1.111)$$

and we find that the x_n that minimize the $< |\underline{\mathbf{e}}_N|^2 >$ satisfy the normal equations

$$< \underline{\mathbf{a}}_m^H \underline{\mathbf{y}} > = \sum_{n=1}^{N} < \underline{\mathbf{a}}_m^H \underline{\mathbf{a}}_n > x_n \ , m = 1, 2, \ldots N \qquad (1.112)$$

1.4   LMS Solutions via the Singular Value Decomposition

1.4.1   Basic Theory Underlying the SVD

In many applications where discrete representations leading to the normal equations of the form (1.80) arise, the expansion vectors are not orthogonal. In fact in many cases, especially when dealing with experimentally derived data, they could be nearly linearly dependent. Solutions based on a direct inversion of the Gram matrix could then be strongly corrupted by roundoff errors and turn out to be nearly useless. A much more powerful approach in such cases is to solve the LMS problem with the aid of a matrix representation known as the singular value decomposition (SVD). In the following we provide a brief account of this technique.

With $\mathbf{A}$ a complex-valued $M \times N$ matrix $(M \geq N)$ we form the $(M + N) \times (M + N)$ matrix

$$\mathbf{W} = \begin{bmatrix} \mathbf{0}_{MM} & \mathbf{A} \\ \mathbf{A}^H & \mathbf{0}_{NN} \end{bmatrix}. \qquad (1.113)$$

It is easy to show that $\mathbf{W}$ is Hermitian for any MXN matrix it can be diagonalized by a unitary transformation comprised of $N + M$ vectors $\mathbf{q}_j$ satisfying the eigenvalue problem

$$\mathbf{W}\mathbf{q}_j = \sigma_j \mathbf{q}_j, \ j = 1, 2, 3, \ldots N + M, \qquad (1.114)$$

where the eigenvalues σ_j are necessarily real. If we partition $\mathbf{q}_j$ to read

$$\mathbf{q}_j = \begin{bmatrix} \mathbf{u}_j \\ \mathbf{v}_j \end{bmatrix} \qquad (1.115)$$

with $\mathbf{u}_j$ and $\mathbf{v}_j$ having dimensions M and N, respectively, (1.114) is equivalent to the following two matrix equations:

$$\mathbf{A}\mathbf{v}_j = \sigma_j \mathbf{u}_j, \qquad (1.116a)$$
$$\mathbf{A}^H \mathbf{u}_j = \sigma_j \mathbf{v}_j. \qquad (1.116b)$$

Multiplying both sides of (1.116a) from the left by $\mathbf{A}^H$ and replacing $\mathbf{A}^H\mathbf{u}_j$ with (1.116b) gives

$$\mathbf{A}^H\mathbf{A}\mathbf{v}_j = \sigma_j^2\mathbf{v}_j. \tag{1.117}$$

Similarly, multiplication of both sides of (1.116b) from the left by $\mathbf{A}$ and a substitution from (1.116a) results in

$$\mathbf{A}\mathbf{A}^H\mathbf{u}_j = \sigma_j^2\mathbf{u}_j. \tag{1.118}$$

Equations (1.117) and (1.118) are eigenvalue problems for the two nonnegative definite Hermitian matrices $\mathbf{A}\mathbf{A}^H$ and $\mathbf{A}^H\mathbf{A}$. As is easy to see, these matrices have the same rank which is just the number of linearly independent columns of $\mathbf{A}$. If we denote this number by R ($R \leq N$), then the number of nonzero eigenvalues σ_j^2 in (1.117) and (1.118) is precisely R. Because the matrices in (1.117) and (1.118) are Hermitian, the eigenvectors $\mathbf{v}_j$ and $\mathbf{u}_j$ can be chosen to form two orthonormal sets. Thus

$$\mathbf{v}_j^H\mathbf{v}_k = \delta_{jk}, \; j,k = 1,2,3,\ldots N, \tag{1.119}$$

and

$$\mathbf{u}_j^H\mathbf{u}_k = \delta_{jk}, \; j,k = 1,2,3,\ldots M. \tag{1.120}$$

By virtue of (1.119) we can resolve an $N \times N$ unit matrix $\mathbf{I}_{NN}$ into a sum of projection operators $\mathbf{v}_j\mathbf{v}_j^H$, i.e.,

$$\sum_{j=1}^{N} \mathbf{v}_j\mathbf{v}_j^H = \mathbf{I}_{NN}. \tag{1.121}$$

If we now multiply (1.116a) from the right by $\mathbf{v}_j^H$, sum both sides over all j and take account of (1.121), we obtain what is referred to as the SVD representation of $\mathbf{A}$:

$$\mathbf{A} = \sum_{j=1}^{R} \sigma_j\mathbf{u}_j\mathbf{v}_j^H. \tag{1.122}$$

The σ_j are referred to as the singular values of $\mathbf{A}$. By convention they are chosen as positive. We note that this can always be done in this representation by absorbing the sign in one of the transformation matrices. Since there can be only R nonzero singular values, only R vectors of each set (1.119) and (1.120) enter into the SVD. Nevertheless, the SVD is sometimes written in block form that includes the complete unitary transformations

$$\mathbf{U} = \begin{bmatrix} \mathbf{u}_1 & \mathbf{u}_2 & \mathbf{u}_3 & \cdots & \mathbf{u}_M \end{bmatrix} \tag{1.123}$$

and

$$\mathbf{V} = \begin{bmatrix} \mathbf{v}_1 & \mathbf{v}_2 & \mathbf{v}_3 & \cdots & \mathbf{v}_N \end{bmatrix}, \tag{1.124}$$

whose dimensions are, respectively, $M \times M$ and $N \times N$. In this notation (1.119) and (1.120) and (1.121) may be replaced by

$$\mathbf{U}^H\mathbf{U} = \mathbf{U}\mathbf{U}^H = \mathbf{I}_{MM}, \; \mathbf{V}^H\mathbf{V} = \mathbf{V}\mathbf{V}^H = \mathbf{I}_{NN}. \tag{1.125}$$

We can then rewrite (1.122) as follows:

$$\mathbf{A} = \mathbf{U}\boldsymbol{\Sigma}\mathbf{V}^H, \tag{1.126}$$

where the $M \times N$ matrix $\boldsymbol{\Sigma}$ has the structure

$$\boldsymbol{\Sigma} = \begin{bmatrix} \sigma_1 & 0 & 0 & 0 & 0 & 0 & . & 0 \\ 0 & \sigma_2 & 0 & 0 & 0 & 0 & . & 0 \\ 0 & 0 & . & 0 & 0 & 0 & . & 0 \\ 0 & 0 & 0 & \sigma_R & 0 & 0 & . & 0 \\ 0 & 0 & 0 & 0 & 0 & 0 & . & 0 \\ . & . & . & . & & . & & . \\ 0 & 0 & 0 & 0 & 0 & 0 & 0 & 0 \end{bmatrix}. \tag{1.127}$$

The representation (1.122) is usually referred to as the reduced SVD form. A generally adopted convention is to order the singular values in accordance with

$$\sigma_1 > \sigma_2 > \ldots > \sigma_R.$$

An alternative block matrix representation is obtained by eliding all but the first R columns in $\mathbf{U}$ and $\mathbf{V}$. Thus if we denote the resulting matrices by $\mathbf{U}_R$ and $\mathbf{V}_R$ (whose respective dimensions are $M \times R$ and $N \times R$), we obtain

$$\mathbf{A} = \mathbf{U}_R\boldsymbol{\Sigma}_R\mathbf{V}_R^H, \tag{1.128}$$

where

$$\boldsymbol{\Sigma}_R = \begin{bmatrix} \sigma_1 & 0 & 0 & 0 \\ 0 & \sigma_2 & 0 & 0 \\ 0 & 0 & . & 0 \\ 0 & 0 & 0 & \sigma_R \end{bmatrix}. \tag{1.129}$$

Note that since the columns of $\mathbf{U}_R$ and $\mathbf{V}_R$ are orthonormal, we have

$$\mathbf{U}_R^H\mathbf{U}_R = \mathbf{I}_{RR} \tag{1.130}$$

and

$$\mathbf{V}_R^H\mathbf{V}_R = \mathbf{I}_{RR}. \tag{1.131}$$

On the other hand, $\mathbf{U}_R\mathbf{U}_R^H$ is no longer a unit matrix but an $M \times M$ (rank R) projection operator into the range of $\mathbf{A}$. Similarly, $\mathbf{V}_R\mathbf{V}_R^H$ is an $N \times N$ (rank R) projection operator into the domain of $\mathbf{A}$.

It is perhaps worth noting at this juncture that the approach to the SVD via the two eigenvalue problems (1.117) and (1.118) has been adopted herein mainly for didactic reasons and is not recommended for actual numerical work. In practice the member matrices entering in (1.128) can be computed using much more efficient algorithms. These are described in [8] and we shall not discuss them. For purposes of providing numerical examples we shall content ourselves in using the routines incorporated in MATLAB.

1.4.2   Solutions of the Normal Equations Using the SVD

We now solve the normal equations (1.80) using the SVD. Substituting (1.128) for $\mathbf{A}$ in (1.80) we get

$$\mathbf{A}^H \mathbf{A} \mathbf{x} = \mathbf{V}_R \mathbf{\Sigma}_R \mathbf{U}_R^H \mathbf{U}_R \mathbf{\Sigma}_R \mathbf{V}_R^H \mathbf{x} = \mathbf{V}_R \mathbf{\Sigma}_R^2 \mathbf{V}_R^H \mathbf{x} = \mathbf{V}_R \mathbf{\Sigma}_R \mathbf{U}_R^H \mathbf{y}. \qquad (1.132)$$

Multiplying from the left by $\mathbf{\Sigma}_R^{-2} \mathbf{V}_R^H$ and taking account of (1.131) yield for the last two members of (1.132)

$$\mathbf{V}_R^H \mathbf{x} = \mathbf{\Sigma}_R^{-1} \mathbf{U}_R^H \mathbf{y}. \qquad (1.133)$$

In case $R = N$, we have $\mathbf{V}_N \mathbf{V}_N^H = \mathbf{I}_{NN}$, so that we can solve (1.133) for $\mathbf{x}$ by simply multiplying both sides by $\mathbf{V}_N$. We then obtain the unique solution

$$\mathbf{x} \equiv \mathbf{x}_{LS} = \mathbf{V}_N \mathbf{\Sigma}_N^{-1} \mathbf{U}_N^H \mathbf{y}. \qquad (1.134)$$

When $R < N$, (1.133) possesses $N - R + 1$ solutions. In many physical situations it is reasonable to select the solution vector having the smallest norm. It is easily shown that this is given by

$$\mathbf{x} \equiv \mathbf{x}_{LS}^0 = \mathbf{V}_R \mathbf{\Sigma}_R^{-1} \mathbf{U}_R^H \mathbf{y}. \qquad (1.135)$$

To show this, we first note that $\mathbf{x}_{LS}^0$ satisfies (1.133). Thus substituting (1.135) into the left side of (1.133) and taking account of (1.131) yield

$$\mathbf{V}_R^H \mathbf{x}_{LS}^0 = \mathbf{V}_R^H \mathbf{V}_R \mathbf{\Sigma}_R^{-1} \mathbf{U}_R^H \mathbf{y} = \mathbf{\Sigma}_R^{-1} \mathbf{U}_R^H \mathbf{y}. \qquad (1.136)$$

Second, let $\mathbf{x}_{LS}$ be any other solution of (1.133) and set

$$\mathbf{x}_{LS} = \mathbf{x}_{LS}^0 + \boldsymbol{\xi}. \qquad (1.137)$$

Again from (1.133), $\mathbf{V}_R^H \left(\mathbf{x}_{LS} - \mathbf{x}_{LS}^0 \right) = 0$, so that $\mathbf{V}_R^H \boldsymbol{\xi} = 0$, or, which is the same thing,

$$\boldsymbol{\xi}^H \mathbf{V}_R = 0. \qquad (1.138)$$

Multiplying (1.135) from the left by $\boldsymbol{\xi}^H$ and taking account of (1.138) we find $\boldsymbol{\xi}^H \mathbf{x}_{LS}^0 = 0$. Hence

$$\| \mathbf{x}_{LS} \|^2 = \left\| \mathbf{x}_{LS}^0 + \boldsymbol{\xi} \right\|^2 = \left\| \mathbf{x}_{LS}^0 \right\|^2 + \| \boldsymbol{\xi} \|^2 \geq \left\| \mathbf{x}_{LS}^0 \right\|^2, \qquad (1.139)$$

which proves that $\mathbf{x}_{LS}^0$ is the solution of (1.133) with the smallest norm.

The matrix $\mathbf{A}^{\tilde{}1} \equiv \mathbf{V}_R \mathbf{\Sigma}_R^{-1} \mathbf{U}_R^H$ is called the Moore–Penrose pseudoinverse of $\mathbf{A}$. Introducing this notation into (1.135) we write

$$\mathbf{x}_{LS}^0 = \mathbf{A}^{\tilde{}1} \mathbf{y}. \qquad (1.140)$$

Note that

$$\mathbf{A}^{\tilde{}1} \mathbf{A} = \mathbf{V}_R \mathbf{V}_R^H, \qquad (1.141)$$

so that the right side reduces to a unit matrix only if $\mathbf{A}^{-1}$ has full rank, in which case $\mathbf{A}^{-1}$ is the true left inverse of $\mathbf{A}$. (A right inverse does not exist as long as $M > N$). The general solution of (1.133) is given by (1.140) plus any of the $N - R$ possible solutions $\boldsymbol{\xi}_\ell$ of the homogeneous set

$$\mathbf{V}_R \mathbf{V}_R^H \boldsymbol{\xi}_\ell = \mathbf{0}, \tag{1.142}$$

i.e.,

$$\mathbf{x}_{LS} = \mathbf{A}^{-1} \mathbf{y} + \boldsymbol{\xi}_\ell, \; \ell = 1, 2, \dots N - R. \tag{1.143}$$

Let us now represent the overdetermined system in the form

$$\mathbf{y} = \mathbf{A}\mathbf{x} + \mathbf{e}_R, \tag{1.144}$$

where by analogy with (1.73) $\mathbf{e}_R$ represents the residual error. If $\mathbf{x}$ is chosen in accordance with (1.143), then (1.144) becomes

$$\mathbf{y} = \mathbf{U}_R \left(\mathbf{U}_R^H \mathbf{y} \right) + \mathbf{e}_R, \tag{1.145}$$

i.e., we obtain an expansion of $\mathbf{y}$ in terms of R orthonormal basis vectors (columns of $\mathbf{U}_R$) plus the residual vector $\mathbf{e}_R$ whose norm has been minimized. We find

$$
\begin{aligned}
\|\mathbf{e}_R\|^2 &= \varepsilon_{N\min} = \mathbf{y}^H \mathbf{y} - \mathbf{y}^H \mathbf{A} \mathbf{x}_{LS} \\
&= \mathbf{y}^H \mathbf{y} - \mathbf{y}^H \mathbf{U}_R \boldsymbol{\Sigma}_R \mathbf{V}_R^H \left[\mathbf{V}_R \boldsymbol{\Sigma}_R^{-1} \mathbf{U}_R^H \mathbf{y} + \boldsymbol{\xi}_\ell \right] \\
&= \mathbf{y}^H \mathbf{y} - \mathbf{y}^H \mathbf{U}_R \mathbf{U}_R^H \mathbf{y},
\end{aligned}
\tag{1.146}
$$

which is identical for all $\mathbf{x}_{LS}$ in (1.143), and, in particular, for the minimum norm solution is given directly by the Moore–Penrose pseudoinverse in (1.140).

The minimum norm least squares solution (1.140) ignores contributions from the vector $\mathbf{y}$ not in the subspace defined by the columns of $\mathbf{U}_R$. For if, $\mathbf{y}^c$ is such a vector, then $\mathbf{U}_R^H \mathbf{y}^c = 0$. This is, of course, beneficial if $\mathbf{y}^c$ happens to comprise only noise (of either numerical or physical origin) for then $\mathbf{U}_R$ acts as an ideal noise suppression filter. If, on the other hand, $\mathbf{y}^c$ includes a substantial signal component, we introduce errors. The only way to capture such a signal component is to enlarge the set of basis vectors. Clearly, a net benefit would accrue from such an enlargement only if the additional signal contribution exceeds the noise contribution. Thus, ideally, the signal subspace should be known a priori. In practice this is rarely the case. The usual approach is to construct (on the basis of physical theory and/or empirical data) a nominally full rank matrix $\mathbf{A}$ and set to zero all but the first R dominant singular values so that R is taken as the estimate of the "effective" dimension of the signal subspace and the columns of $\mathbf{U}_R$ as its basis. This semi-empirical procedure is only meaningful if the singular value plot versus the index exhibits a more or less sharp threshold, so that R can be identified with little ambiguity.

1.4.3   Signal Extraction from Noisy Data

In the following we illustrate the use of the SVD in the extraction of signals from data corrupted by additive random noise. We consider a model wherein the signal to be detected consists of a weighted sum of known waveforms and the objective is to determine the weighting factors. The measured data are the sum

$$y(t_m) \equiv y_m = \sum_{k=1}^{N} x_k \, f_k(t_m) + n(t_m) \ , \ m = 1, 2, \ldots M,$$

wherein the $f_k(t)$ are known functions, x_k the unknown weighting factors. $n(t)$ represents the random noise and the t_m the time instances at which the data are recorded. The number of samples M is larger (preferably much larger) than N. For our purposes the sampling time interval $t_m - t_{m-1}$ is not important but in practice would probably be governed the signal bandwidth. We shall base our estimate on a single record taken over some fixed time interval. We use the SVD to estimate $\mathbf{x}^T = [x_1 \ x_2 \ x_3 \ldots x_N]$ using the measured data vector $\mathbf{y}^T = [y_1 \ y_2 \ y_3 \ldots x_N]$ and the $M \times N$ matrix $A_{mk} = f_k(t_m)$. The estimated weighting factors are given by

$$\mathbf{x} = \mathbf{A}^{\sim 1} \mathbf{y}$$

taking into account only of the dominant singular values.

We illustrate this estimation technique by the following example. Consider the data vector $\mathbf{y}$ with components

$$y(m) = (m - 1)^2 + n\,(m) \ ; 1 \leq m \leq 64,$$

where $(m - 1)^2$ may be taken as the desired signal and $n(m)$ is a random noise disturbance. Let us model this disturbance by setting

$$n(m) = 3,000 rand(m),$$

wherein $rand(m)$ is a pseudorandom sequence uniformly distributed between 1 and -1. We shall attempt to extract the signal from the noisy data by approximating the data vector in the LMS sense with the polynomial

$$yest(m) = a_0 + a_1 m + a_2 m^2 \ ; \ 1 \leq m \leq 64.$$

To relate this to our previous notation, we have

$$\mathbf{A} = \begin{bmatrix} 1 & 1 & 1 \\ 1 & 2 & 4 \\ 1 & 3 & 9 \\ . & . & . \\ 1 & 64 & 4096 \end{bmatrix}, \ \mathbf{x} = \begin{bmatrix} a_0 \\ a_1 \\ a_2 \end{bmatrix}$$

and we seek a vector $\mathbf{x}$ that provides the best LMS fit of $\mathbf{Ax}$ to $\mathbf{y}$. The solution is given at once by multiplying the Moore–Penrose pseudoinverse by the data vector as in (1.140). By using the MATLAB SVD routine we find

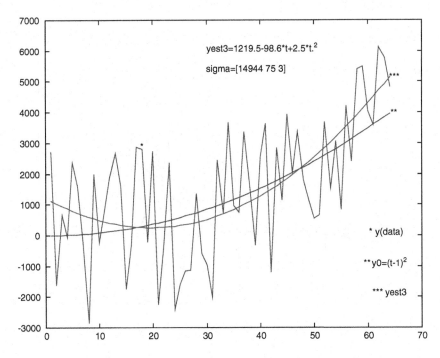

Figure 1.7: LMS fit of polynomial to noisy data (R = 3)

$$\boldsymbol{\Sigma}_3 = \begin{bmatrix} 14944 & 0 & 0 \\ 0 & 75 & 0 \\ 0 & 0 & 3 \end{bmatrix},$$

while the pseudoinverse gives

$$\mathbf{x}^T = \begin{bmatrix} 1219.5 & -98.6 & 2.5 \end{bmatrix}.$$

The resulting polynomial fit, $yest3$, together with the original data and the signal $y0$ are plotted in Fig. 1.7.

We observe that the estimated signal approximates the uncorrupted (noise free) signal much better than does the original data. Still the reconstruction is not very satisfactory, particularly for low indices. Nevertheless this is the best one can achieve with a straightforward LMS procedure. It should be evident that a direct solution of the normal equations would yield identical results. We have, however, not yet exploited the full potential of the SVD approach. We note that the given $\mathbf{A}$ matrix is ill conditioned, the ratio of the maximum to the minimum singular value being $14,944/3 \approx 5,000$. Thus we should expect an improved signal estimate by removing the lowest singular value and employing the truncated SVD with

$$\boldsymbol{\Sigma}_2 = \begin{bmatrix} 14944 & 0 \\ 0 & 75 \end{bmatrix}.$$

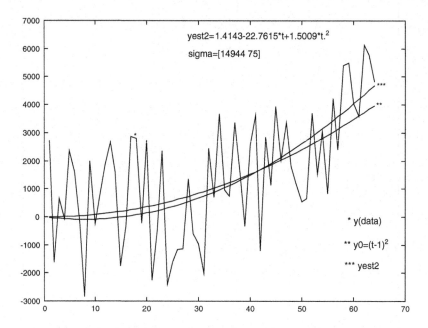

Figure 1.8: LMS fit of polynomial to noisy data (R = 2)

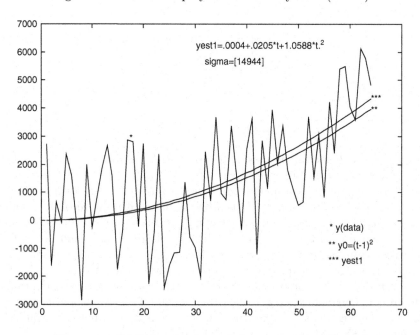

Figure 1.9: LMS fit of polynomial to noisy data (R = 1)

The recomputation of the pseudoinverse with $R = 2$ yields

$$\mathbf{x}^T = \begin{bmatrix} -1.4143 & -22.7615 & 1.5009 \end{bmatrix},$$

and the resulting polynomial fit, $yest2$, together with the original data and the signal $y0$ are plotted in Fig. 1.8.

Clearly the removal of the lowest singular value resulted in an improved signal estimate. Indeed, we can improve the estimate even further by retaining only the one dominant singular value. For this case the polynomial coefficients are

$$\mathbf{x}^T = \begin{bmatrix} .0004 & .0205 & 1.0588 \end{bmatrix},$$

and the results are plotted in Fig. 1.9.

1.4.4   The SVD for the Continuum

Let us now apply the SVD representation to the LMS approximation problem within the realm of the continuum. We note that for any finite set of expansion functions the equivalent algebraic problem is overdetermined, since formally $M = \infty$. If we think of the expansion functions $\phi_n(t)$ as an infinite-dimensional column vector, the matrix $\mathbf{A}$ assumes the form

$$\mathbf{A} = \begin{bmatrix} \phi_1(t) & \phi_2(t) & \cdots & \phi_N(t) \end{bmatrix}. \tag{1.147}$$

To obtain the SVD representation in this case we first solve the eigenvalue problem for $\mathbf{v}_j$ in (1.117). In view of (1.147) we have

$$\mathbf{A}^H \mathbf{A} = \mathbf{G}, \tag{1.148}$$

where $\mathbf{G}$ is the Gram matrix defined by (1.33). Thus for the continuum (1.117) reads

$$\mathbf{G}\mathbf{v}_j = \sigma_j^2 \mathbf{v}_j \; ; \; j = 1, 2, \ldots R, \tag{1.149}$$

where again $R \leq N$. The (formally) infinite-dimensional vectors $\mathbf{u}_j$ then follow from (1.116a):

$$\mathbf{u}_j \equiv u_j(t) = \frac{1}{\sigma_j} \sum_{\ell=1}^{N} \phi_\ell(t) v_{\ell j} \; ; \; j = 1, 2, \ldots R, \tag{1.150}$$

where $v_{\ell j}$ is the ℓ-th element of $\mathbf{v}_j$. If the chosen $\phi_n(t)$ are strictly linearly independent, then, of course, $R = N$. However, it may turn out that some of the singular values σ_j are very small (indicative of quasi linear dependence). In such cases only R of the functions $u_j(t)$ are retained corresponding to the indices of the R dominant contributors. If the eigenvectors $\mathbf{v}_j$ in (1.149) are normalized to unity, the $u_j(t)$ are automatically orthonormal, just as in the discrete case in (1.120). We can show this directly using (1.150). Thus using the inner product notation for the continuum we have

$$
\begin{aligned}
(u_j, u_k) &= \left(\frac{1}{\sigma_j} \sum_{\ell=1}^{N} \phi_\ell v_{\ell j} \, , \, \frac{1}{\sigma_k} \sum_{p=1}^{N} \phi_p v_{pk} \right) \\
&= \frac{1}{\sigma_j \sigma_k} \sum_{\ell=1}^{N} v_{\ell j}^* \sum_{p=1}^{N} (\phi_\ell, \phi_p) \, v_{pk} = \frac{1}{\sigma_j \sigma_k} \sum_{\ell=1}^{N} v_{\ell j}^* \sum_{p=1}^{N} G_{\ell p} v_{pk} \\
&= \frac{1}{\sigma_j \sigma_k} \sum_{\ell=1}^{N} \sigma_k^2 v_{\ell j}^* v_{\ell k} = \frac{\sigma_k}{\sigma_j} \sum_{\ell=1}^{N} v_{\ell j}^* v_{\ell k} \\
&= \delta_{jk} \, , \, j = 1, 2, \ldots R.
\end{aligned}
\tag{1.151}
$$

The SVD representation of $\mathbf{A}$ follows from (1.122). If we denote the matrix elements of $\mathbf{A}$ by A_{tn} (1.122) is equivalent to

$$
A_{tn} \equiv \phi_n(t) = \sum_{j=1}^{R} \sigma_j u_j(t) v_{nj}^* \, ; \, a \le t \le b, \, 1 \le n \le N.
\tag{1.152}
$$

An alternative form not involving the singular values explicitly follows from a replacement of $u_j(t)$ in the preceding by (1.150):

$$
A_{tn} = \sum_{\ell=1}^{N} \sum_{j=1}^{R} \phi_\ell(t) v_{\ell j} v_{nj}^* \, ; \, a \le t \le b, \, 1 \le n \le N.
\tag{1.153}
$$

The corresponding elements of the Moore–Penrose pseudoinverse are

$$
A_{nt}^{-1} = \sum_{j=1}^{R} \sigma_j^{-1} v_{nj} u_j(t)^* \, ; \, 1 \le n \le N, \, a \le t \le b .
\tag{1.154}
$$

The LMS minimum norm solution to the normal equations (1.77) now reads

$$
\hat{f}_n = \sum_{j=1}^{R} \sigma_j^{-1} v_{nj} (u_j, f).
\tag{1.155}
$$

Replacing u_j in this expression by (1.150) yields the following:

$$
\hat{f}_n = \sum_{\ell=1}^{N} \lambda_{\ell n} (\phi_\ell, f) ,
\tag{1.156}
$$

where

$$
\lambda_{\ell n} = \sum_{j=1}^{R} \sigma_j^{-2} v_{nj} v_{\ell j}^*.
\tag{1.157}
$$

In the special case of N orthonormal expansion functions $R = N$, the σ_j are all unity so that $\lambda_{\ell n} = \delta_{\ell n}$. As expected, in this case (1.156) reduces to (1.41) (with $Q_n = 1$).

Using the coefficients in (1.155)

$$
\begin{aligned}
\sum_{n=1}^{N} \hat{f}_n \phi_n \left(t \right) &= \sum_{n=1}^{N} \left(\sum_{j=1}^{R} \sigma_j^{-1} v_{nj} (u_j, f) \right) \phi_n \left(t \right) \\
&= \sum_{j=1}^{R} (u_j, f) \sum_{n=1}^{N} \sigma_j^{-1} v_{nj} \phi_n \left(t \right) \\
&= \sum_{j=1}^{R} (u_j, f) u_j \left(t \right),
\end{aligned}
$$

wherein the last step we used (1.150). Hence

$$
f\left(t \right) = \sum_{j=1}^{R} \left(u_j, f \right) u_j \left(t \right) + e_N \left(t \right), \tag{1.158}
$$

which is the expansion (1.145) for the continuum, wherein (u_j, f) are the expansion coefficients and $e_N(t)$ the residual error with minimum norm.

The basis set represented by (1.150) is particularly useful when one is faced with the problem of constructing a basis for a set of functions $\phi_n(t)$ $n = 1, 2, 3, \ldots N$ only a small number of which are linearly independent. This situation is found, for example, in digital communications where it is frequently advantageous to use modulation formats involving a set of linearly dependent waveforms. It is usually easy to deduce an appropriate basis for the waveforms for small N (e.g., by using symmetry arguments). For large N (and particularly for $N \gg R$) or whenever symmetry arguments and simple deductions fail, the algorithm embodied in (1.149) and (1.150) provides an alternative and systematic approach.

1.4.5 Frames

We form the sum $\sum_{n=1}^{N} \phi_n(t) \phi_n^*(t')$ and substitute for each $\phi_n(t)$ and $\phi_n^*(t')$ the corresponding expansion from (1.152). Upon taking account of the orthogonality $\sum_{n=1}^{N} v_{nj}^* v_{nk} = \delta_{jk}$ we obtain the following identity

$$
\sum_{n=1}^{N} \phi_n \left(t \right) \phi_n^* \left(t' \right) = \sum_{j=1}^{R} \sigma_j^2 u_j \left(t \right) u_j^* \left(t' \right). \tag{1.159}
$$

Multiplying both sides by $f^*(t)$ and $f(t')$ and integrating with respect to both variables give

$$
\sum_{n=1}^{N} |(f, \phi_n)|^2 = \sum_{j=1}^{R} \sigma_j^2 |(f, u_j)|^2.
$$

With $A = \min\{\sigma_j^2\}$ and $B = \max\{\sigma_j^2\}$; $1 \leq j \leq R$ the preceding implies the following inequality:

$$A \sum_{j=1}^{R} |(f, u_j)|^2 \leq \sum_{n=1}^{N} |(f, \phi_n)|^2 \leq B \sum_{j=1}^{R} |(f, u_j)|^2. \qquad (1.160)$$

Since the (f, u_j) are coefficients in an orthonormal expansion, (1.41) applies with the replacements $Q_n \to 1$, and $\hat{f}_n \to (u_n, f)$ and $\varepsilon_{N\min} \to \varepsilon_{R\min}$. Solving (1.158) for $\sum_{n=1}^{R} |(f, u_n)|^2$ and noting that $\varepsilon_{N\min} \equiv \varepsilon_{R\min}$ (in accordance with the SVD decomposition the minimized LMS error obtained with the N expansions functions ϕ_n is the same as that with the R orthonormal u_n) and substituting into (1.160) we obtain

$$A[(f, f) - \varepsilon_{N\min}] \leq \sum_{n=1}^{N} |(f, \phi_n)|^2 \leq B[(f, f) - \varepsilon_{N\min}]. \qquad (1.161)$$

When the N expansion functions represent $f(t)$ with zero LMS error they are said to constitute a frame. Thus for a frame the preceding inequality reads

$$A(f, f) \leq \sum_{n=1}^{N} |(f, \phi_n)|^2 \leq B(f, f). \qquad (1.162)$$

We should note that thus far no assumptions have been made about the nature of the expansion functions. Let us now assume that they are of unit norm. If they are also orthogonal, then the Gram matrix is an $N \times N$ unit matrix and $A = B = 1$. A frame with $A = B$ is said to be tight. Thus an orthogonal set of linearly independent functions constitutes a tight frame. However tight frames can result even when the functions are linearly dependent. For example, consider the following set of four functions: $\phi_1, \phi_1, \phi_2, \phi_2$ with $(\phi_1, \phi_2) = \rho$. We have $N = 4$ and $R = 2$. By assumption all functions have unit norm so that the Gram matrix is

$$\begin{bmatrix} 1 & 1 & \rho & \rho \\ 1 & 1 & \rho & \rho \\ \rho^* & \rho^* & 1 & 1 \\ \rho^* & \rho^* & 1 & 1 \end{bmatrix}.$$

The two nonzero eigenvalues are found to be $2(1 + |\rho|) \equiv B$ and $2(1 - |\rho|) \equiv A$. We see that whenever ϕ_1 and ϕ_2 are orthogonal the frame becomes tight with $A = B = 2$. We might suspect that this factor of two is indicative of the fact that we have twice as many orthogonal functions at our disposal than the minimum required to span the space of dimension $R = 2$. This is actually the case. To see that this rule applies generally consider the following set of R groups each comprised of K identical functions

$$\phi_1 \cdots \phi_1 \phi_2 \cdots \phi_2 \phi_3 \cdots \phi_3 \cdots \cdots \phi_R \cdots \phi_R,$$

that are assumed be orthonormal. It is not hard to see that the corresponding Gram matrix is comprised entirely of R $K \times K$ matrices of ones along the main

diagonal and that a typical eigenvector corresponding to nonzero eigenvalues
has the structure

$$\mathbf{v}^T = [\ 0 \quad \dots \quad 0 \ 1 \quad \dots \quad 1 \ 0 \quad \dots \quad 0 \ ],$$

wherein the j-th block of K ones lines up with the j-th sub matrix of ones.
As a result the R nonzero eigenvalues are all identical and numerically equal
to K. Thus we have again a tight frame with $A = B = K$, indicating a K-th
order redundancy in our choice of the orthonormal basis. Note that both a tight
frame and the numerical correspondence of the constant A with the redundancy
require orthogonality of the linearly independent subset of the N expansion
functions. Also a tight frame requires that the redundancy be uniformly ap-
portioned among the basis functions. For example, the set $\phi_1, \phi_2, \phi_2, \phi_2, \phi_3, \phi_3$
is not a tight frame because the eigenvalues of the associated Gram matrix are
$1, 3,$ and 2. To transform this set into a frame we could, e.g., supplement it with
two additional ϕ_1 and one ϕ_3 resulting in $A = B = 3$.

1.4.6   Total Least Squares

In the approach to the LMS minimization of $\| \mathbf{Ax} - \mathbf{y} \|^2$ using the SVD there
is a tacit assumption that the $M \times N$ data matrix $\mathbf{A}$ is known exactly. In
practice this is not always the case. In the following we present a modified LMS
technique that takes account of possible uncertainties in the data matrix. The
development follows closely the ideas introduced by Golub and Van Loan [9].

The problem may be formulated as follows. Denoting by $\mathbf{a}_i^H$ the i-th row
of $\mathbf{A}$, and by y_i the i-th element of $\mathbf{y}$ we interpret the minimization of (1.146)
as the problem of finding a vector $\mathbf{x}$ whose projection on $\mathbf{a}_i$ is close to the
prescribed y_i for $i = 1, 2 \ldots M$. We now seek an $N \times 1$ vector $\hat{\mathbf{a}}_i$ such that the
distance between the $1 \times (N + 1)$ vector $\mathbf{c}_i^T$ defined by

$$\mathbf{c}_i^T = [\hat{\mathbf{a}}_i^H \ \hat{\mathbf{a}}_i^H \mathbf{x}] \tag{1.163}$$

and the vector $[\mathbf{a}_i^H \ y_i]$ defined entirely by the data, is minimized for a given i.
This minimization allows for the possibility that both the data matrix $\mathbf{A}$ and
the data vector $\mathbf{y}$ may not be consistently specified (in the sense that y_i may
not be well approximated by $\mathbf{a}_i^H \mathbf{x}$ for *any* $\mathbf{x}$) To find $\hat{\mathbf{a}}_i$ we set

$$\psi_i(\mathbf{x}) = \min \left[\| \ \mathbf{a}_i^H - \hat{\mathbf{a}}_i^H \|^2 + \left| y_i - \hat{\mathbf{a}}_i^H \mathbf{x} \right|^2 \right] \tag{1.164}$$

and equate the variational derivative to zero. Thus

$$\delta \left[\| \ \mathbf{a}_i^H - \hat{\mathbf{a}}_i^H \|^2 + \left| y_i - \hat{\mathbf{a}}_i^H \mathbf{x} \right|^2 \right] = - [\mathbf{a}_i^H - \hat{\mathbf{a}}_i^H] \, \delta \hat{\mathbf{a}}_i - \delta \hat{\mathbf{a}}_i^H [\mathbf{a}_i - \hat{\mathbf{a}}_i]$$
$$- [y_i - \hat{\mathbf{a}}_i^H \mathbf{x}] \, \mathbf{x}^H \delta \hat{\mathbf{a}}_i - \delta \hat{\mathbf{a}}_i^H \mathbf{x} [y_i^* - \mathbf{x}^H \hat{\mathbf{a}}_i] = 0 \tag{1.165}$$

Since the increments $\delta \hat{\mathbf{a}}_i$ and $\delta \hat{\mathbf{a}}_i^H$ are arbitrary, the coefficients of each must
vanish independently. This yields for the coefficients of $\delta \hat{\mathbf{a}}_i^H$

$$\hat{\mathbf{a}}_i - \mathbf{a}_i = \mathbf{x} \left[y_i^* - \mathbf{x}^H \hat{\mathbf{a}}_i \right] \tag{1.166}$$

and a similar (but redundant equation) for the coefficients of $\delta \hat{\mathbf{a}}_i$. Moving the terms containing $\hat{\mathbf{a}}_i$ to the left of the equality sign gives

$$\left[\mathbf{I}_{NN} + \mathbf{x}\mathbf{x}^H\right] \hat{\mathbf{a}}_i = \mathbf{a}_i + y_i^* \mathbf{x} \tag{1.167}$$

One can readily verify that

$$\left[\mathbf{I}_{NN} + \mathbf{x}\mathbf{x}^H\right]^{-1} = \mathbf{I}_{NN} - \mathbf{x}\mathbf{x}^H / (1 + \beta) \tag{1.168}$$

where $\beta = \mathbf{x}^H \mathbf{x}$. With the aid of the above relationship one readily finds

$$\hat{\mathbf{a}}_i = \left[\mathbf{I}_{NN} - \frac{\mathbf{x}\mathbf{x}^H}{1+\beta}\right] \mathbf{a}_i + \frac{y_i^*}{1+\beta}\mathbf{x} \tag{1.169}$$

Next we compute $\psi_i(\mathbf{x})$ by using (1.169) in (1.164). This gives

$$
\begin{aligned}
\|\hat{\mathbf{a}}_i - \mathbf{a}_i\|^2 &= \frac{1}{(1+\beta)^2} \left[-\left(\mathbf{a}_i^H \mathbf{x}\right) \mathbf{x}^H + y_i \mathbf{x}^H\right] \left[-\mathbf{x}\left(\mathbf{x}^H \mathbf{a}_i\right) + y_i^* \mathbf{x}\right] \\
&= \frac{1}{(1+\beta)^2} \left[\begin{array}{c} \left|\mathbf{a}_i^H \mathbf{x}\right|^2 \beta - y_i \beta \left(\mathbf{x}^H \mathbf{a}_i\right) - y_i^* \beta \left(\mathbf{x}^H \mathbf{a}_i\right)^* \\ + \beta |y_i|^2 \end{array}\right] \tag{1.170}
\end{aligned}
$$

$$
\begin{aligned}
\left|y_i - \hat{\mathbf{a}}_i^H \mathbf{x}\right|^2 &= \frac{1}{(1+\beta)^2} \left[y_i^* - \left(\mathbf{x}^H \mathbf{a}_i\right)\right] \left[y_i - \left(\mathbf{x}^H \mathbf{a}_i\right)^*\right] \\
&= \frac{1}{(1+\beta)^2} \left[\begin{array}{c} |y_i|^2 - y_i \left(\mathbf{x}^H \mathbf{a}_i\right) - y_i^* \left(\mathbf{x}^H \mathbf{a}_i\right)^* \\ + \left|\mathbf{a}_i^H \mathbf{x}\right|^2 \end{array}\right]. \tag{1.171}
\end{aligned}
$$

Adding (1.170) and (1.171) we obtain

$$\psi_i(\mathbf{x}) = \frac{\left|\mathbf{a}_i^H \mathbf{x} - y_i\right|^2}{1+\beta}. \tag{1.172}$$

The vector $\mathbf{x}$ is yet to be determined. It is the vector that minimizes $\sum_{i=1}^{M} \psi_i(\mathbf{x})$, or, equivalently

$$\alpha = \min_{\mathbf{x}} \left\{ \frac{\|\mathbf{A}\mathbf{x} - \mathbf{y}\|^2}{1 + \mathbf{x}^H \mathbf{x}} \right\}. \tag{1.173}$$

Taking the variational derivative with respect to $\mathbf{x}$ of the quantity in brackets and setting it to zero we have

$$\frac{[\mathbf{A}\mathbf{x} - \mathbf{y}]^H \mathbf{A}\delta\mathbf{x} - \delta\mathbf{x}^H \mathbf{A}^H [\mathbf{A}\mathbf{x} - \mathbf{y}]}{1 + \mathbf{x}^H \mathbf{x}} - \frac{\|\mathbf{A}\mathbf{x} - \mathbf{y}\|^2 \left[\delta\mathbf{x}^H \mathbf{x} + \mathbf{x}^H \delta\mathbf{x}\right]}{[1 + \mathbf{x}^H \mathbf{x}]^2} = 0$$

Equating the coefficients of $\delta \mathbf{x}$ to zero yields the following:

$$[\mathbf{A}\mathbf{x} - \mathbf{y}]^H \left[\mathbf{A}(1+\beta) - (\mathbf{A}\mathbf{x} - \mathbf{y})\mathbf{x}^H\right] = 0. \tag{1.174}$$

By taking the Hermitian transpose of both sides we get the equivalent form

$$\left[\mathbf{A}^H (1+\beta) - \mathbf{x}\left(\mathbf{x}^H \mathbf{A}^H - \mathbf{y}^H\right)\right] [\mathbf{A}\mathbf{x} - \mathbf{y}] = 0. \tag{1.175}$$

Scalar multiplication of both sides from the left by $\mathbf{x}^H$ gives rise to $\mathbf{x}^H \mathbf{A}^H \beta$ as well as its negative so that these terms cancel and the resulting expression simplifies to

$$\left[\mathbf{x}^H \mathbf{A}^H + \beta \mathbf{y}^H \right] \left[\mathbf{A}\mathbf{x} - \mathbf{y} \right] = \mathbf{0}. \tag{1.176}$$

Taking the Hermitian transpose of this expression results in

$$\left[\mathbf{A}\mathbf{x} - \mathbf{y} \right]^H \left[\mathbf{A}\mathbf{x} + \beta \mathbf{y} \right] = \mathbf{0}. \tag{1.177}$$

We will now use the last result to modify (1.174) so that it can be solved for $\mathbf{x}$. For this purpose we write the term $\mathbf{A}\mathbf{x} - \mathbf{y}$ appearing within the square bracket in (1.174) as follows:

$$\mathbf{A}\mathbf{x} - \mathbf{y} = \mathbf{A}\mathbf{x} + \beta \mathbf{y} - (1 + \beta)\mathbf{y}. \tag{1.178}$$

This modifies (1.174) to read

$$\left[\mathbf{A}\mathbf{x} - \mathbf{y} \right]^H \left[\mathbf{A}\,(1 + \beta) - (\mathbf{A}\mathbf{x} + \beta \mathbf{y})\,\mathbf{x}^H + (1 + \beta)\mathbf{y}\mathbf{x}^H \right] = \mathbf{0}. \tag{1.179}$$

By virtue of (1.177) multiplication of the middle term in brackets by $\left[\mathbf{A}\mathbf{x} - \mathbf{y} \right]^H$ gives zero and a subsequent division by $1 + \beta$ reduces (1.179) to

$$\left[\mathbf{A}\mathbf{x} - \mathbf{y} \right]^H \left[\mathbf{A} + \mathbf{y}\mathbf{x}^H \right] = \mathbf{0}. \tag{1.180}$$

We propose to solve the preceding equation for $\mathbf{x}$ using the SVD. For this purpose we define an $M \times (N + 1)$ augmented data matrix by adjoining the data vector $\mathbf{y}$ to the N columns of $\mathbf{A}$ as follows:

$$\mathbf{C} \equiv [\mathbf{A}\ \mathbf{y}]. \tag{1.181}$$

Equation (1.180) can then be written in the following partitioned form:

$$\left[\ \mathbf{x}^H \quad -1\ \right] \mathbf{C}^H \mathbf{C} \left[\begin{array}{c} \mathbf{I}_{NN} \\ \mathbf{x}^H \end{array} \right] = \mathbf{0}. \tag{1.182}$$

Using the SVD decomposition

$$\mathbf{C} = \mathbf{U}\mathbf{\Sigma}\mathbf{V}^H,$$

where $\mathbf{U}$ and $\mathbf{V}$ have, respectively, dimensions $M \times (N+1)$ and $(N+1) \times (N+1)$, we have

$$\mathbf{C}^H \mathbf{C} = \mathbf{V}\mathbf{\Sigma}\mathbf{V}^H. \tag{1.183}$$

To facilitate subsequent manipulations we partition $\mathbf{V}$ and $\mathbf{\Sigma}$ as follows:

$$\mathbf{V} = \left[\begin{array}{cc} \mathbf{V}_{11} & \mathbf{v}_{12} \\ \mathbf{v}_{21} & v_{22} \end{array} \right], \quad \mathbf{\Sigma} = \left[\begin{array}{cc} \mathbf{\Sigma}_1 & 0 \\ 0 & \sigma_{22} \end{array} \right] \tag{1.184}$$

where $\mathbf{V}_{11}, \mathbf{v}_{12}, \mathbf{v}_{21}$ are, respectively, $N \times N$, $N \times 1, 1 \times N$, v_{22} and σ_{22} are scalars, and

$$\mathbf{\Sigma}_1 = diag\left[\sigma_1\ \sigma_2\ \sigma_3\ \ldots \sigma_N \right]. \tag{1.185}$$

Since $\mathbf{V}$ is unitary,

$$\begin{bmatrix} \mathbf{V}_{11}^H & \mathbf{v}_{21}^H \\ \mathbf{v}_{12}^H & v_{22}^* \end{bmatrix} \begin{bmatrix} \mathbf{V}_{11} & \mathbf{v}_{12} \\ \mathbf{v}_{21} & v_{22} \end{bmatrix} = \begin{bmatrix} \mathbf{I}_{NN} & \mathbf{0} \\ \mathbf{0} & 1 \end{bmatrix} \tag{1.186}$$

which provides the following constraints:

$$\mathbf{V}_{11}^H \mathbf{V}_{11} + \mathbf{v}_{21}^H \mathbf{v}_{21} = \mathbf{I}_{NN}, \tag{1.187a}$$

$$\mathbf{v}_{12}^H \mathbf{V}_{11} + v_{22}^* \mathbf{v}_{21} = \mathbf{0}, \tag{1.187b}$$

$$\mathbf{V}_{11}^H \mathbf{v}_{12} + \mathbf{v}_{21}^H v_{22} = \mathbf{0}, \tag{1.187c}$$

$$\mathbf{v}_{12}^H \mathbf{v}_{12} + v_{22}^* v_{22} = 1. \tag{1.187d}$$

Substituting (1.184) into (1.183) and then into (1.182) we obtain the following:

$$\left[\mathbf{x}^H \mathbf{V}_{11} - \mathbf{v}_{12}\right] \mathbf{\Sigma}_1 \left[\mathbf{V}_{11}^H + \mathbf{v}_{21}^H \mathbf{x}^H\right] + \sigma_{22} \left[\mathbf{x}^H \mathbf{v}_{12} - v_{22}\right] \left[\mathbf{v}_{12}^H + v_{22}^* \mathbf{x}^H\right] = \mathbf{0}. \tag{1.188}$$

Let us assume that $v_{22} \neq 0$. Then (1.188) is satisfied by

$$\mathbf{x} \equiv \mathbf{x}_{TLS} = -\frac{\mathbf{v}_{12}}{v_{22}} \tag{1.189}$$

since this reduces the last factor in (1.188) to zero while the first factor on the left gives

$$\mathbf{x}^H \mathbf{V}_{11} - \mathbf{v}_{12} = -\frac{\mathbf{v}_{12}^H \mathbf{V}_{11}}{v_{22}^*} - \mathbf{v}_{12} = -\frac{1}{v_{22}^*} \left[\mathbf{V}_{11}^H \mathbf{v}_{12} + \mathbf{v}_{21}^H v_{22} \right],$$

which vanishes by virtue of (1.187c).

Equation (1.189) is the total least squares solution of Golub and Van Loan. When $v_{22} = 0$, the TLS solution does not exist. When k singular values are identical (say the last k) then, because the SVD of $\mathbf{C}$ would in that case not be affected by a permutation of the corresponding columns of $\mathbf{V}$, any of the last k columns of $\mathbf{V}$ can serve as the vector $\left[\mathbf{v}_{12}^T \quad v_{22}\right]$ for constructing the TLS solution. In such cases, just as was done in the LMS solution, a reasonable choice is a vector with the smallest norm. Because the matrix is unitary it is clear that this prescription is equivalent to choosing that column of $\mathbf{V}$ for which

$$\max_{N-k+2 \leq \ell \leq N+1} |V(N+1, \ell)| \tag{1.190}$$

The residual error α in (1.173) when written in terms of the SVD of $\mathbf{C}$ reads

$$\alpha = \frac{\mathbf{w}^H \mathbf{V} \mathbf{\Sigma} \mathbf{V}^H \mathbf{w}}{\mathbf{w}^H \mathbf{w}}, \tag{1.191}$$

where $\mathbf{w} = [\mathbf{x}_{TLS} \quad -1]$. Using (1.189) and (1.187d) we obtain

$$\mathbf{w}^H \mathbf{w} = 1/|v_{22}|^2. \tag{1.192}$$

Also

$$\mathbf{V}^H\mathbf{w} = \begin{bmatrix} \mathbf{V}_{11}^H & \mathbf{v}_{21}^H \\ \mathbf{v}_{12}^H & v_{22}^* \end{bmatrix} \begin{bmatrix} \mathbf{x}_{TLS} \\ -1 \end{bmatrix}$$

$$= \begin{bmatrix} -\mathbf{V}_{11}^H\mathbf{v}_{12}/v_{22} & -\mathbf{v}_{21}^H \\ -\mathbf{v}_{12}^H\mathbf{v}_{12}/v_{22} & -v_{22}^* \end{bmatrix} = \begin{bmatrix} \mathbf{O} \\ -1/v_{22} \end{bmatrix} \qquad (1.193)$$

where we have made use of (1.187b) and (1.187d). Taking account of the definition of $\boldsymbol{\Sigma}$ in (1.184), (1.191) reduces to

$$\alpha = \sigma_{22}. \qquad (1.194)$$

Let us suppose that $\mathbf{A}$ has full rank N (and $M > N$). If the rank of $\mathbf{C}$ is also N, then $\mathbf{y}$ is in the subspace spanned by the columns of $\mathbf{A}$ and the system $\mathbf{Ax} = \mathbf{y}$ has a unique solution, which must be identical with the LMS solution. Since the ranks of $\mathbf{C}$ and $\mathbf{A}$ are identical, $\sigma_{22} = 0$ so that the TLS solution is also the same. A different situation arises when the rank of $\mathbf{C}$ is $N+1$ so that $\sigma_{22} \neq 0$. Then $\mathbf{y}$ is not in the subspace spanned by the columns of $\mathbf{A}$ and $\mathbf{Ax} = \mathbf{y}$ has no solutions. However, as long as there is a nonzero projection of the vector $\mathbf{y}$ into the range of $\mathbf{A}$ there exists a nontrivial LMS solution. The TLS solution also exists and is distinct from the LMS solution. Clearly, the two solutions must approach each other when $\sigma_{22} \to 0$.

1.4.7   Tikhonov Regularization

We have shown that when the rank an MXN (M>N) matrix $\mathbf{A}$ is less than N, the solution of the normal equations (1.80) for $\mathbf{x}$ can be obtained using the SVD. An alternative approach is the so-called Tikhonov regularization [24]. The corresponding LMS problem is formulated as follows: instead of seeking an $\mathbf{x}$ that minimizes $\varepsilon = \|\mathbf{Ax} - \mathbf{y}\|^2$ one seeks instead an $\mathbf{x}$ that minimizes

$$\varepsilon = \|\mathbf{Ax} - \mathbf{y}\|^2 + \alpha^2 \|\mathbf{x}\|^2 \qquad (1.195)$$

for a fixed real α (Tikhonov regularization parameter). Using the variational approach (as, e.g., in the preceding subsection) one finds that the required $\mathbf{x}$ must satisfy

$$\left(\mathbf{A}^H\mathbf{A} + \alpha^2\mathbf{I}_N\right)\mathbf{x} = \mathbf{A}^H\mathbf{y}, \qquad (1.196)$$

where $\mathbf{I}_N$ is an NXN unit matrix. Evidently the inverse in (1.196) exists even for a singular $\mathbf{A}^H\mathbf{A}$ provided $\alpha \neq 0$ so that

$$\mathbf{x} = \left(\mathbf{A}^H\mathbf{A} + \alpha^2\mathbf{I}_N\right)^{-1}\mathbf{A}^H\mathbf{y}. \qquad (1.197)$$

Let us now represent $\mathbf{A}$ in terms of its SVD. Thus with $rank(\mathbf{A}) = R \leq N$, we set

$$\mathbf{A} = \mathbf{U}_R\boldsymbol{\Sigma}_R\mathbf{V}_R^H$$

and substitute in (1.196) to obtain

$$\left(\mathbf{V}_R\boldsymbol{\Sigma}_R^2\mathbf{V}_R^H + \alpha^2\mathbf{I}_N\right)\mathbf{x} = \mathbf{V}_R\boldsymbol{\Sigma}_R\mathbf{U}_R^H\mathbf{y} \qquad (1.198)$$

To solve (1.198) for $\mathbf{x}$ we introduce a unitary NXN matrix $\mathbf{V}$ in which $\mathbf{V}_R$ forms the first R columns. Clearly (1.198) is then equivalent to

$$\mathbf{V} \begin{bmatrix} \boldsymbol{\Sigma}_R^2 + \alpha^2 \mathbf{I}_R & 0 \\ 0 & \alpha^2 \mathbf{I}_{N-R} \end{bmatrix} \mathbf{V}^H \mathbf{x} = \mathbf{V} \begin{bmatrix} \boldsymbol{\Sigma}_R \\ 0 \end{bmatrix} \mathbf{U}_R^H \mathbf{y}, \qquad (1.199)$$

where $\mathbf{I}_R$ and $\mathbf{I}_{N-R}$ are, respectively, RXR and $(N-R)X(N-R)$ unit matrices. Multiplying (1.199) from the left first by $\mathbf{V}^H$, then by the inverse of the diagonal matrix, and finally by $\mathbf{V}$, yields

$$\mathbf{x} = \mathbf{V}_R \begin{bmatrix} \frac{\sigma_1}{\sigma_1^2 + \alpha^2} & 0 & 0 \\ 0 & . & 0 \\ 0 & 0 & \frac{\sigma_R}{\sigma_R^2 + \alpha^2} \end{bmatrix} \mathbf{U}_R^H \mathbf{y}. \qquad (1.200)$$

Note that with $\alpha = 0$ we obtain the usual SVD solution. As we know, for sufficiently small singular values (large matrix condition numbers) such a solution could be noisy. As we have seen, the recommended remedy in such cases is to trim the matrix, i.e., eliminate sufficiently small singular values and thus reduce R. From (1.200) we observe that the Tikhonov approach provides an alternative: instead of eliminating the troublesome small singular values, one can introduce a suitable value of α to reduce their influence. This is equivalent to constructing a new matrix with a smaller condition number. Thus, if the condition number of $\mathbf{A}$ is $1/\sigma_{\min}$, the condition number of the new matrix becomes $\sigma_{\min} / (\sigma_{\min}^2 + \alpha^2)$.

1.5   Finite Sets of Orthogonal Functions

1.5.1   LMS and Orthogonal Functions

When the expansion functions are orthogonal the Gram matrix in (1.33) is diagonal so that the expansion coefficients are given by

$$\hat{f}_n = Q_n^{-1} (\phi_n, f) \; ; \; n = 1, 2, 3, \ldots N, \qquad (1.201)$$

where Q_n is the normalization factor in (1.25). We note that once the coefficient $\hat{f}_n$ has been determined for some n, say $n = k$, it remains final, i.e., it does not need to be updated no matter how many more coefficients one decides to calculate for $n > k$. This is definitely not the case for a nonorthogonal system. Thus, suppose we decided to increase our function set from N to $N+1$. Then for a nonorthogonal system all the coefficients have to be updated, but not so for an orthogonal system. This property has been referred to that of "finality," and is closely related to what mathematicians refer to as completeness, to be discussed in 1.7.1. Upon substituting (1.201) in (1.78) we obtain

$$\varepsilon_{N\,\min} = (f, f) - \sum_{n=1}^{N} Q_n \left| \hat{f}_n \right|^2, \qquad (1.202a)$$

and since $\varepsilon_{N\min} \geq 0$ we have

$$(f, f) \geq \sum_{n=1}^{N} Q_n \left| \hat{f}_n \right|^2, \tag{1.202b}$$

which is usually referred to as the Bessel inequality.[7]

1.5.2 Trigonometric Functions

Perhaps the best known orthogonal function set is comprised of the trigonometric functions $\cos(nt)$ and $\sin(nt); n = 0, 1, 2, \ldots N$. The LMS approximation to a signal $f(t)$ in the interval $0 < t \leq 2\pi$ in terms of these functions reads

$$f(t) \sim \hat{f}_0 + \sum_{n=1}^{N} \hat{f}_n^{(e)} \cos(nt) + \sum_{n=1}^{N} \hat{f}_n^{(o)} \sin(nt). \tag{1.203}$$

Using the notation

$$\phi_n^{(e)}(t) = \cos(nt), \tag{1.204a}$$
$$\phi_n^{(o)}(t) = \sin(nt), \tag{1.204b}$$

for $n \geq 1$, we have over the interval $0 \leq t \leq 2\pi$

$$\left(\phi_n^{(e)}, \phi_m^{(e)} \right) = \pi \delta_{nm}, \tag{1.205a}$$
$$\left(\phi_n^{(o)}, \phi_m^{(o)} \right) = \pi \delta_{nm}, \tag{1.205b}$$
$$\left(\phi_n^{(e)}, \phi_m^{(o)} \right) = 0, \tag{1.205c}$$

as is easily demonstrated by direct integration. Thus the $2N$ functions (1.204) together with the constant 1 comprise an orthogonal set of $2N + 1$ functions within the specified interval. In view of (1.201) we have $Q_n = \pi$ for $n > 1$ while $Q_0 = 2\pi$ so that the expansion coefficients are

$$\hat{f}_0 = (2\pi)^{-1} (1, f), \tag{1.206a}$$
$$\hat{f}_n^{(e)} = (\pi)^{-1} \left(\phi_n^{(e)}, f \right), \tag{1.206b}$$
$$\hat{f}_n^{(o)} = (\pi)^{-1} \left(\phi_n^{(o)}, f \right). \tag{1.206c}$$

Since the sinusoids are periodic with period 2π, the orthogonality relationships in (1.205) hold over any interval of 2π duration. For example,

$$\left(\phi_n^{(e)}, \phi_m^{(e)} \right) \equiv \int_{\tau}^{2\pi+\tau} \cos(nt) \cos(mt) \, dt = \pi \delta_{nm}, \tag{1.207}$$

---

[7]Formulas (1.201) and (1.202) assume a more esthetically pleasing form if we assume an orthonormal set for then $Q_n = 1$. Even though this can always be realized by simply dividing each expansion function by $\sqrt{Q_n}$, it is not customary in applied problems and we shall generally honor this custom.

wherein τ is an arbitrary reference value. Thus if a function $f(t)$ were to be specified over the interval $\tau < t \le 2\pi + \tau$, the expansion coefficients (1.206) would be given by inner products over this interval. For example,

$$\hat{f}_n^{(e)} = (\pi)^{-1}\left(\phi_n^{(e)}, f\right) = (\pi)^{-1}\int_{\tau}^{2\pi+\tau} \cos{(nt)}\, f\,(t)\, dt. \tag{1.208}$$

We note that the right side of (1.203) is always periodic over any 2π interval. This does not imply that $f(t)$ itself need to be periodic since the coefficients in this expansion have been chosen to approximate the given function only within the specified finite interval; the behavior of the sum outside this interval need not bear any relation to the function $f(t)$. Thus the fact that the approximation to the function yields a periodic extension outside the specified interval is to be taken as a purely algebraic property of sinusoids. Of course, in special cases it may turn out that the specified function is itself periodic over 2π. Under such special circumstances the periodic extension of the approximation would also approximate the given function and (1.205) would hold for $|t| < \infty$.

The restriction to the period of length 2π is readily removed. For example, suppose we choose to approximate $f(t)$ within the interval $-T/2 < t \le T/2$. Then the expansion functions (1.206) get modified by a replacement of the argument t by $2\pi/T$, i.e.,

$$\phi_n^{(e)}(t) = \cos(2\pi nt/T), \tag{1.209a}$$
$$\phi_n^{(o)}(t) = \sin(2\pi nt/T), \tag{1.209b}$$

while the expansion coefficients are given by

$$\hat{f}_0 = (T)^{-1}(1, f), \tag{1.210a}$$
$$\hat{f}_n^{(e)} = 2\,(T)^{-1}\left(\phi_n^{(e)}, f\right), \tag{1.210b}$$
$$\hat{f}_n^{(o)} = 2\,(T)^{-1}\left(\phi_n^{(o)}, f\right). \tag{1.210c}$$

In applications it is frequently more convenient to employ instead of (1.203) an expansion in terms of the complex exponential functions

$$\phi_n(t) = \cos(2\pi nt/T) + i\sin(2\pi nt/T) = e^{i2\pi nt/T}. \tag{1.211}$$

The LMS approximation to $f(t)$ then assumes the symmetrical form

$$f(t) \sim \sum_{n=-N}^{n=N} \hat{f}_n\phi_n(t). \tag{1.212}$$

The three orthogonality statements (1.205) can now be merged into the single relationship

$$(\phi_n, \phi_m) \equiv \int_{-T/2}^{T/2} \phi_n^*(t)\,\phi_m(t)\, dt = T\delta_{nm}, \tag{1.213}$$

and the expansion coefficients in (1.212) become

$$\hat{f}_n = (T)^{-1} (\phi_n, f).$$ (1.214)

The Bessel inequality in (1.202b) in the present case reads

$$(f, f) = \int_{-T/2}^{T/2} |f(t)|^2 \, dt \geq T \sum_{n=-N}^{n=N} \left| \hat{f}_n \right|^2.$$ (1.215)

As will be shown in Sect. 1.2.1, for piecewise differentiable functions the LMS error can always be made to approach zero as $N \to \infty$ in which case the inequality in (1.215) becomes an equality so that

$$(f, f) = \int_{-T/2}^{T/2} |f(t)|^2 \, dt = T \sum_{n=-\infty}^{n=\infty} \left| \hat{f}_n \right|^2.$$ (1.216)

This relationship is usually referred to as Parseval theorem. The corresponding infinite series is the Fourier series. Its convergence properties we shall study in detail in Chap. 2.

1.5.3   Orthogonal Polynomials [1]

As an additional illustration of LMS approximation of signals by orthogonal functions we shall consider orthogonal polynomials. Although they do not play as prominent a role in signal analysis as trigonometric functions they nevertheless furnish excellent concrete illustrations of the general principles discussed in the preceding. Orthogonal polynomials are of importance in a variety of areas of applied mathematics (for example in constructing algorithms for accurate numerical integration) and are the subject of an extensive technical literature. In the following we shall limit ourselves to extremely brief accounts of only three types of orthogonal polynomials: those associated with the names of Legendre, Laguerre, and Hermit.

Legendre Polynomials

Legendre Polynomials are usually introduced in connection with the defining (Legendre) differential equation. An alternative, and for present purposes, more suitable approach is to view them simply as polynomials that have been constrained to form an orthogonal set in the interval $-1 \leq t \leq 1$. Using the Gram–Schmidt orthogonalization procedure in 1.2.5 the linearly independent set

$$1, t, t^2, \ldots t^{N-1}$$ (1.217)

can be transformed into the set of polynomials

$$P_n(t) = \sum_{\ell=0}^{\hat{n}} \frac{(-1)^\ell [2(n-\ell)]!}{2^n \ell! (n-\ell)! (n-2\ell)!} t^{n-2\ell}, \quad n = 0, 1, 2, \ldots N-1,$$ (1.218)

where

$$\hat{n} = \begin{cases} n/2 \; ; \; n \text{ even,} \\ (n-1)/2 \; ; \; n \text{ odd,} \end{cases}$$

with the orthogonality relationship

$$(P_n, P_m) \equiv \int_{-1}^{1} P_n(t) P_m(t) \, dt = \frac{2}{2n+1} \delta_{nm}. \tag{1.219}$$

Thus the normalization factor in (1.26) is in this case

$$Q_n = 2/(2n+1), \tag{1.220}$$

so that the coefficients in the LMS approximation of $f(t)$ by the first N Legendre Polynomials are

$$\hat{f}_n = Q_n^{-1}(P_n, f). \tag{1.221}$$

The LMS approximation is then

$$f(t) \sim \sum_{n=0}^{N-1} \hat{f}_n P_n(t). \tag{1.222}$$

The Bessel inequality in (1.202b) now reads

$$(f, f) \equiv \int_{-1}^{1} |f(t)|^2 \, dt \geq \sum_{n=0}^{N-1} Q_n \left| \hat{f} \right|^2. \tag{1.223}$$

Again as in the case of trigonometric functions the limiting form $N \to \infty$ leads to a zero LMS error for a large class of functions, in particular for functions that are piecewise differentiable within the expansion interval. Equation (1.223) is then satisfied with equality. This relationship may be referred to as Parseval's theorem by analogy with Fourier Series.

It should be noted that there is no need to restrict the expansion interval to $-1 \leq t \leq 1$ for the simple substitution

$$t = \frac{2}{b-a}\left[t' - \frac{b+a}{2} \right]$$

transforms this interval into the interval $a \leq t' \leq b$ in terms of the new variable t'. Thus the expansion (1.222) can be adjusted to apply to any finite interval.

There are many different series and integral representations for Legendre Polynomials, the utility of which lies primarily in rather specialized analytical investigations. Here we limit ourselves to one additional relationship, the so-called Rodrigue's formula, which is

$$P_n(t) = \frac{1}{2^n n!} \frac{d^n}{dt^n} \left(t^2 - 1 \right)^n. \tag{1.224}$$

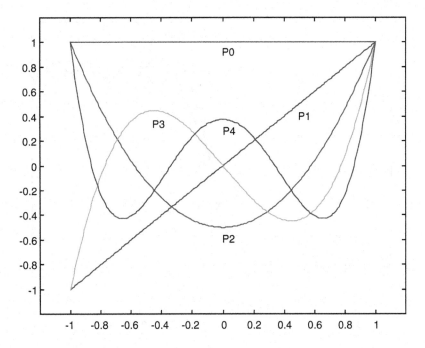

Figure 1.10: Plots of the first five Legendre polynomials

This recursion formula is frequently taken as the basic definition of Legendre Polynomials and represents a convenient way of generating these functions provided n is not too large. For example, the first five of these polynomials are

$$P_0(t) = 1, \tag{1.225a}$$

$$P_1(t) = t, \tag{1.225b}$$

$$P_2(t) = \frac{3}{2}t^2 - \frac{1}{2}, \tag{1.225c}$$

$$P_3(t) = \frac{5}{2}t^3 - \frac{3}{2}t, \tag{1.225d}$$

$$P_4(t) = \frac{35}{8}t^4 - \frac{30}{8}t^2 + \frac{3}{8}. \tag{1.225e}$$

Plots of (1.225) are shown in Fig. 1.10.

As an example suppose we obtain the LMS approximation to the function $3/2 + sign(t)$ by a finite number of Legendre polynomials. Figure 1.11 shows the manner in which the partial sum approximates the given function as the number of (odd) Legendre Polynomials is increased from 4 to 6 and to 11. We note that the approximation at $t = 0$ yields in all cases the arithmetic mean of the function at the discontinuity, i.e.,

$$\frac{f(0^+) + f(0^-)}{2} = 3/2.$$

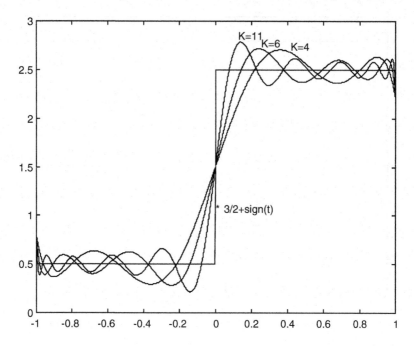

Figure 1.11:   LMS approximation of a step discontinuity by Legendre Polynomials

We shall show later that this is a general property of LMS convergence at a step discontinuity: the partial sums always give the arithmetic mean of the function values at the right and the left side of the discontinuity. In addition, we note that as the number of expansion functions is increased, the oscillatory overshoot does not appear to diminish. We shall meet this behavior again in connection with the study of Fourier series and Fourier integrals, where it is usually referred to as the Gibbs Phenomenon.

Laguerre Polynomials

Suppose we attempt to approximate the function $f(t)$ with an orthogonal set of functions over a semi-infinite interval. It would appear that these expansion functions cannot be polynomials, since the square of a polynomial is not integrable over a semi-infinite interval. Polynomials can still be used, if one is willing to introduce a suitable weighting factor into the definition of the inner product, as described in 1.2.5, to render the inner product convergent. Thus suppose $L_n(t)$ is an n-th order polynomial and we choose e^{-t} as the weighting factor. Clearly the norm defined by

$$\|L_n(t)\| = \sqrt{\int_0^\infty e^{-t}L_n^2(t)\,dt} \qquad (1.226)$$

converges irrespective of the nature of the polynomial.   Polynomials $L_n(t)$ orthogonal over $(0, \infty)$ with an exponential weighting factor, viz.,

$$\int_0^\infty e^{-t} L_n(t) L_m(t) dt = Q_n \delta_{nm},$$ (1.227)

are called Laguerre polynomials. With the Gram–Schmidt orthogonalization such polynomials can be generated from the set $1, t, t^2, \ldots t^{N-1}$. The fundamental definitions are as follows:

$$L_n(t) = \sum_{\ell=0}^n \frac{n!(-1)^\ell}{\ell!} \binom{n}{\ell} t^\ell$$ (1.228)

and

$$Q_n = (n!)^2.$$ (1.229)

The equivalent of the Rodriguez formula for Laguerre polynomials reads

$$L_n(t) = e^t \frac{d^n}{dt^n} \left(t^n e^{-t} \right).$$ (1.230)

Expansions in Laguerre polynomials follow from the normal equations with the inner product as in (1.82) with $a = 0, b = \infty, \phi_n = L_n$, and $\psi = e^{-t/2}$. Using the orthogonality properties (1.227) the expansion coefficients are given by

$$\hat{f}_n = Q_n^{-1} \left(e^{-t} L_n, f \right)$$ (1.231)

and the expansion (1.83) reads

$$f(t) = \sum_{n=0}^{N-1} \hat{f}_n L_n(t) + e_{\psi, N-1} e^{t/2}.$$ (1.232)

This expression suggests that a reasonable alternative definition of the error might be $e_{\psi, N-1} e^{t/2}$. However then

$$\int_0^\infty \left| e_{\psi, N-1} e^{t/2} \right|^2 dt = \int_0^\infty \left| f(t) - f^{(N)}(t) \right|^2 dt,$$ (1.233)

so that $e_{\psi, N-1}$ must decrease at infinity at least exponentially for the integral just to converge. Indeed, the square of this renormalized MS error is always greater or equal to

$$\| e_{\psi, N-1} \|^2 = \int_0^\infty e^{-t} \left| f(t) - f^{(N)}(t) \right|^2 dt,$$ (1.234)

which is just the definition of the norm employed in minimization process leading to the normal equations. The Bessel inequality for these expansion functions reads:

$$\int_0^\infty e^{-t} |f(t)|^2 dt \geq \sum_{n=0}^{N-1} Q_n \left| \hat{f}_n \right|^2.$$ (1.235)

Hermite Polynomials

When a function is to be approximated over an infinite time interval $-\infty < t < \infty$ by polynomials a reasonable weighting factor is $\psi = e^{-t^2/2}$. The polynomials that yield orthogonality in this case are called Hermite polynomials and are defined by

$$H_n\left(t\right) = \left(-1\right)^n e^{t^2} \frac{d^n}{dt^n} e^{-t^2}. \tag{1.236}$$

One can show that

$$\int_{-\infty}^{\infty} e^{-t^2} H_n\left(t\right) H_m(t) dt = Q_n \delta_{nm}, \tag{1.237}$$

with

$$Q_n = 2^n n! \sqrt{\pi}. \tag{1.238}$$

In view of the orthogonality properties (1.237) the expansion coefficients are

$$\hat{f}_n = Q_n^{-1}\left(e^{-t^2} H_n, f\right).$$

1.6   Singularity Functions

1.6.1   The Delta Function

Definition of the Delta Function

In dealing with finite-dimensional vectors the Kronecker symbol δ_{nm}, (1.27), is frequently employed to designate a unit matrix. In that case the n-th component of any M-dimensional vector $\mathbf{x}$ can be represented by

$$x_n = \sum_{m=1}^{M} \delta_{nm} x_m. \tag{1.239}$$

What we now seek is an analogue of (1.239) when the vector comprises a nondenumerably infinite set of components, i.e., we wish to define an identity transformation in function space. Since the sum then becomes an integral, this desideratum is equivalent to requiring the existence of a function $\delta\left(t\right)$ with the property

$$f\left(t\right) = \int_a^b \delta\left(t - t'\right) f\left(t'\right) dt'. \tag{1.240}$$

In order to serve as an identity transformation, this expression should hold for all functions of interest to us that are defined over the interval $a < t < b$. Unfortunately it is not possible to reconcile the required properties of $\delta\left(t\right)$ with the usual definition of a function and any known theory of integration for which (1.240) could be made meaningful. In other words, if we restrict ourselves to conventional definitions of a function, an identity transformation of the kind we seek does not exist. A rigorous mathematical theory to justify (1.240) is

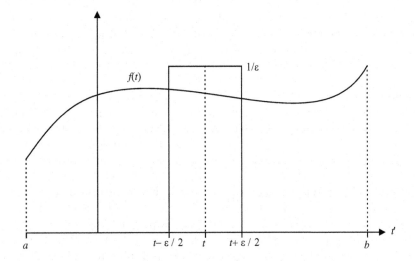

Figure 1.12: Geometrical relationship of $f(t)$ and the pulse function

available in terms of the so-called theory of distributions. Even a brief discussion of this theory would take us too far afield. Instead, we shall follow the more conventional approach and justify the notation (1.240) in terms of a limiting process. For this purpose consider the pulse function defined by

$$K_\epsilon (t) = \left\{ \begin{array}{l} \frac{1}{\epsilon}; |t| \leq \epsilon/2, \\ 0; |t| > \epsilon/2. \end{array} \right. \tag{1.241}$$

Let us next choose an $f(t)$ that is continuous and differentiable everywhere in $a < t < b$ and compute the integral

$$I_\epsilon (t) = \int_a^b K_\epsilon (t - t') f(t') \, dt'. \tag{1.242}$$

It is not hard to see that this is equivalent to

$$I_\epsilon (t) = \frac{1}{\epsilon} \int_{t-\epsilon/2}^{t+\epsilon/2} f(t') \, dt' \tag{1.243}$$

for $a + \epsilon/2 < t < b - \epsilon/2$. Thus the integral is just the sliding arithmetic average of $f(t)$ over a continuum of intervals of length ϵ and may be given the geometrical interpretation as shown in Fig. 1.12.

Clearly, for sufficiently small ϵ the $f(t')$ within the interval encompassed by the pulse function can be approximated by $f(t)$ as closely as desired. Hence formally we can write

$$f(t) = \lim_{\epsilon \to 0} \int_a^b K_\epsilon (t - t') f(t') \, dt'. \tag{1.244}$$

Thus, instead of (1.240), we have established an identity transformation which entails a limiting process. It is tempting to pull the limiting process into the integrand of (1.244) and establish a formal identity with (1.244) by defining

$$\delta\left(t - t'\right) = \lim_{\epsilon \to 0} K_\epsilon\left(t - t'\right). \tag{1.245}$$

a definition which does not actually represent a mathematically valid limiting process. Nevertheless, because of notational convenience and custom we shall use it and similar expressions to "define" the Dirac delta function $\delta\left(t\right)$ but only with the firm understanding that this is merely a mnemonic rule for a limiting process such as (1.244). Of course we could, as is sometimes done, consider (1.245) as representing a "function" which is everywhere zero with the exception of a single point where it becomes infinite. Such a definition is difficult to apply consistently in many physical situations where the delta function provides a very useful shorthand notational tool. Although we may still think physically of the delta function as a sort of tall spike, it is analytically more fruitful to think in terms of a limiting process such as (1.244), especially since the physical significance of a delta function almost invariably manifests itself under an integral sign so that there is no need to ask for the actual value taken on by the function.

The pulse function is not the only way in which to obtain limiting forms of a delta function. In fact there is an infinite set of kernels $K_\Omega\left(t, t'\right)$ with the property that for a continuous function integrable in a, b the following limiting form holds:

$$f(t) = \lim_{\Omega \to \Omega_0} \int_a^b K_\Omega\left(t, t'\right) f\left(t'\right) dt'; \quad a < t < b. \tag{1.246}$$

The parameter Ω_0 is usually either zero or infinity. Every kernel for which the preceding limit exists will be said to define a delta function. Note that this limit is taken with the kernel under the integral sign. What about the limit of the kernel itself, i.e., $\lim K_\Omega\left(t, t'\right)$ as $\Omega \to \Omega_0$? It is not hard to see that as long as $t \neq t'$ this limit is zero. However for $t = t'$ the limit does not exist. Nevertheless, we shall employ the customary notation and define the delta function by

$$\delta\left(t - t'\right) = \lim_{\Omega \to \Omega_0} K_\Omega\left(t, t'\right), \tag{1.247}$$

which is to be understood merely as a shorthand notation for the more elaborate statement (1.246). Implicit in this definition is the interval a, b over which the identity transformation (1.240) is defined.

A general property of delta function kernels is that they are symmetric, i.e.,

$$K_\Omega\left(t, t'\right) = K_\Omega\left(t', t\right). \tag{1.248}$$

In view of (1.247) this means that $\delta\left(\tau\right) = \delta\left(-\tau\right)$ so that for symmetric kernels the delta function may be regarded as an even function. It may turn out that the

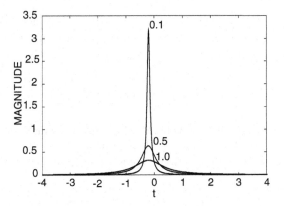

Figure 1.13: Delta function kernel in Eq. (1.249)

kernel itself is a function of the difference of t and t', i.e., $K_\Omega(t, t') = K_\Omega(t - t')$ as in the case for the pulse function in Fig. 1.12. Under these circumstances (1.247) may be replaced by

$$\delta(t) = \lim_{\Omega \to \Omega_0} K_\Omega(t).$$

Limiting Forms Leading to Delta Functions

Besides the pulse function another limiting form of the delta function is the following:

$$\delta(t) = \lim_{\Omega \to 0} \frac{\Omega}{\pi\left[t^2 + \Omega^2\right]}, \tag{1.249}$$

which holds over the infinite time interval, i.e., $a = -\infty$ and $b = \infty$, with $\Omega_0 = 0$. To prove this, we have to carry out the limiting process

$$I(t) \equiv \lim_{\Omega \to 0} \int_{-\infty}^{\infty} \frac{\Omega}{\pi\left[(t - t')^2 + \Omega^2\right]} f(t')\, dt', \tag{1.250}$$

and establish that $I(t) = f(t)$. Plots of the kernel in (1.250) are shown in Fig. 1.13 for several values of Ω.

We note that this kernel tends to zero everywhere except at the solitary point at which the integration variable t' approaches t, where the ratio $\Omega/\left[(t-t')^2+\Omega^2\right]$ becomes indeterminate. In order to estimate the contribution from the neighborhood of this special point we subdivide the integration interval into three parts, as follows:

$$\int_{-\infty}^{t-\epsilon/2}\{\}\,dt' + \int_{t-\epsilon/2}^{t+\epsilon/2}\{\}\,dt' + \int_{t+\epsilon/2}^{\infty}\{\}\,dt',\qquad(1.251)$$

where ϵ is a positive quantity which we may choose as small as desired. Since $t' \neq t$ in the first and third integrals, their limiting values as $\Omega \longrightarrow 0$ are zero. Therefore the entire contribution must come from the middle term alone, i.e.,

$$I(t) = \lim_{\Omega\to 0}\int_{t-\epsilon/2}^{t+\epsilon/2}\frac{\Omega}{\pi\left[(t-t')^2+\Omega^2\right]}f(t')\,dt'.\qquad(1.252)$$

Since $f(t)$ is continuous, we can choose a sufficiently small ϵ so that $f(t')$ may be regarded as a constant within the integration interval and set equal to $f(t)$. Thus, upon factoring $f(t)$ out of the integrand and changing the variable of integration to $x = (t'-t)/\Omega$, (1.252) leads to the final result

$$I = f(t)\lim_{\Omega\to 0}\int_{-\epsilon/2\Omega}^{\epsilon/2\Omega}\frac{dx}{\pi\left[x^2+1\right]} = f(t)\int_{-\infty}^{\infty}\frac{dx}{\pi\left[x^2+1\right]} = f(t),\qquad(1.253)$$

wherein the last step account has been taken of the indefinite integral $\int\left[1+x^2\right]^{-1}dx = \arctan(x)$.

Yet another limiting form for the delta function, which will be shown to play the central role in the theory of the Fourier integral, is the following:

$$\delta(t) = \lim_{\Omega\to\infty}\frac{\sin(\Omega t)}{\pi t}\qquad(1.254)$$

with t again defined over the entire real line. To prove (1.254) we need an intermediate result known as the Riemann–Lebesgue Lemma [16] (RLL) which for our purposes may be phrased as follows: Given a continuous and piecewise differentiable function $g(t)$ we define

$$\Psi_1(\omega,\alpha) = \int_{\alpha}^{\infty}g(t)e^{-i\omega t}dt,\qquad(1.255a)$$

$$\Psi_2(\omega,\beta) = \int_{-\infty}^{\beta}g(t)e^{-i\omega t}dt,\qquad(1.255b)$$

where α and β are real constants with $\alpha > \beta$. If

$$\int_{\alpha}^{\infty}|g(t)|\,dt < \infty,\text{ and}\qquad(1.256a)$$

$$\int_{-\infty}^{\beta}|g(t)|\,dt < \infty,\qquad(1.256b)$$

then the following limits hold:

$$\lim_{\omega \to \infty} \Psi_1(\omega, \alpha) \longrightarrow 0, \tag{1.257a}$$

$$\lim_{\omega \to \infty} \Psi_2(\omega, \beta) \longrightarrow 0. \tag{1.257b}$$

To prove (1.257a) we choose a constant $\tau > \alpha$ and write (1.255a) as the sum of two parts as follows:

$$\Psi_1(\omega, \alpha) = \hat{\Psi}_1(\omega, \alpha, \tau) + \int_\tau^\infty g(t) e^{-i\omega t} dt, \tag{1.258}$$

where $\hat{\Psi}_1(\omega, \alpha, \tau) = \int_\alpha^\tau g(t) e^{-i\omega t} dt$. For any finite τ we have the bound

$$|\Psi_1(\omega, \alpha)| \leq \left| \hat{\Psi}_1(\omega, \alpha, \tau) \right| + \int_\tau^\infty |g(t)|\, dt. \tag{1.259}$$

Integrating $\int_\alpha^\tau g(t) e^{-i\omega t} dt$ by parts, we get

$$\hat{\Psi}_1(\omega, \alpha, \tau) = \frac{g(\tau) e^{-i\omega\tau}}{-i\omega} - \frac{g(\alpha) e^{-i\omega\alpha}}{-i\omega} - \frac{1}{-i\omega} \int_\alpha^\tau \frac{dg(t)}{dt} e^{-i\omega t} dt. \tag{1.260}$$

Since $g(t)$ is bounded, the first two terms on the right of (1.260) approach zero for sufficiently large ω. Also, since $g(t)$ is assumed continuous and piecewise differentiable, its derivative is integrable over any finite interval. Hence as $\omega \to \infty$ last term on the right of (1.260) tends to zero as well so that

$$\lim_{\omega \to \infty} \hat{\Psi}_1(\omega, \alpha, \tau) \longrightarrow 0. \tag{1.261}$$

Since in accordance with (1.256a) $g(t)$ is absolutely integrable the second term on the right of (1.259) can be made arbitrarily small by simply choosing a sufficiently large τ. Thus since the right side of the inequality in (1.259) can be made arbitrarily small, its left side must also approach zero. This proves assertion (1.257a). By an identical argument one can prove (1.257b).

From the preceding proof it should be evident that, in particular,

$$\lim_{\omega \to \infty} \int_a^b g(t) e^{-i\omega t} \longrightarrow 0, \tag{1.262}$$

where the only requirement is that $g(t)$ be finite and piecewise differentiable over the finite interval a, b. Also, it is easy to see that assertions (1.257) and (1.262) remain valid when $e^{-i\omega t}$ is replaced by $\sin(\omega t)$ or $\cos(\omega t)$.

Returning to the proof of (1.254), we have to evaluate

$$I(t) \equiv \lim_{\Omega \to \infty} \int_{-\infty}^\infty \frac{\sin[\Omega(t - t')]}{\pi(t - t')} f(t')\, dt' \tag{1.263}$$

and show that $I(t) = f(t)$. This is more readily accomplished by regarding the integrand in (1.263) as a product of $\sin[\Omega(t - t')]$ and the function

$$g(t, t') = f(t')/\pi(t - t'). \tag{1.264}$$

If we suppose that the latter is absolutely integrable with respect to t' in the sense of (1.256) we can apply the RLL as long as we exclude the point $t' = t$ for there $g(t, t')$ becomes infinite for $f(t') \neq 0$. We again isolate this special point by breaking up the integration interval into three nonoverlapping segments, just as we did in (1.251):

$$
\begin{aligned}
I(t) &= \lim_{\Omega \to \infty} \int_{-\infty}^{t-\epsilon/2} g(t, t') \sin[\Omega(t - t')] \, dt' \\
&\quad + \lim_{\Omega \to \infty} \int_{t-\epsilon/2}^{t+\epsilon/2} g(t, t') \sin[\Omega(t - t')] \, dt' \\
&\quad + \lim_{\Omega \to \infty} \int_{t+\epsilon/2}^{\infty} g(t, t') \sin[\Omega(t - t')] \, dt'.
\end{aligned}
\tag{1.265}
$$

In view of (1.264) and (1.256) for functions $f(t)$ satisfying

$$
\int_{\alpha}^{\infty} \frac{|f(t')|}{t'} \, dt' < \infty \,, \qquad \int_{-\infty}^{\beta} \frac{|f(t')|}{t'} \, dt' < \infty
\tag{1.266}
$$

the first and the third integrals on the right of (1.265) vanish in accordance with the RLL for any $\epsilon > 0$. Hence we need concern ourselves only with the middle integral, i.e.,

$$
I(t) = \lim_{\Omega \to \infty} \int_{t-\epsilon/2}^{t+\epsilon/2} \frac{\sin[\Omega(t - t')]}{\pi(t - t')} f(t') \, dt'.
\tag{1.267}
$$

Again we choose a sufficiently small ϵ so that $f(t')$ may be approximated by $f(t)$ as closely as desired and write

$$
I(t) = f(t) \lim_{\Omega \to \infty} \int_{t-\epsilon/2}^{t+\epsilon/2} \frac{\sin[\Omega(t - t')]}{\pi(t - t')} \, dt'.
$$

The limit on the right is evaluated by first changing the variable of integration from t' to $\tau = \Omega(t' - t)$ to obtain

$$
\lim_{\Omega \to \infty} \int_{t-\epsilon/2}^{t+\epsilon/2} \frac{\sin[\Omega(t - t')]}{\pi(t - t')} \, dt' = \lim_{\Omega \to \infty} \int_{-\Omega\epsilon/2}^{\Omega\epsilon/2} \frac{\sin \tau}{\pi \tau} \, d\tau = \frac{1}{\pi} \int_{-\infty}^{\infty} \frac{\sin \tau}{\tau} \, d\tau.
\tag{1.268}
$$

The last integral converges to a known value, i.e.,

$$
\int_{-\infty}^{\infty} \frac{\sin \tau}{\tau} \, d\tau = \pi,
$$

so that we obtain $I(t) = f(t)$, thus proving (1.254), or, equivalently,

$$
f(t) = \lim_{\Omega \to \infty} \int_{-\infty}^{\infty} \frac{\sin[\Omega(t - t')]}{\pi(t - t')} f(t') \, dt'.
\tag{1.269}
$$

There are other limiting forms that lead to delta functions which we shall discuss in the sequel. From the foregoing we note that such limiting forms evidently require not only the specification of the time interval in question but also the class of admissible functions. For example, in the three cases we have considered, the limiting process required that the functions be continuous and sectionally differentiable throughout the time interval over which the delta function property (i.e., (1.246)) is to hold. The modification needed to include functions with step discontinuities will be discussed in the next subsection. For the case of the Fourier kernel, (1.254), an additional constraint on the growth of the function at infinity, (1.266), had to be imposed. We shall discuss this and other constraints on the growth at infinity in connection with Fourier Transforms in Chap. 2.

1.6.2 Higher Order Singularity Function

Consider now a function $f(t)$ with a sectionally continuous and differentiable first derivative. Then the fundamental limiting form (1.246) should apply to the derivative and we may write

$$\frac{df(t)}{dt} = \lim_{\Omega \to \Omega_0} \int_a^b K_\Omega(t, t') \frac{df(t')}{dt'} dt'; \quad a < t < b. \tag{1.270}$$

If we assume a differentiable kernel the preceding may be integrated by parts and written in the form

$$\frac{df(t)}{dt} = \lim_{\Omega \to \Omega_0} \left[K_\Omega(t, t') f(t') \big|_a^b + \int_a^b -\frac{dK_\Omega(t, t')}{dt'} f(t') dt' \right].$$

Now $\lim_{\Omega \to \Omega_0} K_\Omega(t, t') = 0$ for $t \neq t'$ so that if t does not coincide with the integration limits the preceding is equivalent to

$$-\frac{df(t)}{dt} = \lim_{\Omega \to \Omega_0} \int_a^b \frac{dK_\Omega(t, t')}{dt'} f(t') dt'. \tag{1.271}$$

Just as we abbreviated the limiting process (1.246) by the introduction of the delta function in (1.240) that selects $f(t')$ at a single point $t' = t$ we can abbreviate the limiting process in (1.271) by introducing the symbolic function $\delta^{(1)}(t)$ that selects the negative derivative of $f(t')$ at $t' = t$, i.e.,

$$-\frac{df(t)}{dt} = \int_a^b \delta^{(1)}(t' - t) f(t') dt'. \tag{1.272}$$

This new symbolic function, referred to as the "doublet", can be formally considered a derivative of the delta function. This interpretation follows from the following integration by parts:

$$\int_a^b \frac{d\delta(t' - t)}{dt'} f(t') dt' = \delta(t' - t) f(t') \big|_a^b - \int_a^b \delta(t' - t) \frac{df(t')}{dt'} dt' = -\frac{df(t)}{dt}. \tag{1.273}$$

Comparing the first integral in the preceding with the integral in (1.272) we readily identify $d\delta\,(t'-t)\,/dt'$ with $\delta^{(1)}\,(t'-t)$. Note that if instead of the delta function we were to insert the doublet $\delta^{(1)}\,(t'-t)$ in the first integral in (1.273) the integration by parts would yield $d^2 f(t)/dt^2$. This suggests, by analogy with (1.272), that we interpret $d\delta^{(1)}\,(t'-t)\,/dt' \equiv \delta^{(2)}\,(t'-t)$ as a "higher order" singularity function which, when multiplied by $f(t')$, selects upon integration its second derivative at $t = t'$. Clearly by repeating this procedure we can generate singularity functions of increasingly higher order and thus generalize (1.272) to[8]

$$(-1)^k \frac{d^k f(t)}{dt^k} = \int_a^b \delta^{(k)}\,(t'-t)\,f\,(t')\,dt' \quad ;k = 1,2,3,\ldots \tag{1.274}$$

wherein the k-th order singularity function $\delta^{(k)}\,(t)$ is defined (in a purely formal way) as the k-th order derivative of a delta function, i.e.,

$$\delta^{(k)}\,(t) = \frac{d^k \delta\,(t)}{dt^k}. \tag{1.275}$$

These singularity functions and the attendant operations are to be understood as purely formal shorthand notation whose precise meaning, just as the meaning of the delta function itself, must be sought in the underlying limiting forms of suitable kernels. Fortunately in most applications these detailed limiting processes need not be considered explicitly and one is usually content with the formal statements (1.274) and (1.275).

As an example of a limiting form leading to a doublet consider the kernel defined by the triangle function

$$K_\Omega\,(t-t') = \begin{cases} \frac{2}{\Omega}\left(1 - \frac{|t-t'|}{\Omega/2}\right); |t-t'| < \Omega/2, \\ 0; |t-t'| \geq \Omega/2. \end{cases} \tag{1.276}$$

A plot of this function and its derivative is shown in Fig. 1.14

By repeating the limiting arguments employed in conjunction with the pulse function, (1.243), one readily finds that for a continuous and differentiable function $f\,(t)$

$$\lim_{\Omega \to 0} \int_{-\infty}^{\infty} K_\Omega\,(t)\,f\,(t)\,dt = f\,(0)$$

so that (1.276) defines a delta function kernel. On the other hand, employing the derivative depicted in Fig. 1.14 we obtain

$$\lim_{\Omega \to 0} \int_{-\infty}^{\infty} \frac{dK_\Omega\,(t)}{dt} f\,(t)\,dt = \lim_{\Omega \to 0} \{[f\,(-\Omega/2) - f\,(\Omega/2)]\,(\Omega/2)\,(4/\Omega^2)\}$$

$$= \frac{f\,(-\Omega/2) - f\,(\Omega/2)}{\Omega/2} = -\frac{df\,(t)}{dt}\Big|_{t=0} \tag{1.277}$$

in agreement with formula (1.271) (with $\Omega_0 = 0$ and $t = 0$).

---

[8]The reader may be puzzled by the change in the argument from $t-t'$, which we employed in the preceding subsection for the delta function, to $t'-t$. We do this to avoid distinguishing between the derivative with respect to the argument and with respect to t' in the definition of $\delta^{(k)}\,(t)$.

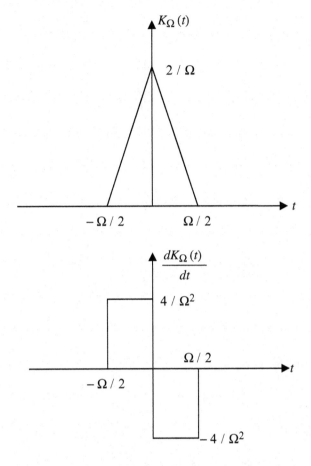

Figure 1.14: Triangle delta function kernel and its derivative

1.6.3 Idealized Signals

In this as well as in subsequent sections repeated use will be made of several idealized signals. We list them here for ready reference.

1. The unit step function

$$U(t) = \begin{cases} 1; t > 0, \\ 0; t < 0. \end{cases} \tag{1.278a}$$

2. Rectangular pulse

$$p_T(t) = \begin{cases} 1; |t| < T, \\ 0; |t| > T. \end{cases} \tag{1.278b}$$

3. The sign function

$$sign(t) = \begin{cases} 1; t > 0, \\ -1 : t < 0. \end{cases} \tag{1.278c}$$

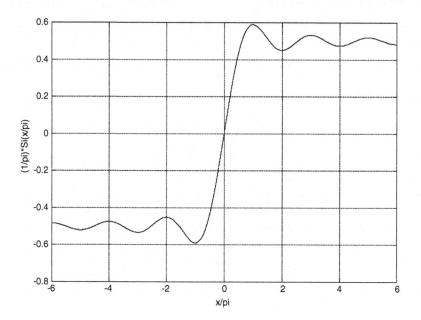

Figure 1.15: Sine integral function

4. The triangular pulse

$$q_T(t) = \begin{cases} 1 - \frac{|t|}{T}; & |t| < T, \\ 0; & |t| > T. \end{cases} \tag{1.278d}$$

5. The sine integral function

$$\mathrm{Si}(t) = \int_0^t \frac{\sin\tau}{\tau} d\tau. \tag{1.278e}$$

This function, plotted in Fig. 1.15, plays an important role in the theory of Fourier series and integrals. The damped oscillations reach their maximum (minimum) at $x = \pm\pi$ of ± 1.859 and decay asymptotically at $\pm\infty$ where $\mathrm{Si}(\pm\infty) = \pm\pi/2$

6. Derivative of the unit step function

   In the following we shall obtain the derivative of a unit step as a limiting form of the derivative of the function defined by

$$U_\epsilon(t) = [1 - q_\epsilon(t + \epsilon/2)]\, U(t + \epsilon/2)$$

   shown plotted in Fig. 1.16.

   From the figure we note that $dU_\epsilon(t)/dt = (1/\epsilon)\, p_{\epsilon/2}(t) = K_\epsilon(t)$, as defined in (1.241). Consequently, for any function $f(t)$ continuous and differentiable at the origin,

$$\lim_{\epsilon \longrightarrow 0} \int_{-\infty}^{\infty} \frac{dU_\epsilon(t)}{dt} f(t)\, dt = f(0),$$

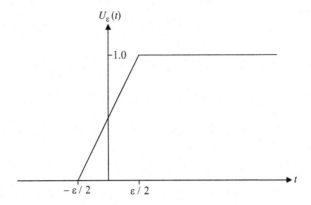

Figure 1.16: Step function with finite rise time

so that we make the formal identification

$$\frac{dU(t)}{dt} = \delta(t).$$

(1.279)

1.6.4   Representation of Functions with Step Discontinuities

The most general analogue signal of interest to us is represented by a piecewise differentiable function with a denumerable set of step discontinuities. If we denote the smooth portion of the signal (i.e., defined as continuous and differentiable) by $f_s(t)$ and the points of discontinuity by t_k, such a function $f(t)$ may be represented as follows:

$$f(t) = f_s(t) + \sum_k \left[f(t_k^+) - f(t_k^-) \right] U(t - t_k),$$

(1.280)

where $f(t_k^+) - f(t_k^-)$ represents the function increment (step) at the discontinuity point t_k as indicated in Fig. 1.17. Even though (1.280)

$$\frac{df(t)}{dt} = \frac{df_s(t)}{dt} + \sum_k \left[f(t_k^+) - f(t_k^-) \right] \delta(t - t_k),$$

(1.281)

where $\frac{df_s(t)}{dt}$ is an ordinary derivative of the smooth part of the function.

We have shown that limiting forms of certain kernels lead to delta function properties provided the signal is smooth (i.e., continuous and differentiable). With the aid of the representation (1.279) we can examine these limiting properties for signals with step discontinuities. For example, consider the limit with the Fourier kernel in (1.269). After substituting (1.279) in (1.269) we may take account of the fact that the limiting process involving $f_s(t)$ follows along the

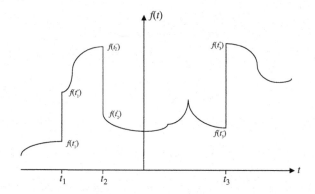

Figure 1.17: A piecewise smooth signal

lines leading to (1.269), i.e., the corresponding limit is just $f_s(t)$. Hence the limiting form for the discontinuous function can therefore be written as follows:

$$I(t) = f_s(t) + \sum_k \left[f\left(t_k^+\right) - f\left(t_k^-\right) \right] \lim_{\Omega \to \infty} \int_{t_k}^{\infty} \frac{\sin\left[\Omega\left(t - t'\right)\right]}{\pi\left(t - t'\right)} dt'. \qquad (1.282)$$

Changing the integration variable to $\tau = \Omega\left(t - t'\right)$, the last integral becomes

$$\lim_{\Omega \to \infty} \int_{t_k}^{\infty} \frac{\sin\left[\Omega\left(t - t'\right)\right]}{\pi\left(t - t'\right)} dt' = \lim_{\Omega \to \infty} \frac{1}{\pi} \int_{-\infty}^{\Omega(t - t_k)} \frac{\sin\tau}{\tau} d\tau = \begin{cases} 1 \; ; \; t > t_k, \\ \frac{1}{2} \; ; \; t = t_k, \\ 0 \; ; \; t < t_k. \end{cases}$$
$$(1.283)$$

In order to see what happens when the last result is substituted in (1.282) let us temporarily assume that we have only one discontinuity, say at $t = t_1$. Then

$$I(t) = \begin{cases} f_s(t) + f\left(t_1^+\right) - f\left(t_1^-\right) \; ; \; t > t_1, \\ f_s(t) + \frac{1}{2}\left[f\left(t_1^+\right) - f\left(t_1^-\right)\right] \; ; \; t = t_1, \\ f_s(t) \; ; \; t < t_1. \end{cases} \qquad (1.284)$$

Note that at the discontinuity $I(t) = I(t_1) = f_s(t_1) + \frac{1}{2}\left[f\left(t_1^+\right) - f\left(t_1^-\right)\right] = f\left(t_1^-\right) + \frac{1}{2}\left[f\left(t_1^+\right) - f\left(t_1^-\right)\right] = \frac{1}{2}\left[f\left(t_1^+\right) + f\left(t_1^-\right)\right]$, i.e., the representation converges to the arithmetic mean of the two function values. On the other hand, for all other values of t, $I(t)$ converges to $f(t)$. Clearly the same argument applies to any number of discontinuities. We can therefore write the general result as follows:

$$\lim_{\Omega \to \infty} \int_{-\infty}^{\infty} \frac{\sin\left[\Omega\left(t - t'\right)\right]}{\pi\left(t - t'\right)} f\left(t'\right) dt' = \frac{1}{2}\left[f\left(t^+\right) + f\left(t^-\right)\right], \qquad (1.285)$$

where the symbols t^+ and t^- represent time values "just after" and "just before" the discontinuity at t. When the function is continuous at t then

$f(t^+) = f(t^-) = f(t)$ and (1.285) reduces to the old formula (1.269). It turns out that the limiting form (1.285) is not restricted to the Fourier kernel but is a general property of delta function kernels. Thus for any piecewise differentiable function with step discontinuities the fundamental delta function property (1.240) becomes

$$\frac{1}{2}\left[f(t^+) + f(t^-)\right] = \int_a^b \delta(t - t')\, f(t')\, dt'. \tag{1.286}$$

1.6.5 Delta Function with Functions as Arguments

In applications the delta function appears frequently with an "ordinary" function in the argument, e.g., $\delta\left[g(t)\right]$. A typical case might involve another function $f(t)$ in the integrand in the form

$$I = \int_a^b f(t)\, \delta\left[g(t)\right] dt. \tag{1.287}$$

Assuming that $g(t)$ is differentiable we can replace the argument of the delta function by a new variable $\tau = g(t)$ and follow the standard procedure for changing the variable of integration to obtain

$$I = \int_{g(a)}^{g(b)} f\left[g^{-1}(\tau)\right] \frac{\delta(\tau)}{\left|\dfrac{dg(t)}{dt}\right|_{t=g^{-1}(\tau)}} d\tau, \tag{1.288}$$

where symbol $g^{-1}(\tau)$ represents the inversion of $g(t)$, i.e., a solution of $g(t) = \tau$ for t. Clearly if $g(t) \neq 0$ anywhere within the interval of integration, then $I = 0$. On the other hand, if

$$g(t_k) = 0\ ;\ a < t_k < b;\ k = 1, 2, 3, \ldots L, \tag{1.289}$$

the delta function $\delta(\tau)$ will give a contribution at each t_k so that (1.288) evaluates to

$$I = \sum_{k=1}^{L} \frac{f(t_k)}{\left|\dfrac{dg(t)}{dt}\right|_{t=t_k}}. \tag{1.290}$$

Clearly we obtain the same result if in (1.287) we simply set

$$\delta\left[g(t)\right] = \sum_{k=1}^{L} \frac{\delta(t - t_k)}{\left|\dfrac{dg(t)}{dt}\right|_{t=t_k}}. \tag{1.291}$$

Thus (1.290) may be taken as the formal definition of $\delta\left[g(t)\right]$.

As an example, consider the integral

$$I = \int_a^b \frac{1}{t^4 + 1} \delta(t^2 - 3t + 2)\, dt. \tag{1.292}$$

The roots of $t^2 - 3t + 2$ are $t_1 = 1$ and $t_2 = 2$ so that (1.291) gives

$$\delta(t^2 - 3t + 2) = \delta\,(t - 1) + \delta\,(t - 2)\,.$$

The value of I depends on whether none, one, or both roots fall within the integration limits. Thus

$$I = \begin{cases} 0 \;; a < 1, b < 1 \text{ or } a > 2, b > 2 \text{ or } 1 < a < 2, 1 < b < 2, \\ \frac{1}{2} \;;\; a < 1, 1 < b < 2, \\ \frac{1}{17} \;; 1 < a < 2, b > 2, \\ \frac{1}{17} + \frac{1}{2} \;; a < 1, b > 2. \end{cases} \tag{1.293}$$

If a root happens to coincide with one of the integration limits, one obtains a situation analogous to integrating a discontinuous function so that (1.287) applies. Thus in the preceding example with $a = 1$ and $1 < b < 2$ one obtains $I = 1/4$. Note that to obtain a nonzero contribution requires that $g\,(t)$ have real roots. For example,

$$I = \int_a^b \frac{1}{t^4 + 1} \delta(t^2 + 1) dt = 0$$

for all (real) choices of a and b since $t^2 + 1 = 0$ possesses no real roots.

1.7   Infinite Orthogonal Systems

1.7.1   Deterministic Signals

Thus far we considered the LMS approximation of signals only in terms of a finite number of expansion functions. In a finite, N-dimensional, vector space we know that the LMS error can always be made to approach zero by simply choosing N linearly independent basis vectors. In fact in this case not only would the LMS error be zero but the actual error would also be zero. The analogous situation in an infinite-dimensional vector space requires an infinite number of expansion functions. However, unlike in the finite-dimensional space, the LMS error may approach zero without at the same time the actual error tending to zero. In the former case we speak of convergence in the mean, in the latter of pointwise convergence. Thus

$$\lim_{N \to \infty} \int_a^b |\psi(t)|^2 \left| f\,(t) - f^N\,(t) \right|^2 dt = 0 \tag{1.294}$$

for convergence in the mean and

$$\lim_{N \to \infty} f^N\,(t) = f\,(t) \tag{1.295}$$

for pointwise convergence. While it is obvious that (1.295) implies (1.294), the converse is in general not true. We shall call the infinite set of expansion functions $\phi_n\,(t)$ complete if for every piecewise differentiable function $f\,(t)$ (1.294) holds while (1.295) holds at all points of continuity [5, 21].

Assume that the set $\phi_n(t)$ is orthonormal in a, b with a weighting factor $w(t) = |\psi(t)|^2$, i.e.,

$$(\phi_m, \phi_n) = \int_a^b w(t)\, \phi_m^*(t)\, \phi_n(t)\, dt = \delta_{nm}. \tag{1.296}$$

We now select the $\hat{f}_n$ in

$$f^N(t) = \sum_{n=1}^N \hat{f}_n \phi_n(t) \tag{1.297}$$

such that $(f - f^N, f - f^N)$ is minimized. Recall that this is accomplished with

$$\hat{f}_n = (\phi_n, f) = \int_a^b w(t')\, \phi_n^*(t')\, f(t')\, dt'. \tag{1.298}$$

We now substitute this $\hat{f}_n$ in (1.297) to obtain

$$f^N(t) = \int_a^b f(t')\, w(t') \sum_{n=1}^N \phi_n(t)\, \phi_n^*(t')\, dt'. \tag{1.299}$$

Now (1.295) implies that

$$\lim_{N \to \infty} \int_a^b f(t')\, w(t') \sum_{n=1}^N \phi_n(t)\, \phi_n^*(t')\, dt' = f(t). \tag{1.300}$$

If we now refer to our fundamental definition of the delta function in (1.246), we identify in the above $N = \Omega$, $\Omega_0 = \infty$, and

$$K_N(t, t') = w(t') \sum_{n=1}^N \phi_n(t)\, \phi_n^*(t'), \tag{1.301}$$

then (1.300) implies that

$$\lim_{N \to \infty} w(t') \sum_{n=1}^N \phi_n(t)\, \phi_n^*(t') = \delta(t - t'), \tag{1.302}$$

or, alternatively,

$$\frac{\delta(t - t')}{w(t')} = \sum_{n=1}^{\infty} \phi_n(t)\, \phi_n^*(t'). \tag{1.303}$$

which is the statement of completeness at points of continuity.

In addition, whenever (1.294) holds (but not necessarily (1.295)), we have

$$\int_a^b w(t)\, |f(t)|^2\, dt = \sum_{n=1}^{\infty} \left| \hat{f}_n \right|^2 \tag{1.304}$$

a relationship known in the special case of Fourier analysis as Parseval's formula. It follows either by direct substitution of (1.297) in (1.294) or by taking the limiting form of the Bessel inequality (1.202b). When the expansion functions are not of unit norm but instead $(\phi_m, \phi_n) = Q_n \delta_{nm}$, (1.304) becomes

$$\int_a^b w(t) |f(t)|^2 \, dt = \sum_{n=1}^{\infty} Q_n \left| \hat{f}_n \right|^2. \tag{1.305}$$

From the foregoing we see that delta function kernels arise naturally in conjunction with expansions in orthonormal functions. Equation (1.303) (or the formally equivalent statements (1.300) or (1.302)) is referred to as the completeness relation for the orthonormal set $\phi_n(t)$ with weighting factor $w(t)$. It can be shown that at points of discontinuity the right side of (1.300) should be replaced by $\frac{1}{2}[f(t^+) + f(t^-)]$ with the LMS error tending to zero in accordance with (1.294). In Chap. 2 we shall demonstrate this explicitly for Fourier series and the Fourier integral.

1.7.2 Stochastic Signals: Karhunen–Loeve Expansion*

Suppose we approximate a stochastic process $\underline{f}(t)$ by the sum

$$\underline{f}^N(t) = \sum_{n=1}^{N} \underline{\hat{f}}_n \phi_n(t), \tag{1.306}$$

where the $\phi_n(t)$ are deterministic expansion functions and the coefficients $\underline{\hat{f}}_n$ are random variables. The error in the approximation, $\underline{e}_N(t) = \underline{f}(t) - \underline{f}^N(t)$, is a stochastic process and will differ for each sample function (realization) of $\underline{f}(t)$. Similarly the MS error $(\underline{e}_N, \underline{e}_N)$ is a random variable. Selecting the $\underline{\hat{f}}_n$ so as to minimize the MS error for each sample function of the stochastic process leads, just as in the deterministic case, to the projection theorem $(\phi_m, \underline{e}_N) = 0$ so that the $\underline{\hat{f}}_n$ satisfy the stochastic normal equations

$$(\phi_m, \underline{f}) = \sum_{n=1}^{N} (\phi_m, \phi_n) \underline{\hat{f}}_n.$$

Denoting the inverse of the Gram matrix by $\{G_{nm}\}^{-1}$ the solution for the random variables $\underline{\hat{f}}_n$ reads

$$\underline{\hat{f}}_n = \sum_{m=1}^{N} \{G_{nm}\}^{-1} (\phi_m, \underline{f})$$

For each sample function of $\underline{f}(t)$ the LMS error is

$$
\begin{aligned}
(\underline{e}_N, \underline{e}_N) &= (\underline{f}, \underline{f}) - \sum_{n=1}^{N} \hat{\underline{f}}_n (\underline{f}, \phi_n) \\
&= (\underline{f}, \underline{f}) - \sum_{n=1}^{N} \sum_{m=1}^{N} \{G_{nm}\}^{-1} (\phi_m, \underline{f}) (\underline{f}, \phi_n). \quad (1.307)
\end{aligned}
$$

Denoting by $R_{ff}(t, t') = < \underline{f}(t)\underline{f}^*(t') >$ the autocorrelation function of the process we compute the ensemble average of both sides of (1.307) to obtain

$$
\begin{aligned}
\varepsilon_{N \min} &= < (\underline{e}_N, \underline{e}_N) >= \int_a^b R_{ff}(t, t)\, dt \\
&\quad - \sum_{n=1}^{N} \sum_{m=1}^{N} \{G_{nm}\}^{-1} \int_a^b dt \int_a^b dt' \phi_m^*(t') \phi_n(t) R_{ff}(t', t). \quad (1.308)
\end{aligned}
$$

In the preceding no assumption other than linear independence has been made about the expansion functions. We now assume that they are eigenfunctions of the integral equation

$$
\int_a^b dt' R_{ff}(t, t')\phi_n(t') = \lambda_n \phi_n(t) \ , \ n = 1, 2, \ldots \quad (1.309)
$$

It can be shown that the properties of the autocorrelation function guarantee the existence of an infinite set of orthogonal functions satisfying (1.309) and the completeness relationship

$$
\sum_{n=1}^{\infty} \phi_n(t) \phi_n^*(t') = \delta(t - t'). \quad (1.310)
$$

The eigenvalues λ_n are positive for all finite n and converge to zero as n tends to infinity. To obtain an expansion of the autocorrelation function in terms of its eigenfunctions we multiply both sides of (1.309) by $\phi_n^*(t'')$ and sum over n. This yields a delta function under the integral sign so that upon integration we get (after replacing t'' with t') the expansion

$$
R_{ff}(t, t') = \sum_{n=1}^{\infty} \lambda_n \phi_n(t) \phi_n^*(t'). \quad (1.311)
$$

The expansion functions are orthogonal so that with unit normalization the expansion coefficients in (1.306) are

$$
\hat{\underline{f}}_n = \int_a^b \phi_n^*(t) \underline{f}(t)\, dt. \quad (1.312)
$$

The correlation matrix elements for these coefficients are given by the ensemble averages

$$< \hat{\underline{f}}_n \hat{\underline{f}}_m^* >= \int_a^b \phi_n^*(t)\, dt \int_a^b dt' R_{ff}(t,t') \phi_m(t') \qquad (1.313)$$

and using (1.309) we get

$$< \hat{\underline{f}}_n \hat{\underline{f}}_m^* >= \lambda_n \int_a^b \phi_n^*(t)\, \phi_m(t) = \lambda_n \delta_{nm}. \qquad (1.314)$$

This shows that when the expansion functions in (1.306) are chosen to be eigenfunctions of the autocorrelation function the expansion coefficients are uncorrelated.

Let us compute the average LMS error when a stochastic process is approximated by such an expansion. For this purpose we may use (1.308) with $\{G_{nm}\}^{-1} = \delta_{nm}$. Taking account of (1.309) we get as $N \to \infty$

$$\varepsilon_{\infty \min} = \int_a^b R_{ff}(t,t)dt - \sum_{n=1}^{\infty} \lambda_n. \qquad (1.315)$$

But in view of (1.311) $\int_a^b R_{ff}(t,t)dt = \sum_{n=1}^{\infty} \lambda_n$ so that $\varepsilon_{\infty \min} = 0$.
The LMS expansion of the stochastic process $\underline{f}(t)$

$$\underline{f}(t) \sim \sum_{n=1}^{\infty} \hat{\underline{f}}_n \phi_n(t) \qquad (1.316)$$

in which the expansion functions are eigenfunctions of the autocorrelation function is known as the Karhunen–Loeve [18] expansion and finds extensive applications in statistical signal analysis. As shown above, it converges in the sense that

$$\lim_{N \to \infty} < \int_a^b \left| \underline{f}(t) - \sum_{n=1}^{N} \hat{\underline{f}}_n \phi_n(t) \right|^2 dt > \to 0. \qquad (1.317)$$

Problems

1. Prove that the p-norm defined by (1.20) satisfies postulates (1.14a), (1.14b) and (1.14c).

2. Show that the functions $\phi_1(t) = t\sqrt{3/2}$ and $\phi_2(t) = (3t^2 - 1)\sqrt{5/8}$ are orthonormal over $-1 \le t \le 1$ and find the Euclidean distance between the signals $f(t)$ and $g(t)$,

$$\begin{aligned} f(t) &= 0.5\phi_1(t) + \phi_2(t), \\ g(t) &= -\phi_1(t) + 0.5\phi_2(t). \end{aligned}$$

3. Approximate $f(t) = \sin t$ within the interval $0, \pi$ in the LMS sense by the function sets $1, t, t^2$ and $1, t, t^2, t^3, t^4$. Compute the corresponding LMS errors and plot $f^3(t)$ and $f^5(t)$ together with $f(t)$ on the same set of axes.

4. Generate a data set $y[n]$ as follows:

$$y[n] = (2n+1)^2 + 12000 * \text{rand}(n), \quad 1 \le n \le 64,$$

where $\text{rand}(n)$ is a random sequence of 64 numbers uniformly distributed between -1 and 1. Find the coefficients a_0, a_1, a_2 such that

$$y_{est}[n] \equiv a_0 + a_1 n + a_2 n^2$$

represents an LMS fit to the data. Carry out the coefficient evaluation using:

(a) The Normal Equations.

(b) The SVD retaining (i) all three singular values (ii) two dominant singular values, and (iii) one dominant singular value.

(c) Plot the $y_{est}[n]$ obtained in (a) and (b) together with the data set as well as the "signal" $y_0[n] \equiv (2n+1)^2$ on the same set of axes.

5. Prove that the three functions $e^{-t}, te^{-t}, t^2 e^{-t}$ are linearly independent in the interval $0, \infty$ and transform them into an orthonormal set using

(a) The Gram–Schmidt procedure.

(b) The orthonormal functions defined by (1.150).

(c) Repeat (a) and (b) with the addition of the fourth function e^{-2t} to the set.

6. Find an orthonormal set defined in the interval $0, \pi$ corresponding to the following three functions : $\sin t, t \sin t, 2\sin t + 3t \sin t$ using:

(a) Eq. (1.56)

(b) The Gram–Schmidt orthogonalization procedure

7. For the $N \times 4$ matrix $(N > 4)$ $\mathbf{A}$ defined by

$$\mathbf{A} = [\mathbf{a}_1 \mathbf{a}_2 \mathbf{a}_2 \mathbf{a}_1],$$

where $\mathbf{a}_1$ and $\mathbf{a}_2$ are two N-dimensional orthogonal column vectors, with $\mathbf{a}_1^H \mathbf{a}_1 = 2$ and $\mathbf{a}_2^H \mathbf{a}_2 = 3$. Find

(a) The SVD of $\mathbf{A}$

(b) The Moore–Penrose pseudoinverse of $\mathbf{A}$

(c) Given an arbitrary N-dimensional column vector $\mathbf{y}$ use the result of (b) to solve $\mathbf{A}\mathbf{x} \sim \mathbf{y}$ in the LMS sense for $\mathbf{x}$.

(d) Find the LMS error in (c)

8. A set of data x_n, y_n, $n = 1, 2, 3, \ldots N$ is to fitted to the parabola

$$y = ax^2 + b$$

in the LMS sense with the correspondence $x_n \Leftrightarrow x$ and $y_n \Leftrightarrow y$.

   (a) Write down the normal equations
   (b) Solve for a and b
   (c) Find the LMS error

9. The function

$$f(t) = 1 - |t|$$

is to be approximated in the LMS sense in the interval $-1 \leq t \leq 1$ by the polynomial

$$\sum_{n=0}^{2} \hat{f}_n t^n$$

   (a) Find the coefficients $\hat{f}_n$
   (b) Find the LMS error

10. Prove the following (Hadamard) inequality:

$$\det[(\phi_n, \phi_m)] \leq \prod_{k=1}^{N} (\phi_k, \phi_k).$$

11. Using the three M-dimensional column vectors b_1, b_2, b_3, we form the $MX9$ matrix B,

$$B = \begin{bmatrix} b_1 & b_1 & b_1 & b_1 & b_2 & b_2 & b_2 & b_3 & b_3 \end{bmatrix},$$

Assuming $M \geq 9$ and that b_1, b_2, b_3 are orthonormal determine the SVD of B.

12. In Problem 11 show that the column vectors of B constitute a frame in the three-dimensional subspace spanned by b_1, b_2, b_3. What are the frame bounds A and B? Using the three vectors b_1, b_2, b_3 explain how one could modify the matrix B to obtain a tight frame.

13. The signal

$$f(t) = \cos(2\pi t/T)$$

is to be approximated in the interval $0 \leq t \leq T$ in the LMS sense by the N functions defined by

$$\phi_n(t) = p_\delta (t - n\delta),$$

where $\delta = T/N$ and

$$p_\delta(t) = \begin{cases} 1/\sqrt{\delta}; & 0 \leq t \leq \delta, \\ 0 \; ; otherwise. \end{cases}$$

(a) Denoting the partial sum by

$$f^N(t) = \sum_{n=0}^{N-1} \hat{f}_n \phi_n(t)$$

find the coefficients $\hat{f}_n$.

(b) Compute the LMS error as a function of T and N.

(c) Prove that the LMS error approaches 0 as $N \longrightarrow \infty$.

14. With

$$\begin{aligned}
\phi_n^{(e)}(t) &= \cos\left[2\pi nt/(b-a)\right] \\
\phi_n^{(o)}(t) &= \sin\left[2\pi nt/(b-a)\right]
\end{aligned}$$

and $n = 1, 2, 3, \ldots$ prove directly the orthogonality relations

$$\begin{aligned}
\int_a^b \phi_n^{(e)}(t)\,\phi_m^{(e)}(t) &= \frac{b-a}{2}\delta_{nm}, \\
\int_a^b \phi_n^{(o)}(t)\,\phi_m^{(o)}(t) &= \frac{b-a}{2}\delta_{nm}, \\
\int_a^b \phi_n^{(e)}(t)\,\phi_m^{(o)}(t) &= 0.
\end{aligned}$$

15. Consider the set of polynomials $\phi_k = t^{k-1}, k = 1, 2, \ldots N$. With the inner product defined by

$$(\phi_k, \phi_j) = \int_{-\infty}^{\infty} \phi_k(t)\,\phi_j(t)\,e^{-t^2}\,dt,$$

employ the Gram–Schmidt orthogonalization procedure to show that the resulting orthogonal functions are proportional to the Hermite polynomials in (1.236).

16. Prove the following:

(a) $\delta(t-t') = \lim\limits_{a\to\infty} \frac{\sin^2[a(t-t')]}{\pi a(t-t')^2}$ for all piecewise differentiable functions satisfying (1.266)

(b) $\delta(t-t') = \lim\limits_{a\to 0} \frac{1}{\sqrt{2\pi a^2}}e^{-\frac{(t-t')^2}{2a^2}}$ for all piecewise differentiable and bounded functions.

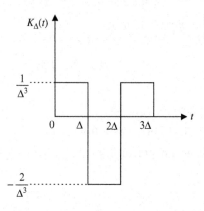

17. Evaluate the following integrals

   (a) $\int_1^4 \frac{\delta(t^2-2)}{t^2+1} dt$

   (b) $\int_{-2\pi}^{2\pi} \frac{\delta[\sin(2t)]}{1+\cos^2(2t)} dt$

   (c) $\int_{-1}^1 \frac{\delta(t^2+1)}{2+t} dt$

18. Prove that every real square integrable function (i.e., one satisfying $\int_{-\infty}^{\infty} f(t)^2 dt < \infty$) also satisfies (1.266).

19. For the kernel defined by $K_\Delta(t) = \frac{1}{\Delta^3} p_{\Delta/2}(t-\Delta/2) - \frac{2}{\Delta^3} p_{\Delta/2}(t-3\Delta/2) + \frac{1}{\Delta^3} p_{\Delta/2}(t-5\Delta/2)$ and plotted in the following sketch prove that for a piecewise continuous function $f(t)$ that is twice differentiable

$$\lim_{\Delta \to 0} \int_{-\infty}^{\infty} K_\Delta(t) f(t) dt = f''(0),$$

   i.e., that $\lim_{\Delta \to 0} K_\Delta(t) = \delta^{(2)}(t)$

Chapter 2

Fourier Series and Integrals with Applications to Signal Analysis

Perhaps the most important orthogonal functions in engineering applications are trigonometric functions. These were briefly discussed in 1.5.2 as one example of LMS approximation by finite orthogonal function sets. In this chapter we reexamine the LMS approximation problem in terms of infinite trigonometric function sets. When the approximating sum converges to the given function we obtain a Fourier Series; in case of a continuous summation index (i.e., an integral as in (1.92) the converging approximating integral is referred to as a Fourier Integral.

2.1 Fourier Series

2.1.1 Pointwise Convergence at Interior Points for Smooth Functions

We return to 1.5.2 and the LMS approximation of $f(t)$ within the interval $-T/2 < t < T/2$ by $2N+1$ using complex exponentials as given by (1.211). The approximating sum reads

$$f^N(t) = \sum_{n=-N}^{N} \hat{f}_n e^{i2\pi nt/T}, \tag{2.1}$$

while for the expansion coefficients we find from (1.214)

$$\hat{f}_n = \frac{1}{T} \int_{-T/2}^{T/2} f(t) e^{-i2\pi nt/T} dt. \tag{2.2}$$

W. Wasylkiwskyj, *Signals and Transforms in Linear Systems Analysis*,
DOI 10.1007/978-1-4614-3287-6_2, © Springer Science+Business Media, LLC 2013

Upon substituting (2.2) in (2.1) and interchanging summation and integration we obtain

$$f^N(t) = \int_{-T/2}^{T/2} f(t') K_N(t - t') \, dt', \qquad (2.3)$$

where

$$K_N(t - t') = \frac{1}{T} \sum_{n=-N}^{N} e^{i2\pi n(t-t')/T} = \sum_{n=-N}^{N} \left(\frac{1}{\sqrt{T}} e^{i2\pi nt/T} \right) \left(\frac{1}{\sqrt{T}} e^{i2\pi nt'/T} \right)^*,$$
$$(2.4)$$

and as shown in the following, approaches a delta function at points of continuity as N approaches infinity. The last form highlights the fact that this kernel can be represented as a sum of symmetric products of expansion functions in conformance with the general result in (1.301) and (1.302). Using the geometrical series sum formula we readily obtain

$$K_N(t - t') = \frac{\sin\left[2\pi(N + 1/2)(t - t')/T\right]}{T \sin\left[\pi(t - t')/T\right]}, \qquad (2.5)$$

which is known as the Fourier series kernel. As is evident from (2.4) this kernel is periodic with period T and is comprised of an infinite series of regularly spaced peaks each similar to the a-periodic sinc function kernel encountered in (1.254). A plot of $T\,K_N(\tau)$ for $N = 5$ as a function of $(t - t')/T \equiv \tau/T$ is shown in Fig. 2.1. The peak value attained by $T\,K_N(\tau)$ at $\tau/T = 0, \pm 1, \pm 2,$

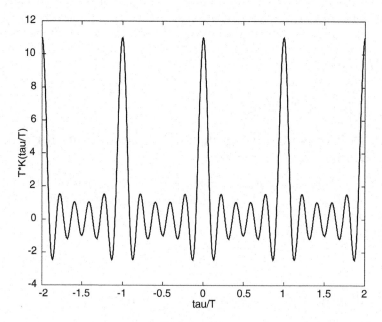

Figure 2.1: Fourier series kernel (N=5)

is in general $2N + 1$, as may be verified directly with the aid of (2.4) or (2.5). As the peaks of these principal lobes grow in proportion with N their widths diminish with increasing N. In fact we readily find directly from (2.5) that the peak-to-first null lobe width is $\Delta\tau = T/(2N + 1)$. We note that $\Delta\tau K_N$ $(\pm kT) = 1$, so that the areas under the principal lobes should be on the order of unity for sufficiently large N. This suggests that the infinite series of peaks in Fig. 2.1 should tend to an infinite series of delta functions as the number N increases without bound. This is in fact the case. To prove this we must show that for any piecewise differentiable function defined in any of the intervals $(k - 1/2)T < t < (k + 1/2)T$, $k = 0, \pm 1, \ldots$ the limit

$$\lim_{N \longrightarrow \infty} \int_{(k-1/2)T}^{(k+1/2)T} f\left(t'\right) K_N \left(t - t'\right) dt' = \frac{1}{2} \left[f\left(t^+\right) + f\left(t^-\right)\right] \qquad (2.6)$$

holds. Of course, because of the periodicity of the Fourier Series kernel it suffices if we prove (2.6) for $k = 0$ only. The proof employs steps very similar to those following (1.263) except that the constraint on the behavior of $f(t)$ at infinity, (1.266), presently becomes superfluous since the integration interval is finite. Consequently the simpler form of the RLL given by (1.262) applies. As in (1.267) we designate the limit by $I(t)$ and write

$$I\left(t\right) = \lim_{N \longrightarrow \infty} \int_{-T/2}^{T/2} g\left(t, t'\right) \sin\left[2\pi\left(N + 1/2\right)\left(t - t'\right)/T\right] dt', \qquad (2.7)$$

where by analogy with (1.264) we have defined the function

$$g\left(t, t'\right) = \frac{f\left(t'\right)}{T \sin\left[\pi\left(t - t'\right)/T\right]}. \qquad (2.8)$$

In (2.7) we may identify the large parameter $2\pi\left(N + 1/2\right)/T$ with ω in (1.262) and apply the RLL provided we again exclude the point $t = t'$ where $g(t, t')$ becomes infinite. We then proceed as in (1.265) to obtain

$$\begin{aligned} I\left(t\right) =\; & \lim_{N \longrightarrow \infty} \int_{-T/2}^{t-\epsilon/2} g\left(t, t'\right) \sin\left[2\pi\left(N + 1/2\right)\left(t - t'\right)/T\right] dt' \\ & + \lim_{N \longrightarrow \infty} \int_{t-\epsilon/2}^{t+\epsilon/2} g\left(t, t'\right) \sin\left[2\pi\left(N + 1/2\right)\left(t - t'\right)/T\right] dt' \\ & + \lim_{N \longrightarrow \infty} \int_{t+\epsilon/2}^{T/2} g\left(t, t'\right) \sin\left[2\pi\left(N + 1/2\right)\left(t - t'\right)/T\right] dt', \qquad (2.9) \end{aligned}$$

where ϵ is an arbitrarily small positive number. Let us first assume that $f(t')$ is smooth, i.e., piecewise differentiable and continuous. In that case then $g(t, t')$ has the same properties provided $t \neq t'$. This is true for the function in the integrand of the first and third integral in (2.9). Hence by the RLL these vanish so that $I(t)$ is determined solely by the middle integral

$$I\left(t\right) = \lim_{N \longrightarrow \infty} \int_{t-\epsilon/2}^{t+\epsilon/2} f\left(t'\right) \frac{\sin\left[2\pi\left(N + 1/2\right)\left(t - t'\right)/T\right]}{T \sin\left[\pi\left(t - t'\right)/T\right]} dt'. \qquad (2.10)$$

Since ϵ is arbitrarily small, $f(t')$ can be approximated as closely as desired by $f(t)$ and therefore factored out of the integrand. Also for small ϵ the $\sin[\pi(t - t')/T]$ in the denominator can be replaced by its argument. With these changes (2.10) becomes

$$I(t) = f(t) \lim_{N \longrightarrow \infty} \int_{t-\epsilon/2}^{t+\epsilon/2} \frac{\sin[2\pi(N+1/2)(t - t')/T]}{\pi(t - t')} dt'. \qquad (2.11)$$

The final evaluation becomes more transparent when the integration variable is changed from t' to $x = 2\pi(N+1/2)(t' - t)/T$ which transforms (2.11) into

$$I(t) = f(t) \lim_{N \longrightarrow \infty} \frac{1}{\pi} \int_{-\epsilon\pi(N+1/2)/T}^{\epsilon\pi(N+1/2)/T} \frac{\sin x}{x} dx = f(t) \frac{1}{\pi} \int_{-\infty}^{\infty} \frac{\sin x}{x} dx = f(t). \qquad (2.12)$$

This establishes the delta function character of the Fourier series kernel. Equivalently, we have proven that for any smooth function $f(t)$ the Fourier series

$$f(t) = \sum_{n=-\infty}^{\infty} \hat{f}_n e^{i2\pi nt/T} \qquad (2.13)$$

with coefficients given by (2.2) converges in the interval $-T/2 < t < T/2$.

2.1.2 Convergence at Step Discontinuities

Note that in the preceding limiting argument we have excluded the endpoints of the interval, i.e., we have shown convergence only in the open interval. In fact, as we shall shortly see, pointwise convergence can in general not be achieved by a Fourier series at $t = \pm T/2$ even for a function with smooth behavior in the open interval. It turns out that convergence at the endpoints is intimately related to convergence at a step discontinuity, to which we now turn our attention. Thus suppose our function possesses a finite number of step discontinuities in the (open) interval under consideration. We can then represent it as a sum comprising a smooth function $f_s(t)$ and a sum of step functions as in (1.280). In order not to encumber the development with excessive notation we confine the discussion to one typical discontinuity, say at $t = t_1$, and write

$$f(t) = f_s(t) + \left[f(t_1^+) - f(t_1^-)\right] U(t - t_1). \qquad (2.14)$$

The Fourier coefficients follow from (2.2) so that

$$\hat{f}_n = \frac{1}{T} \int_{-T/2}^{T/2} f_s(t) e^{-i2\pi nt/T} dt + \frac{\left[f(t_1^+) - f(t_1^-)\right]}{T} \int_{t_1}^{T/2} e^{-i2\pi nt/T} dt \qquad (2.15)$$

and substitution in (2.1) yields the partial sum

$$f^N(t) = \int_{-T/2}^{T/2} f_s(t') \frac{\sin[2\pi(N+1/2)(t - t')/T]}{T \sin[\pi(t - t')/T]} dt' + \left[f(t_1^+) - f(t_1^-)\right] \lambda_N(t), \qquad (2.16)$$

where

$$\lambda_N(t) = \int_{t_1}^{T/2} \frac{\sin\left[2\pi\left(N+1/2\right)\left(t-t'\right)/T\right]}{T\sin\left[\pi\left(t-t'\right)/T\right]} dt' . \qquad (2.17)$$

The limiting form of the first integral on the right of (2.16) as $N \longrightarrow \infty$ has already been considered so that

$$\lim_{N\to\infty} f^N(t) = f_s(t) + \left[f\left(t_1^+\right) - f\left(t_1^-\right)\right] \lim_{N\to\infty} \lambda_N(t) \qquad (2.18)$$

and only the last limit introduces novel features. Confining our attention to this term we distinguish three cases: the interval $-T/2 < t < t_1$, wherein $t' \neq t$ so that the RLL applies, the interval $t_1 < t < T/2$, and the point of discontinuity $t = t_1$. In the first case $\lambda_N(t)$ approaches zero. In the second case we divide the integration interval into three subintervals as in (2.9). Proceeding in identical fashion we find that $\lambda_N(t)$ approaches unity. For $t = t_1$ we subdivide the integration interval into two subintervals as follows:

$$\begin{aligned}
\lambda_N(t_1) &= \int_{t_1}^{t_1+\epsilon/2} \frac{\sin\left[2\pi\left(N+1/2\right)\left(t_1-t'\right)/T\right]}{T\sin\left[\pi\left(t_1-t'\right)/T\right]} dt' \\
&+ \int_{t_1+\epsilon/2}^{T/2} \frac{\sin\left[2\pi\left(N+1/2\right)\left(t_1-t'\right)/T\right]}{T\sin\left[\pi\left(t_1-t'\right)/T\right]} dt', \qquad (2.19)
\end{aligned}$$

where again ϵ is an arbitrarily small positive quantity. In the second integral $t' \neq t_1$ so that again the RLL applies and we obtain zero in the limit. Hence the limit is given by the first integral which we compute as follows:

$$\begin{aligned}
\lim_{N\to\infty} \lambda_N(t_1) &= \lim_{N\to\infty} \int_{t_1}^{t_1+\epsilon/2} \frac{\sin\left[2\pi\left(N+1/2\right)\left(t_1-t'\right)/T\right]}{T\sin\left[\pi\left(t_1-t'\right)/T\right]} dt' \\
&= \lim_{N\to\infty} \int_0^{\frac{\pi(2N+1)\epsilon}{2T}} \frac{\sin x}{\pi\left(2N+1\right)\sin\frac{x}{2N+1}} dx \\
&= \lim_{N\to\infty} \int_0^{\frac{\pi(2N+1)\epsilon}{2T}} \frac{\sin x}{\pi x} dx = \int_0^{\infty} \frac{\sin x}{\pi x} dx = \frac{1}{2}. \quad (2.20)
\end{aligned}$$

Summarizing the preceding results we have

$$\lim_{N\to\infty} \lambda_N(t) = \begin{cases} 0 \; ; & -T/2 < t < t_1, \\ 1/2 \; ; & t = t_1, \\ 1 \; ; & t_1 < t < T/2. \end{cases} \qquad (2.21)$$

Returning to (2.18) and taking account of the continuity of $f_s(t)$ we have the final result

$$\lim_{N\to\infty} f^N(t_1) = \frac{1}{2}\left[f\left(t_1^+\right) + f\left(t_1^-\right)\right]. \qquad (2.22)$$

Clearly this generalizes to any number of finite discontinuities within the expansion interval. Thus, for a piecewise differentiable function with step discontinuities the Fourier series statement (2.13) should be replaced by

$$\frac{1}{2}\left[f\left(t^+\right) + f\left(t^-\right)\right] = \sum_{-\infty}^{\infty} \hat{f}_n e^{i2\pi nt/T}. \qquad (2.23)$$

Although the limiting form (2.23) tells us what happens when the number of terms in the series is infinite, it does not shed any light on the behavior of the partial approximating sum for finite N. To assess the rate of convergence we should examine (2.17) as a function of t with increasing N. For this purpose let us introduce the function

$$\text{Si}\, s(x, N) = \int_0^x \frac{\sin[(N+1/2)\theta]}{2\sin(\theta/2)}\, d\theta \tag{2.24}$$

so that the dimensionless parameter x is a measure of the distance from the step discontinuity ($x = 0$). The integrand in (2.24) is just the sum $(1/2)\sum_{n=-N}^{n=N}\exp(-in\theta)$ which we integrate term by term and obtain the alternative form

$$\text{Si}\, s(x, N) = \frac{x}{2} + \sum_{n=1}^{N}\frac{\sin(nx)}{n}. \tag{2.25}$$

Note that for any N the preceding gives $\text{Si}\, s(\pi, N) = \pi/2$. As $N \to \infty$ with $0 < x < \pi$ this series converges to $\pi/2$. A plot of (2.25) for $N = 10$ and $N = 20$ is shown in Fig. 2.2. For larger values of N the oscillatory behavior of $\text{Si}\, s(y, N)$

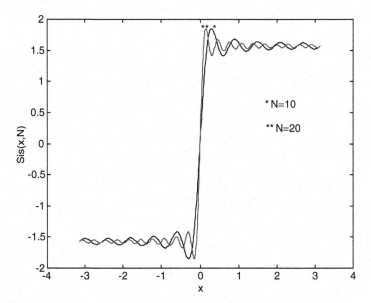

Figure 2.2: FS convergence at a step discontinuity for N=10 and N=20

damps out and the function approaches the asymptotes $\pm\pi/2$ for $y \neq 0$. Note that as N is increased the peak amplitude of the oscillations does not diminish but migrates toward the location of the step discontinuity, i.e., $y = 0$. The numerical value of the overshoot is ± 1.852 or about 18% above (below) the positive (negative) asymptote. When expressed in terms of (2.25), (2.17) reads

$$\lambda_N(t) = \frac{1}{\pi}\text{Si}\, s[(T/2 - t)2\pi/T, N] - \frac{1}{\pi}\text{Si}\, s[(t_1 - t)2\pi/T, N]. \tag{2.26}$$

Taking account of the limiting forms of (2.25) we note that as long as $t < T/2$ in the limit as $N \to \infty$ the contribution from the first term on the right of (2.26) approaches $1/2$, while the second term tends to $-1/2$ for $t < t_1$, $1/2$ for $t > t_1$ and 0 for $t = t_1$, in agreement with the limiting forms enumerated in (2.21).

Results of sample calculations of $\lambda_N(t)$ (with $t_1 = 0$) for $N = 10$, 20, and 50 are plotted in Fig. 2.3. Examining these three curves we again observe that increasing N does not lead to a diminution of the maximum amplitude of the

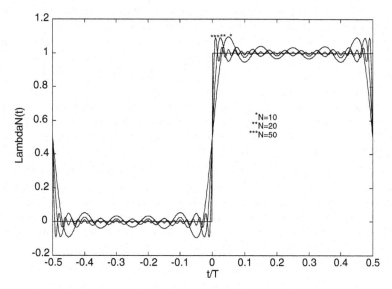

Figure 2.3: Convergence at a step discontinuity

oscillations. On the contrary, except for a compression of the timescale, the oscillations for $N = 50$ have essentially the same peak amplitudes as those for $N = 10$ and in fact exhibit the same overshoot as in Fig. 2.2. Thus N appears to enter into the argument in (2.22) merely as scaling factor of the abscissa, so that the magnitude of the peak overshoot appears to persist no matter how large N is chosen. The reason for this behavior can be demonstrated analytically by approximating (2.24) for large N. We do this by first changing the variable of integration in (2.24) to $y = (N + 1/2)\theta$ to obtain

$$\mathrm{Si}\, s(x, N) = \int_0^{(N+1/2)x} \frac{\sin y}{(2N+1)\sin[y/(2N+1)]}\, dy. \qquad (2.26^*)$$

Before proceeding with the next algebraic step we note that as $N \to \infty$ the numerator in (2.24) will be a rapidly oscillating sinusoid so that its contributions to the integral will mutually cancel except for those in the neighborhood of small θ. In terms of the variables in (2.26*) this means that for large N the argument $y/(2N+1)$ of the sine function will remain small. In that case we may replace the sine by its argument which leads to the asymptotic form

$$\mathrm{Si}[s(N, x)] \sim \mathrm{Si}[(N+1/2)x], \qquad (2.26^{**})$$

where $\mathrm{Si}(z)$ is the sine integral function defined in (1.278e) and plotted in Fig. 1.15. If we use this asymptotic form in (2.26), we get

$$\lambda_N(t) = \frac{1}{\pi} \mathrm{Si}[(N+1/2)(T/2-t)2\pi/T] - \frac{1}{\pi} \mathrm{Si}[(N+1/2)(t_1-t)2\pi/T],$$

which shows directly that N enters as a scaling factor of the abscissa. Thus as the number of terms in the approximation becomes infinite the oscillatory behavior in Fig. 2.3 compresses into two vanishingly small time intervals which in the limit may be represented by a pair of infinitely thin spikes at $t = 0^+$ and $t = 0^-$. Since in the limit these spikes enclose zero area we have here a direct demonstration of convergence in the mean (i.e., the LMS error rather than the error itself tending to zero with increasing N). This type of convergence, characterized by the appearance of an overshoot as a step discontinuity is approached, is referred to as the Gibbs phenomenon, in honor of Willard Gibbs, one of the America's greatest physicists. Gibbs phenomenon results whenever an LMS approximation is employed for a function with step discontinuities and is by no means limited to approximations by sinusoids (i.e., Fourier series). In fact the numerical example in Fig. 1.11 demonstrates it for Legendre Polynomials.

Another aspect of the Gibbs phenomenon worth mentioning is that it affords an example of nonuniform convergence. For as we have seen $\lim N \to \infty$ $\lambda_N(t_1) \to 1/2$. On the other hand, the limit approached when N is allowed to approach infinity first and the function subsequently evaluated at t as it is made to approach t_1 (say, through positive values) is evidently unity. Expressed in symbols, these two alternative ways of approaching the limit are

$$\lim_{N\to\infty}\lim_{t\to t_1^+} \lambda_N(t) = 1/2, \qquad (2.26***a)$$

$$\lim_{t\to t_1^+}\lim_{N\to\infty} \lambda_N(t) = 1. \qquad (2.26***b)$$

In other words, the result of the limiting process depends on the order in which the limits are taken, a characteristic of nonuniform convergence. We can view (2.26***) as a detailed interpretation of the limiting processes implied in the Fourier series at step discontinuities which the notation (2.23) does not make explicit.

2.1.3 Convergence at Interval Endpoints

The preceding discussion applies only to convergence properties of the Fourier series within the open interval. To complete the discussion of convergence we must still consider convergence at the interval endpoints $\pm T/2$. We start with the approximate form (2.20) (c.f. Fig. 2.3) which, together with the periodicity of $\lambda_N(t)$ based on the exact form (2.17), gives

$$\lim_{N\to\infty} \lambda_N(\pm T/2) = 1/2.$$

Thus in view of (2.16) we have at the endpoints

$$\lim_{N \to \infty} f^N \left(\pm T/2 \right)$$

$$= \lim_{N \to \infty} \int_{-T/2}^{T/2} f_s \left(t' \right) \frac{\sin \left[2\pi \left(N + 1/2 \right) \left(\pm T/2 - t' \right) /T \right]}{T \sin \left[\pi \left(\pm T/2 - t' \right) /T \right]} dt'$$

$$+ \frac{\left[f \left(t_1^+ \right) + f \left(t_1^- \right) \right]}{2}. \tag{2.27}$$

Since the observation points $\pm T/2$ coincide with the integration limits, the limiting procedure following (2.9) is not directly applicable. Rather than examining the limiting form of the integral in (2.27) directly, it is more instructive to infer the limit in the present case from (2.24) and the periodicity of the Fourier series kernel. This periodicity permits us to increment the integration limits in (2.27) by an arbitrary amount, say τ, provided we replace $f_s \left(t \right)$ by its periodic extension

$$f_s^{ext} \left(t \right) = \sum_{n=-\infty}^{n=\infty} f_s \left(t - nT \right). \tag{2.28}$$

With this extension the endpoints $\pm T/2$ now become the interior points in an infinite sequence of expansion intervals ... $\left(\tau - 3T/2, \tau - T/2 \right)$, $\left(\tau - T/2, \tau + T/2 \right)$ .... These intervals are all of length T and may be viewed as centered at $t = \tau \pm nT$, as may be inferred from Fig. 2.4. We note that unless $f_s \left(T/2 \right) = f_s \left(-T/2 \right)$ the periodic extension of the originally smooth $f_s \left(t \right)$ will have a step discontinuity at the new interior points of the amount $f_s \left(-T/2 \right) - f_s \left(T/2 \right)$. Thus with a suitable shift of the expansion interval and the replacement of

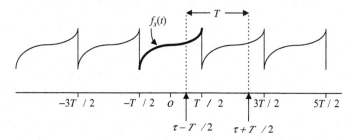

Figure 2.4: Step discontinuity introduced by a periodic extension of $f_s \left(t \right)$

$f_s \left(t' \right)$ by the $f_s^{ext} \left(t' \right)$ in (2.28) we can mimic the limiting process employed following (2.17) without change. Carrying this out we get an identical result at each endpoint, viz., $\left[f_s \left(-T/2 \right) + f_s \left(T/2 \right) \right] /2$. Clearly as far as any "real" discontinuity at an interior point of the original expansion interval is concerned, say at $t = t_1$, its contribution to the limit is obtainable by simply adding the last term in (2.27). Hence

$$\lim_{N \to \infty} f^N \left(\pm T/2 \right) = \frac{f \left(-T/2 \right) + f \left(T/2 \right)}{2}. \tag{2.29}$$

Of course, as in the convergence at an interior discontinuity point, the limit (2.29) gives us only part of the story, since it sidesteps the very important issue of Gibbs oscillations for finite N. A representative example of what happens when the given function assumes different values at the two endpoints is demonstrated by the Fourier expansion of e^{-t} as shown in Fig. 2.5, where the expansion interval is $0, 1$, and 21 terms ($N = 10$) are employed. Clearly the convergence at $t = 0$ and $t = 1$ is quite poor. This should be contrasted with the plot in Fig. 2.6 which shows the expansion of $e^{-|t-1/2|}$ over the same interval

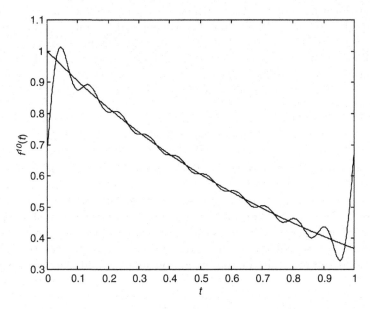

Figure 2.5: Fourier series approximation of e^{-t} with 21 sinusoids

and the same number of expansion functions. When the discontinuity occurs in the interior of the interval, the convergence is also marred by the Gibbs oscillations as illustrated in Fig. 2.7 for the pulse $p_{.5}(t - .5)$, again using 21 sinusoids. Fig. 2.8 shows a stem diagram of the magnitude of the Fourier coefficients $\hat{f}_n$ plotted as a function of ($m = n + 10$, $n = -10, -9, \ldots 11$). Such Fourier coefficients are frequently referred to as (discrete) spectral lines and are intimately related to the concept of the frequency spectrum of a signal as will be discussed in detail in connection with the Fourier integral.

2.1.4 Delta Function Representation

The convergence properties of Fourier series can be succinctly phrased in terms of delta functions. Thus the Fourier series kernel can be formally represented by the statement

$$\lim_{N \to \infty} \frac{\sin\left[2\pi\left(N + 1/2\right)\left(t - t'\right)/T\right]}{T \sin\left[\pi\left(t - t'\right)/T\right]} = \sum_{k=-\infty}^{\infty} \delta\left(t - t' - kT\right). \qquad (2.30)$$

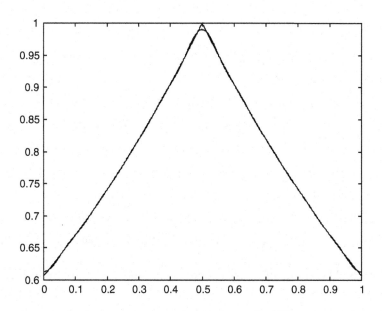

Figure 2.6: Fourier series approximation of $e^{-|t-1/2|}$ with 21 sinusoids

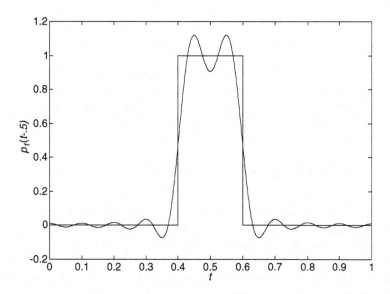

Figure 2.7: Fourier series approximation of a pulse using 21 terms

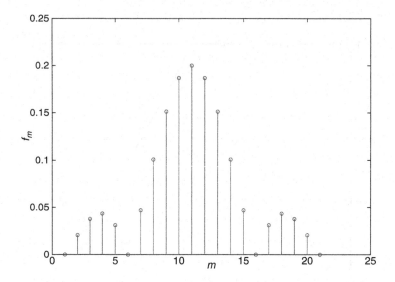

Figure 2.8: Magnitude of Fourier series coefficients for the pulse in Fig. 2.7

Alternatively, we can replace the kernel by the original geometric series and write

$$\sum_{n=-\infty}^{\infty} \left(\frac{1}{\sqrt{T}} e^{i2\pi nt/T}\right) \left(\frac{1}{\sqrt{T}} e^{i2\pi nt'/T}\right)^{*}$$

$$= \frac{1}{T} \sum_{n=-\infty}^{\infty} e^{i2\pi n(t-t')/T} = \sum_{k=-\infty}^{\infty} \delta\left(t - t' - kT\right). \qquad (2.31)$$

These expressions, just as the corresponding completeness statements for general orthogonal sets discussed in 1.7.1, are to be understood as formal notational devices invented for efficient analytical manipulations; their exact meaning is to be understood in terms of the limiting processes discussed in the preceding subsection.

2.1.5   The Fejer Summation Technique

The poor convergence properties exhibited by Fourier series at step discontinuities due to the Gibbs phenomenon can be ameliorated if one is willing to modify the expansion coefficients (spectral lines) by suitable weighting factors. The technique, generally referred to as "windowing," involves the multiplication of the Fourier series coefficients by a suitable (spectral) "window" and summation of the new trigonometric sum having modified coefficients. In general, the new series will not necessarily converge to the original function over the entire interval. The potential practical utility of such a scheme rests on the fact the approximating sum may represent certain features of the given function that are of particular interest better than the original series. This broad subject is treated in detail in books specializing in spectral estimation. Here we merely illustrate the technique with the so-called Fejer summation approach, wherein the modified trigonometric sum actually does converge to the original function. In fact this representation converges uniformly to the given function and thus completely eliminates the Gibbs phenomenon.

The Fejer [16] summation approach is based on the following result from the theory of limits. Given a sequence f^N such that $\lim_{N \to \infty} f^N \to f$ exists, the arithmetic average

$$\sigma_M = \frac{1}{M+1} \sum_{N=0}^{M} f^N \tag{2.32}$$

approaches the same limit as $M \to \infty$, i.e.,

$$\lim_{M \to \infty} \sigma_M \to f. \tag{2.33}$$

In the present case we take for $f^N = f^N(t)$, i.e., the partial Fourier series summation. Thus if this partial sum approaches $f(t)$ as $N \to \infty$, the preceding theorem states that $\sigma_M = \sigma_M(t)$ will also converge to $f(t)$. Since $f^N(t)$ is just a finite sum of sinusoids we should be able to find a closed-form expression for $\sigma_M(t)$ by a geometrical series summation. Thus

$$\sigma_M(t) = \frac{1}{M+1} \{ \hat{f}_0 + \left[\hat{f}_0 + \hat{f}_1 e^{i2\pi t/T} + \hat{f}_{-1} e^{-i2\pi t/T} \right] + \left[\begin{array}{l} \hat{f}_0 + \hat{f}_1 e^{i2\pi t/T} + \hat{f}_2 e^{i2(2\pi t/T)} + \\ \hat{f}_{-1} e^{-i2\pi t/T} + \hat{f}_{-2} e^{-i2(2\pi t/T)} \end{array} \right] + \dots \}.$$

This can be rewritten as follows:

$$\sigma_M(t) = \frac{1}{M+1} \{ (M+1) \hat{f}_0 + M \left(\hat{f}_1 e^{i2\pi t/T} + \hat{f}_{-1} e^{-i2\pi t/T} \right) + (M-1) \left(\hat{f}_2 e^{i2(2\pi t/T)} + \hat{f}_{-2} e^{-i2(2\pi t/T)} \right) + \dots \}$$

$$= \frac{1}{M+1} \{ (M+1) \hat{f}_0 + \sum_{k=1}^{M} \hat{f}_k (M-k+1) e^{ik(2\pi t/T)} + \sum_{k=1}^{M} \hat{f}_{-k} (M-k+1) e^{-ik(2\pi t/T)} \}.$$

After changing the summation index from k to $-k$ in the last sum we get

$$\sigma_M(t) = \sum_{k=-M}^{M} \hat{f}_k \left(1 - \frac{|k|}{M+1} \right) e^{ik(2\pi t/T)}, \tag{2.34}$$

which we now identify as the expansion of the function $\sigma_M(t)$ in terms of $2M + 1$ trigonometric (exponential) functions. We note that expansion coefficients are obtained by multiplying the Fourier series coefficients $\hat{f}_k$ by the triangular spectral window

$$\hat{w}_k\left(M\right) = 1 - \frac{|k|}{M+1} \quad k = 0, \pm 1, \pm 2, \ldots \pm M. \tag{2.35}$$

We can view (2.34) from another perspective if we substitute the integral representation (2.3) of the partial sum $f^N\left(t\right)$ into (2.32) and carry out the summation on the Fourier series kernel (2.5). Thus after setting $\xi = 2\pi(t - t')/T$ we get the following alternative form:

$$\begin{aligned}
\sigma_M\left(t\right) &= \frac{1}{M+1} \int_{-T/2}^{T/2} \sum_{N=0}^{M} \frac{\sin\left[(N + 1/2)\,\xi\right]}{T\sin\left[\xi/2\right]} f\left(t'\right) dt' \\
&= \frac{1}{M+1} \int_{-T/2}^{T/2} \frac{f\left(t'\right) dt'}{T\sin(\xi/2)} \sum_{N=0}^{M} \left(\frac{e^{i(N+1/2)\xi}}{2i} - \frac{e^{-i(N+1/2)\xi}}{2i}\right). \tag{2.36}
\end{aligned}$$

Using the formula

$$\sum_{N=0}^{M} e^{iN\xi} = e^{i\xi M/2}\frac{\sin\left[(M+1)\,\xi/2\right]}{\sin\left[\xi/2\right]}$$

to sum the two geometric series transforms (2.36) into

$$\sigma_M\left(t\right) = \int_{-T/2}^{T/2} \frac{\sin^2\left[(M+1)\,\pi(t-t')/T\right]}{T(M+1)\sin^2\left[\pi(t-t')/T\right]} f\left(t'\right) dt'. \tag{2.37}$$

This representation of $\sigma_M\left(t\right)$ is very much in the spirit of (2.3). Indeed in view of (2.33) $\sigma_M\left(t\right)$ must converge to the same limit as the associated Fourier series. The new kernel function

$$K_M(t - t') = \frac{\sin^2\left[(M+1)\,\pi(t-t')/T\right]}{T(M+1)\sin^2\left[\pi(t-t')/T\right]} \tag{2.38}$$

is called the Fejer kernel and (2.34) the Fejer sum. Just like the Fourier series kernel the Fejer kernel is periodic with period T so that in virtue of (2.33) we may write

$$\lim_{M\to\infty} \frac{\sin^2\left[(M+1)\,\pi(t-t')/T\right]}{T(M+1)\sin^2\left[\pi(t-t')/T\right]} = \sum_{k=-\infty}^{\infty} \delta\left(t - t' - kT\right). \tag{2.39}$$

Alternatively with the aid of limiting arguments similar to those employed in (2.11) and (2.12) one can easily verify (2.39) directly by evaluating the limit in (2.37) as $M \to \infty$.

Figure 2.9 shows the approximation achieved with the Fejer sum (2.34) (or its equivalent (2.37)) for $f\left(t\right) = U\left(t - 0.5\right)$ with 51 sinusoids ($M = 25$). Also shown for comparison is the partial Fourier series sum for the same value of M.

Note that in the Fejer sum the Gibbs oscillations are absent but that the approximation underestimates the magnitude of the jump at the discontinuity. In effect, to achieve a good fit to the "corners" at a jump discontinuity the penalty one pays with the Fejer sum is that more terms are needed than with a Fourier sum to approximate the smooth portions of the function. To get some idea of the rate of convergence to the "corners" plots of Fejer sums for $M = 10, 25, 50$, and 100 are shown in Fig. 2.10, where (for $t > 0.5$) $\sigma_{10}(t) < \sigma_{25}(t) < \sigma_{50}(t) < \sigma_{100}(t)$.

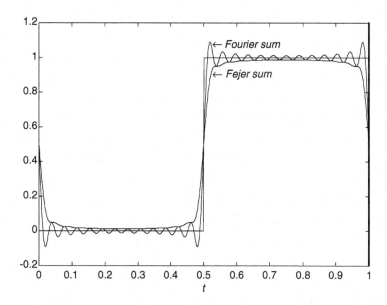

Figure 2.9: Comparison of Fejer and Fourier convergence

In passing we remark that the Fejer sum (2.34) is not a partial Fourier series sum because the expansion coefficients themselves, $\hat{\sigma}_k = \hat{w}_k(M)\hat{f}_k$ are functions of M. Trigonometric sums of this type are not unique. In fact by forming the arithmetic mean of the Fejer sum itself

$$\sigma_M^{(1)}(t) = \frac{1}{M+1}\sum_{N=0}^{M}\sigma_N(t) \tag{2.40}$$

we can again avail ourselves of the limit theorem in (2.32) and (2.33) and conclude that the partial sum $\sigma_M^{(1)}(t)$ must approach $f(t)$ in the limit of large M, i.e.,

$$\lim_{M\to\infty}\sigma_M^{(1)}(t) = f(t). \tag{2.41}$$

For any finite M we may regard $\sigma_M^{(1)}(t)$ as the second-order Fejer approximation. Upon replacing M by N in (2.34) and substituting for $\sigma_N(t)$ we can easily carry

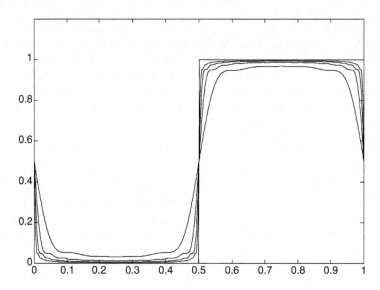

Figure 2.10: Convergence of the Fejer approximation

out one of the sums and write the final result in the form

$$\sigma_M^{(1)}(t) = \sum_{k=-M}^{M} \hat{f}_k \hat{w}_k^{(1)}(M) e^{ik(2\pi t/T)}, \tag{2.42}$$

where

$$\hat{w}_k^{(1)}(M) = \frac{1}{M+1} \sum_{n=1}^{M-|k|+1} \frac{n}{|k|+n}, \quad k = 0, \pm 1, \pm 2, \ldots \pm M \tag{2.43}$$

is the new spectral window. We see that we no longer have the simple linear taper that obtains for the first-order Fejer approximation. Unfortunately this sum does not appear to lend itself to further simplification. A plot of (2.43) in the form of a stem diagram is shown in Fig. 2.11 for $M = 12$. Figure 2.12 shows plots of the first- and second-order Fejer approximations for a rectangular pulse using $M = 25$. We see that the second-order approximation achieves a greater degree of smoothing but underestimates the pulse amplitude significantly more than does the first-order approximation. Apparently to reduce the amplitude error to the same level as achieved with the first-order approximation much larger spectral width (values of M) are required. This is consistent with the concave nature of the spectral taper in Fig. 2.11 which, for the same bandwidth, will tend to remove more energy from the original signal spectrum than a linear taper.

Clearly higher order Fejer approximations can be generated recursively with the formula

$$\sigma_M^{(m)}(t) = \frac{1}{M+1} \sum_{k=0}^{M} \sigma_k^{(m-1)}(t), \tag{2.44a}$$

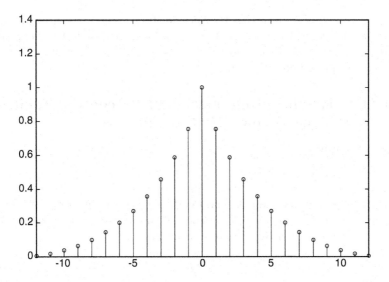

Figure 2.11: Second-order Fejer spectral window

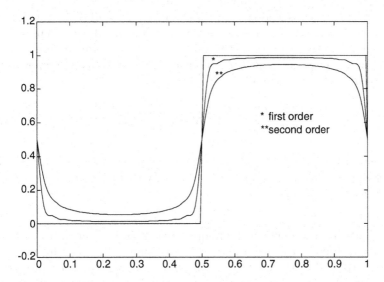

Figure 2.12: First- and second-order Fejer approximations

wherein $\sigma_k^{(0)}(t) \equiv \sigma_k(t)$. It should be noted that Fejer approximations of all orders obey the limiting property

$$\lim_{M \to \infty} \sigma_M^{(m-1)}(t) = \frac{1}{2}[f(t^+) + f(t^-)] \; ; \quad m = 1, 2, 3, \ldots. \tag{2.44b}$$

i.e., at step discontinuities the partial sums converge to the arithmetic average of the given function, just like ordinary Fourier series. The advantage of higher order Fejer approximations is that they provide for a greater degree of smoothing in the neighborhood of step discontinuities. This is achieved at the expense of more expansion terms (equivalently, requiring wider bandwidths) to reach a given level of approximation accuracy.

2.1.6   Fundamental Relationships Between the Frequency and Time Domain Representations

Parseval Formula

Once all the Fourier coefficients of a given function are known they may be used, if desired, to reconstruct the original function. In fact, the specification of the coefficients and the time interval within which the function is defined is, in principle, equivalent to the specification of the function itself. Even though the $\hat{f}_n$ are components of the infinite-dimensional vector

$$\mathbf{f} = [\ldots \hat{f}_n \ldots]^T, \tag{2.45}$$

we can still interpret them as the projections of the signal $f(t)$ along the basis functions $e^{i2\pi nt/T}$ and think of them geometrically as in Fig. 1.3. Because each $\hat{f}_n$ is uniquely associated with a radian frequency of oscillation ω_n, with $\omega_n/2\pi = n/T \ Hz$, $\mathbf{f}$ is said to constitute the frequency domain representation of the signal, and the elements of $\mathbf{f}$ the signal (line) spectrum. A very important relationship between the frequency domain and the time domain representations of the signal is Parseval formula

$$\frac{1}{T} \int_{-T/2}^{T/2} |f(t)|^2 \, dt = \sum_{n=-\infty}^{n=\infty} \left| \hat{f}_n \right|^2. \tag{2.46}$$

This follows as a special case of (1.305) and is a direct consequence of the LMS error in the approximation tending to zero. With $\mathbf{f}' \equiv \sqrt{T}\mathbf{f}$ we can rewrite (2.46) using the notation

$$(f, f) = \|\mathbf{f}'\|^2, \tag{2.47}$$

which states that the norm in the frequency domain is identical to that in the time domain. Since physically the time average on the left of (2.46) may generally be interpreted as the average signal power (or some quantity proportional to it), Parseval formula in effect states that the average power in the time and frequency domains is preserved.

Given the two functions, $f(t)$ and $g(t)$ within the interval $-T/2, T/2$ with Fourier coefficients $\hat{f}_n$ and $\hat{g}_n$, it is not hard to show (problem 2-2) that (2.46) generalizes to

$$\frac{1}{T} \int_{-T/2}^{T/2} f(t) g^*(t) \, dt = \sum_{n=-\infty}^{n=\infty} \hat{f}_n \hat{g}_n^*. \tag{2.48}$$

Time and Frequency Domain Convolution

An important role in linear system analysis is played by the convolution integral. From the standpoint of Fourier series this integral is of the form

$$h\left(t\right) = \frac{1}{T} \int_{-T/2}^{T/2} f\left(\tau\right) g\left(t - \tau\right) d\tau. \tag{2.49}$$

We now suppose that the Fourier series coefficients $\hat{f}_n$ and $\hat{g}_n$ of $f\left(t\right)$ and $g\left(t\right)$, defined within $-T/2, T/2$, are known. What will be the Fourier coefficients $\hat{h}_m$ of $h\left(t\right)$ when expanded in the same interval? The answer is readily obtained when we represent $f\left(\tau\right)$ by its Fourier series (2.13) and similarly $g\left(t - \tau\right)$. Thus

$$
\begin{aligned}
h\left(t\right) &= \frac{1}{T} \int_{-T/2}^{T/2} \sum_{n=-\infty}^{\infty} \hat{f}_n e^{i2\pi n\tau/T} \sum_{m=-\infty}^{\infty} \hat{g}_m e^{i2\pi m(t-\tau)/T} d\tau \\
&= \frac{1}{T} \sum_{m=-\infty}^{\infty} \hat{g}_m e^{i2\pi mt/T} \sum_{n=-\infty}^{\infty} \hat{f}_n \int_{-T/2}^{T/2} e^{i2\pi(n-m)\tau/T} d\tau \\
&= \frac{1}{T} \sum_{m=-\infty}^{\infty} \hat{g}_m e^{i2\pi mt/T} \sum_{n=-\infty}^{\infty} \hat{f}_n T \delta_{nm} \\
&= \sum_{m=-\infty}^{\infty} \hat{g}_m \hat{f}_m e^{i2\pi mt/T} = \sum_{m=-\infty}^{\infty} \hat{h}_m e^{i2\pi mt/T} \tag{2.50}
\end{aligned}
$$

from which we identify $\hat{h}_m = \hat{g}_m \hat{f}_m$. A dual situation frequently arises when we need the Fourier coefficients of the product of the two functions, e.g., $q(t) \equiv f\left(t\right) g\left(t\right)$. Here we can proceed similarly

$$
\begin{aligned}
q(t) &\equiv f\left(t\right) g\left(t\right) \\
&= \sum_{n=-\infty}^{\infty} \hat{f}_n e^{i2\pi nt/T} \sum_{m=-\infty}^{\infty} \hat{g}_m e^{i2\pi mt/T} \\
&= \sum_{n=-\infty}^{\infty} \sum_{m=-\infty}^{\infty} \hat{f}_n \hat{g}_m e^{i2\pi(n+m)t/T} \\
&= \sum_{n=-\infty}^{\infty} \sum_{k=-\infty}^{\infty} \hat{f}_n \hat{g}_{k-n} e^{i2\pi kt/T} \\
&= \sum_{n=-\infty}^{\infty} \left(\sum_{m=-\infty}^{\infty} \hat{f}_m \hat{g}_{n-m} \right) e^{i2\pi nt/T} = \sum_{n=-\infty}^{\infty} \hat{q}_n e^{i2\pi nt/T}, \tag{2.51}
\end{aligned}
$$

where in the last step we identify the Fourier coefficient of $q(t)$ as $\hat{q}_n = \sum_{m=-\infty}^{\infty} \hat{f}_m \hat{g}_{n-m}$ which is a convolution sum formed with the Fourier coefficients of the two functions.

Symmetries

Frequently (but not always) the signal in the time domain will be real. In that case the formula for the coefficients gives

$$\hat{f}_{-n} = \hat{f}_n^*, \tag{2.52}$$

which means that the magnitude of the line spectrum is symmetrically disposed with respect to the index $n = 0$. Simplifications also arise when the signal is either an even or an odd function with respect to $t = 0$. In case of an even function $f(t) = f(-t)$ we obtain

$$\hat{f}_n = \frac{2}{T} \int_0^{T/2} f(t) \cos\left(2\pi nt/T\right) dt \tag{2.53}$$

and since $\hat{f}_{-n} = \hat{f}_n$ the Fourier series reads

$$f(t) = \hat{f}_0 + 2 \sum_{n=1}^{\infty} \hat{f}_n \cos\left(2\pi nt/T\right). \tag{2.54}$$

In case of an odd function $f(t) = -f(-t)$ the coefficients simplify to

$$\hat{f}_n = \frac{-i2}{T} \int_0^{T/2} f(t) \sin\left(2\pi nt/T\right) dt \tag{2.55}$$

and since $\hat{f}_{-n} = -\hat{f}_n$ we have for the Fourier series

$$f(t) = i2 \sum_{n=1}^{\infty} \hat{f}_n \sin\left(2\pi nt/T\right). \tag{2.56}$$

It is worth noting that (2.53-2.54) hold for complex functions in general, independent of (2.52).

2.1.7   Cosine and Sine Series

In our discussion of convergence of Fourier series we noted that whenever a function assumes unequal values at the interval endpoints its Fourier series coverages at either endpoint to the arithmetic mean of the two endpoint values. An illustration of how the approximation manifests itself when finite partial sums are involved may be seen from the plot in Fig. 2.5 for an exponential function. It turns out that these pathological convergence properties can actually be eliminated by a judicious choice of the expansion interval. The approach rests on the following considerations. Suppose function $f(t)$ to be expanded is defined in the interval $0, T$ while the nature of its periodic extension is outside the domain of the problem of interest and, consequently, at our disposal. In that case we may artificially extend the expansion interval to $-T, T$ and define a function over this new interval as $f(|t|)$, as shown in Fig. 2.13. This function is continuous

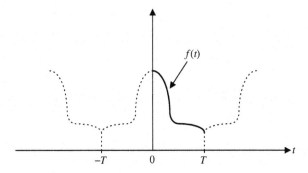

Figure 2.13: Extension of the function for the cosine series

at $t = 0$ and moreover assumes identical values at $-T$ and T. Hence its periodic extension is also continuous at these endpoints which means that its Fourier series will converge uniformly throughout the closed interval $-T, T$ to $f(|t|)$ and, in particular, to the prescribed function $f(t)$ throughout the desired range $0 \le t \le T$. Of course, since $f(|t|)$ is even with respect to $t = 0$, this Fourier series contains only cosine terms. However, because the expansion interval is $2T$ rather than T, the arguments of the expansion functions are $\pi nt/T$ rather than $2\pi nt/T$. Hence

$$
\begin{aligned}
\hat{f}_n &= \frac{1}{2T} \int_{-T}^{T} f(|t|) \cos(\pi nt/T)\, dt \\
&= \frac{1}{T} \int_{0}^{T} f(t) \cos(\pi nt/T)\, dt.
\end{aligned}
\tag{2.57}
$$

The Fourier cosine series reads

$$
\begin{aligned}
f(t) &= \hat{f}_0 + 2 \sum_{n=1}^{\infty} \hat{f}_n \cos(\pi nt/T) \\
&= \sum_{n=0}^{\infty} \hat{f}_n^c \cos(\pi nt/T),
\end{aligned}
\tag{2.58}
$$

where

$$
\hat{f}_n^c = \begin{cases} \frac{1}{T} \int_0^T f(t)\, dt \ ; \ n = 0, \\ \frac{2}{T} \int_0^T f(t) \cos(\pi nt/T)\, dt \ ; n > 0. \end{cases}
\tag{2.59}
$$

The approximation to e^{-t} using a cosine series comprised of 10 terms is plotted in Fig. 2.14. We note a significant improvement in the approximation over that obtained with the conventional partial Fourier series sum in Fig. 2.5, where 21 terms are employed to approximate the same function.

It should be noted that the coefficients of the cosine series (2.59) are nothing more than the solution to the normal equations for the LMS problem phrased in terms of the cosine functions

$$
\phi_n^c(t) = \cos(\pi nt/T), \quad n = 0, 1, 2, \ldots
\tag{2.60}
$$

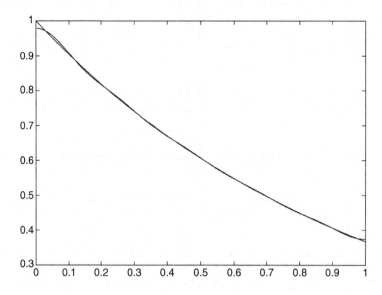

Figure 2.14: Cosine series approximation (N=10)

As may be verified directly, they are orthogonal over the interval $0, T$. In our compact notation this reads

$$(\phi_n^c, \phi_m^c) = (T/\varepsilon_n)\,\delta_{nm},$$

where we have introduced the abbreviation

$$\varepsilon_n = \left\{ \begin{array}{l} 1; n = 0, \\ 2\ ; n > 0, \end{array} \right.$$

which is usually referred to as the Neumann symbol.

The convergence properties of the cosine series at points of continuity and at jump discontinuities within the interval are identical to those of the complete Fourier series from which, after all, the cosine series may be derived. The cosine expansion functions form a complete set in the space of piecewise differentiable functions whose derivatives must vanish at the interval endpoints. This additional restriction arises because of the vanishing of the derivative of $\cos(\pi nt/T)$ at $t = 0$ and $t = T$. In accordance with (1.303), the formal statement of completeness may be phrased in terms of an infinite series of products of the orthonormal expansion functions $\sqrt{\varepsilon_n/T}\phi_n^c(t)$ as follows:

$$\delta(t - t') = \sum_{n=0}^{\infty} \sqrt{\frac{\varepsilon_n}{T}} \cos(\pi nt/T) \sqrt{\frac{\varepsilon_n}{T}} \cos(\pi nt'/T). \qquad (2.61)$$

Sine Series

If instead of an even extension of $f(t)$ into the interval $-T, 0$ as in Fig. 2.13, we employ an odd extension, as in Fig. 2.15, and expand the function $f(|t|)\,sign(t)$ in a Fourier series within the interval $-T, T$, we find that the cosine terms

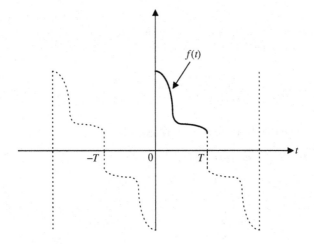

Figure 2.15: Function extension for sine series

vanish and the resulting Fourier series is comprised entirely of sines. Within the original interval $0, T$ it converges to the prescribed function $f(t)$ and constitutes the so-called sine series expansion, to wit,

$$f(t) = \sum_{n=0}^{\infty} \hat{f}_n^s \sin(\pi nt/T),\qquad (2.62)$$

where

$$\hat{f}_n^s = \frac{2}{T}\int_0^T f(t)\sin(\pi nt/T)\,dt.\qquad (2.63)$$

Evidently because the sine functions vanish at the interval endpoints the sine series will necessarily converge to zero there. Since at a discontinuity a Fourier series always converges to the arithmetic mean of the left and right endpoint values, we see from Fig. 2.15 that the convergence of the sine series to zero at the endpoints does not require that the prescribed function also vanishes there. Of course, if this is not the case, only LMS convergence is guaranteed at the endpoints and an approximation by a finite number of terms will be vitiated by the Gibbs effect. A representative illustration of the expected convergence behavior in such cases can be had by referring to Fig. 2.5. For this reason the sine series is to be used only with functions that vanish at the interval endpoints. In such cases convergence properties very similar to those of cosine series are achieved. A case in point is the approximation shown in Fig. 2.6.

The sine expansion functions

$$\phi_n^s(t) = \sin(\pi nt/T),\ n = 1,2,3,\dots\qquad (2.64)$$

possess the orthogonality properties

$$(\phi_n^s, \phi_m^s) = (T/2)\,\delta_{nm};\qquad (2.65)$$

they form a complete set in the space of piecewise differentiable functions that vanish at the interval endpoints. Again the formal statement of this completeness may be summarized by the delta function representation

$$\delta\left(t-t'\right)=\sum_{n=0}^{\infty}\sqrt{\frac{2}{T}}\sin\left(\pi nt/T\right)\sqrt{\frac{2}{T}}\sin\left(\pi nt'/T\right). \tag{2.66}$$

2.1.8   Interpolation with Sinusoids

Interpolation Using Exponential Functions

Suppose $f\left(t\right)$ can be represented exactly by the sum

$$f\left(t\right)=\sum_{n=-N}^{N}c_{n}e^{i\frac{2\pi nt}{T}}\ ;\ 0\leq t\leq T. \tag{2.67}$$

If $f\left(t\right)$ is specified at $M=2N+1$ points within the given interval (2.67) can be viewed as a system of M linear equations for the M unknown coefficients c_{n}. A particularly simple formula for the coefficients results if we suppose that the function is specified on uniformly spaced points within the interval. To derive it we first change the summation index in (2.67) from n to $m=N+n$ to obtain

$$f\left(t\right)=\sum_{m=0}^{2N}c_{m-N}e^{i\frac{2\pi(m-N)t}{T}}. \tag{2.68}$$

With $t=\ell\Delta t$ and $\Delta t=T/M$ (2.68) becomes

$$f\left(\ell\Delta t\right)=\sum_{m=0}^{M-1}c_{m-N}e^{i\frac{2\pi(m-N)\ell}{M}}. \tag{2.69}$$

From the geometric series $\sum_{m=0}^{M-1}e^{im\alpha}=e^{i\alpha(M-1)/2}\sin\left(M\alpha/2\right)/\sin(\alpha/2)$ we readily establish the orthogonality relationship

$$\sum_{\ell=0}^{M-1}e^{i\frac{2\pi\ell(m-k)}{M}}=M\delta_{mk}. \tag{2.70}$$

Upon multiplying both sides of (2.69) by $e^{-i\frac{2\pi\ell k}{M}}$ and summing on ℓ and using (2.70) we obtain the solution for the coefficients

$$c_{m-N}=\frac{1}{M}\sum_{\ell=0}^{M-1}f\left(\ell\Delta t\right)e^{-i\frac{2\pi\ell(m-N)}{M}}. \tag{2.71}$$

Reverting to the index n and $M=2N+1$ the preceding is equivalent to

$$c_{n}=\frac{1}{2N+1}\sum_{\ell=0}^{2N}f\left(\ell\Delta t\right)e^{-i\frac{2\pi\ell n}{2N+1}}. \tag{2.72}$$

On the other hand we know that the solution for c_n in (2.67) is also given by the integral

$$c_n = \frac{1}{T} \int_0^T f(t) e^{-i\frac{2\pi n t}{T}} dt. \tag{2.73}$$

If in (2.72) we replace $1/(2N+1)$ by its equivalent $\Delta t/T$, we can interpret (2.71) as a Riemann sum approximation to (2.73). However we know from the foregoing that (2.72) is in fact an exact solution of (2.69). Thus whenever $f(t)$ is comprised of a finite number of sinusoids the Riemann sum will represent the integral (2.73) exactly provided $2N+1$ is chosen equal to or greater than the number of sinusoids. Evidently, if the number of sinusoids is exactly $2N+1$, the c_n as computed using either (2.73) or (2.72) must be identically zero whenever $|n| > N$. If $f(t)$ is a general piecewise differentiable function, then (2.67) with the coefficients determined by (2.72) provides an interpolation to $f(t)$ in terms of sinusoids. In fact by substituting (2.72) into (2.67) and again summing a geometric series we obtain the following explicit interpolation formula:

$$f(t) = \sum_{\ell=0}^{M-1} f(\ell \Delta t) \frac{\sin\left[\pi \left(\frac{t}{\Delta t} - \ell\right)\right]}{M \sin\left[\frac{\pi}{M} \left(\frac{t}{\Delta t} - \ell\right)\right]}. \tag{2.74}$$

Unlike the LMS approximation problem underlying the classical Fourier series, the determination of the coefficients in the interpolation problem does not require the evaluation of integrals. This in itself is of considerable computational advantage. How do interpolation-type approximations compare with LMS approximations? Figure 2.16 shows the interpolation of e^{-t} achieved with 11 sinusoids while Fig. 2.17 shows the approximation with the same number of sinusoids using the LMS approximation. We note that the fit is comparable in the two cases except at the endpoints where, as we know, the LMS approximation necessarily converges to $(1 + e^{-1})/2$. As the number of terms in the interpolation is increased the fit within the interval improves. Nevertheless, the interpolated function continues to show considerable undamped oscillatory behavior near the endpoints as shown by the plot in Fig. 2.18.

Interpolation Using Cosine Functions

Recalling the improvement in the LMS approximation achieved with the cosine series over the complete Fourier expansion, we might expect a similar improvement in case of interpolation. This turns out actually to be the case. As will be demonstrated, the oscillatory behavior near the endpoints in Fig. 2.18 can be completely eliminated and a substantially better fit to the prescribed function achieved throughout the entire approximating interval using an alternative interpolation that employs only cosine functions, i.e., an interpolation formula based on (2.58) rather than (2.67). In this case we set the interpolation interval to

$$\Delta t = T/(M - 1/2) \tag{2.75}$$

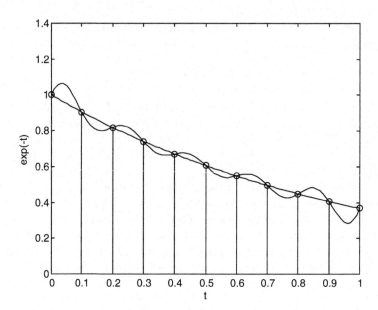

Figure 2.16: Interpolation of e^{-t} using 11 sinusoids

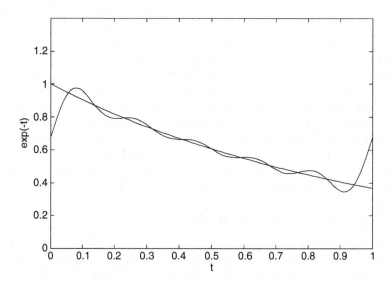

Figure 2.17: LMS approximation to e^{-t} using 11 sinusoids

and with $t = m\Delta t$ in (2.58) we obtain

$$f(m\Delta t) = \sum_{n=0}^{M-1} c_n^c \cos\left[\pi nm/(M-1/2)\right] \quad ; \quad m = 0, 1, 2, \dots M-1, \quad (2.76)$$

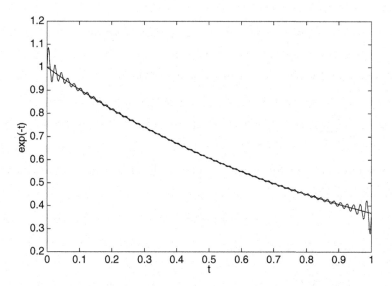

Figure 2.18: Interpolation of e^{-t} using 101 sinusoids

where the c_n^c are the unknown coefficients. The solution for the c_n^c is made somewhat easier if one first extends the definition of $f(m\Delta t)$ to negative indices as in Fig. 2.13 and rewrites (2.76) in terms of complex exponentials. Thus

$$f(m\Delta t) = \sum_{n=-(M-1)}^{M-1} c_n'^c e^{i2\pi nm/(2M-1)} \; ; \quad m = 0, \pm 1, \pm 2, \ldots \pm(M-1), \quad (2.77)$$

where in addition to $f(m\Delta t) = f(-m\Delta t)$ we postulated that $c_n^c = c_{-n}^c$ and defined

$$c_n'^c = \begin{cases} c_0^c \; ; \; n = 0, \\ c_n^c/2 \; ; \; n \neq 0. \end{cases} \quad (2.78)$$

Again using the geometric series sum formula we have the orthogonality

$$\sum_{n=-(M-1)}^{M-1} e^{i2\pi n(m-k)/(2M-1)} = \frac{\sin[\pi(m-k)]}{\sin[\pi(m-k)/(2M-1)]}$$

$$\equiv (2M-1)\delta_{mk} \quad (2.79)$$

with the aid of which the solution for the $c_n'^c$ in (2.78) follows at once:

$$c_n'^c = \frac{1}{2M-1} \sum_{m=-(M-1)}^{M-1} f(m\Delta t) e^{-i2\pi nm/(2M-1)}$$

$$= \frac{1}{2M-1} \sum_{m=0}^{M-1} \varepsilon_m f(m\Delta t) \cos[2\pi nm/(2M-1)]$$

$$= \frac{1}{M - 1/2} \sum_{m=0}^{M-1} (\varepsilon_m/2) f (m\Delta t) \cos [\pi nm/(M - 1/2)].$$

Taking account of (2.78) we obtain the final result

$$c_n^c = \frac{2}{M - 1/2} \sum_{m=0}^{M-1} (\varepsilon_n \varepsilon_m/4) f (m\Delta t) \cos [\pi nm/(M-1/2)]; \quad n = 0, 1, 2, \ldots M-1.$$

$$(2.80)$$

The final interpolation formula now follows through a direct substitution of (2.80) into

$$f(t) = \sum_{n=0}^{M-1} c_n^c \cos (\pi nt/T).$$ (2.81)

After summation over n we obtain

$$f(t) = \frac{1}{M - 1/2} \sum_{m=0}^{M-1} (\varepsilon_m/2) f (m\Delta t) \{1 + k_M (t/\Delta t - m) + k_M (t/\Delta t + m)\},$$

$$(2.82)$$

where

$$k_M (t) = \cos \left(\frac{\pi M}{2M - 1} t \right) \frac{\sin \left[\frac{\pi(M-1))}{2(M-1/2)} t \right]}{\sin \left[\frac{\pi}{2(M-1/2)} t \right]}.$$ (2.83)

Fig. 2.19 shows the interpolation of e^{-t} using 11 cosine functions.

The improvement over the interpolation wherein both sines and cosines were employed, Fig. 2.16, is definitely noticeable. A more important issue with general sinusoids is the crowding toward the interval endpoints as in Fig. 2.18. With the cosine interpolation these oscillations are completely eliminated, as may be seen from the plot in Fig. 2.20.

By choosing different distributions of the locations and sizes of the interpolation intervals the interpolation properties can be tailored to specific classes of functions. Of course, a nonuniform distribution of interpolation intervals will in general not lead to analytically tractable forms of expansion coefficients and will require a numerical matrix inversion. We shall not deal with nonuniform distribution of intervals. There is, however, a slightly different way of specifying a uniform distribution of interpolation intervals from the one we have just considered which is worth mentioning since it leads to formulas for the so-called discrete cosine transform commonly employed in data and image compression work. Using the seemingly innocuous modification of (2.75) to

$$\Delta t = \frac{T}{2M}$$ (2.84)

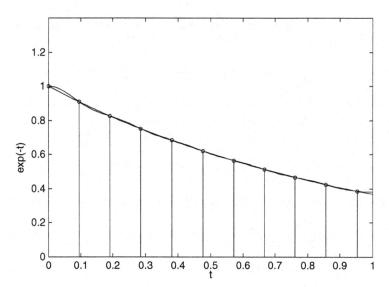

Figure 2.19: Interpolation of e^{-t} with 11 cosine functions

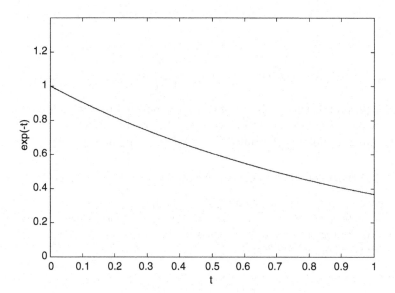

Figure 2.20: Interpolation of e^{-t} with 101 cosine functions

and forcing the first and the last step size to equal $\Delta t/2$ we replace (2.76) by

$$f\left[\Delta t\left(m+1/2\right)\right] = \sum_{n=0}^{M-1} \hat{c}_n^c \cos\left[\pi n\left(2m+1\right)/2M\right] \quad ; \quad m = 0, 1, 2, \ldots M - 1.$$

$$(2.85)$$

With the aid of the geometrical sum formula we can readily verify the orthogonality relationship

$$\sum_{m=0}^{M-1} \cos\left[\pi n\left(2m+1\right)/2M\right] \cos\left[\pi k\left(2m+1\right)/2M\right] = \frac{M}{\varepsilon_n}\delta_{nk} \tag{2.86}$$

with the aid of which we solve for the coefficients in (2.85):

$$\hat{c}_n^c = \frac{\varepsilon_n}{M}\sum_{m=0}^{M-1} f\left[\Delta t\left(m+1/2\right)\right]\cos\left[\pi n\left(2m+1\right)/2M\right]. \tag{2.87}$$

Replacing c_n^c in (2.81) by $\hat{c}_n^c$ of (2.87) yields the interpolation formula

$$f\left(t\right) = \frac{1}{M}\sum_{m=0}^{M-1} f\left[\Delta t\left(m+1/2\right)\right]\left[1 + \hat{k}_M\left(\tau^+\right) + \hat{k}_M\left(\tau^-\right)\right], \tag{2.88}$$

where

$$\tau^+ = t/\Delta t - \left(m+1/2\right) \tag{2.88a}$$
$$\tau^- = t/\Delta t + \left(m+1/2\right) \tag{2.88b}$$

and

$$\hat{k}_M\left(t\right) = \cos(\pi t/2)\frac{\sin\left(\frac{\pi(M-1)}{2M}t\right)}{\sin\left(\frac{\pi}{2M}t\right)}. \tag{2.89}$$

Equation (2.85) together with (2.87) is usually referred to as the discrete cosine transform pair. Here we have obtained it as a by-product along our route toward a particular interpolation formula comprised of cosine functions.

2.1.9   Anharmonic Fourier Series

Suppose we approximate the signal $f(t)$ in the LMS sense by a sum of sinusoids with radian frequencies $\mu_1, \mu_2, \ldots \mu_N$ which are not necessarily harmonically related. Assuming the signal is specified in the interval $a \leq t \leq b$ we write this approximating sum as follows:

$$f(t) \sim \sum_{n=1}^{N} \hat{f}_n\psi_n\left(t\right), \tag{2.90}$$

wherein

$$\psi_n\left(t\right) = A_n \sin\mu_n t + B_n \cos\mu_n t \tag{2.91}$$

and A_n and B_n are suitable normalization constants. It is not hard to show that as long as all the μ_n are distinct the Gram matrix $\Gamma_{nm} = \left(\psi_n, \psi_m\right)$ is nonsingular so that the normal equations yield a unique set of expansion coefficients $\hat{f}_n$. Of course their computation would be significantly simplified if it were possible to choose a sets of radian frequencies μ_n such that the Gram matrix is diagonal, or, equivalently, that the ψ_n are orthogonal over the chosen interval. We know that this is always the case for harmonically related radian frequencies. It turns out that orthogonality also obtains when the radian frequencies are not harmonically related provided they are chosen such that for a given pair of real constants α and β the $\psi_n(t)$ satisfy the following endpoint conditions:

$$\psi_n(a) = \alpha \psi_n'(a), \tag{2.92a}$$
$$\psi_n(b) = \beta \psi_n'(b). \tag{2.92b}$$

To prove orthogonality we first observe that the $\psi_n(t)$ satisfy the differential equation of the harmonic oscillator, i.e.,

$$\frac{d^2 \psi_n}{dt^2} + \mu_n^2 \psi_n = 0, \tag{2.93}$$

where we may regard the ψ_n as an eigenvector and μ_n^2 as the eigenvalue of the differential operator $-d^2\psi_n/dt^2$. Next we multiply (2.93) by ψ_m and integrate the result over $a \leq t \leq b$ to obtain

$$\psi_m \frac{d\psi_n}{dt} \Big|_a^b - \int_a^b \frac{d\psi_m}{dt} \frac{d\psi_n}{dt} dt + \mu_n^2 \int_a^b \psi_m \psi_n dt = 0, \tag{2.94}$$

where the second derivative has been eliminated by an integration by parts. An interchange of indices in (2.94) gives

$$\psi_n \frac{d\psi_m}{dt} \Big|_a^b - \int_a^b \frac{d\psi_n}{dt} \frac{d\psi_m}{dt} dt + \mu_m^2 \int_a^b \psi_n \psi_m dt = 0 \tag{2.95}$$

and subtraction of (2.95) from (2.94) yields

$$\psi_m \frac{d\psi_n}{dt} \Big|_a^b - \psi_n \frac{d\psi_m}{dt} \Big|_a^b = (\mu_n^2 - \mu_m^2) \int_a^b \psi_n \psi_m dt. \tag{2.96}$$

We now observe that substitution of the endpoint conditions (2.92) into the left side of (2.96) yields zero. This implies orthogonality provided we assume that for $n \neq m$ μ_m and μ_n are distinct. For then

$$\int_a^b \psi_n \psi_m dt = 0 \; ; \; n \neq m. \tag{2.97}$$

The fact that the eigenvalues μ_n^2 are distinct follows from a direct calculation. To compute the eigenvalues we first substitute (2.91) into (2.92) which yields the following set of homogeneous algebraic equations:

$$(\sin \mu_n a - \alpha \mu_n \cos \mu_n a) A_n + (\cos \mu_n a + \alpha \mu_n \sin \mu_n a) B_n = 0, \tag{2.98a}$$
$$(\sin \mu_n b - \beta \mu_n \cos \mu_n b) A_n + (\cos \mu_n a + \alpha \mu_n \sin \mu_n a) B_n = 0. \tag{2.98b}$$

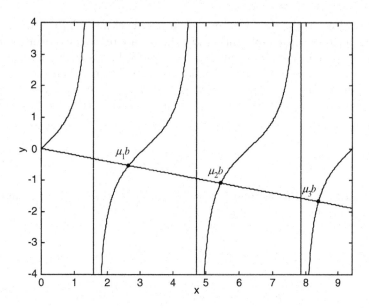

Figure 2.21: Diagram of the transcendental equation $-0.2x = \tan x$

A nontrivial solution for A_n and B_n is only possible if the determinant of the coefficients vanishes. Computing this determinant and setting the result to zero yield the following equation for μ_n :

$$(\beta - \alpha)\mu_n \cos\left[\mu_n(b-a)\right] - (1 + \alpha\beta\mu_n^2) \sin\left[\mu_n(b-a)\right] = 0. \qquad (2.99)$$

This transcendental equation possesses an infinite set of distinct positive simple zeros μ_n. For an arbitrary set of parameters these roots can only be determined numerically. Many standard root finding algorithms are available for this purpose. Generally these are iterative techniques that require a "good" first guess of the root. In case of (2.99) an approximate location to start the iteration can be got from a graphical construction. We illustrate it for $\alpha = 0$ and $a = 0$ in which case (2.99) becomes $\beta\mu_n \cos\mu_n b - \sin\mu_n b = 0$, which is equivalent to

$$\beta\mu_n = \tan(\mu_n b). \qquad (2.100)$$

Defining the nondimensional variable $x = \mu_n b$ in (2.100) we obtain the roots from the intersection of the straight line $y = x\beta/b$ with the curves defined by the various branches of $y = \tan x$ as shown in Fig. 2.21 for $\beta/b = -0.2$. The first three roots expressed in terms of the nondimensional quantities $\mu_1 b, \mu_2 b$ and $\mu_3 b$ may be read off the abscissa. When $\alpha = 0$ and $a = 0$ (2.98a) requires that $B_n = 0$ so that the expansion functions that correspond to the solutions of (2.100) are

$$\psi_n(t) = A_n \sin \mu_n t. \qquad (2.101)$$

Setting $A_n = 1$ we compute the normalization constant

$$Q_n \quad = \quad \int_0^b \sin^2 \mu_n t \, dt = \int_0^b \frac{1 - \cos 2\mu_n t}{2} dt$$

$$= \left(b/2 - \frac{\sin 2\mu_n b}{4\mu_n}\right) = \frac{b}{2}(1 - \frac{\sin \mu_n b \cos \mu_n b}{\mu_n b}).$$

The expansion coefficients in (2.90) for this case are

$$\hat{f}_n = \frac{2}{b}\left(1 - \frac{\sin \mu_n b \cos \mu_n b}{\mu_n b}\right)^{-1} \int_0^b f(t) \sin \mu_n t \, dt. \tag{2.102}$$

From Fig. 2.21 we note that as n increases the abscissas of the points where the straight line intersects the tangent curves approach $\pi/2(2n-1) \approx n\pi$. Hence for large n the radian frequencies of the anharmonic expansion (2.90) are asymptotically harmonic, i.e.,

$$\mu_n \underset{n \sim \infty}{\sim} n\pi/b. \tag{2.103}$$

Taking account of (2.103) in (2.102) we also observe that for large n formula (2.102) represents the expansion coefficient of a sine Fourier series (2.63). Thus the anharmonic character of the expansion appears to manifest itself only for finite number of terms. Hence we would expect that the convergence properties of anharmonic expansions to be essentially the same as harmonic Fourier series.

An anharmonic series may be taken as a generalization of a Fourier series. For example, it reduces to the (harmonic) sine series in (2.62) when $\alpha = \beta = 0$ and when $\alpha = \beta \to \infty$ to the (harmonic) cosine series (2.58), provided $f(a) \neq 0$ and $f(b) \neq 0$. When the endpoint conditions (2.92) are replaced by a periodicity condition we obtain the standard Fourier series.

2.2 The Fourier Integral

2.2.1 LMS Approximation by Sinusoids Spanning a Continuum

Instead of approximating $f(t)$ by a sum of $2N+1$ sinusoids with discrete frequencies $\omega_n = 2\pi n/T$ we now suppose that the frequencies ω span a continuum between $-\Omega$ and Ω. With

$$f^\Omega(t) = \int_{-\Omega}^{\Omega} \hat{f}(\omega) e^{i\omega t} d\omega \tag{2.104}$$

we seek a function $\hat{f}(\omega)$ such that the MS error

$$\varepsilon_\Omega(T) \equiv \int_{-T/2}^{T/2} \left| f(t) - f^\Omega(t) \right|^2 dt \tag{2.105}$$

is minimized. As we know, this minimization leads to the normal equation (1.99), where we identify $\phi(\omega, t) = e^{i\omega t}, a = -T/2, b = T/2$, so that with the aid of (1.100) we obtain

$$\int_{-T/2}^{T/2} f(t) e^{-i\omega t} dt = \int_{-\Omega}^{\Omega} \hat{f}(\omega') \frac{2\sin\left[(\omega - \omega')T/2\right]}{(\omega - \omega')} d\omega'. \tag{2.106}$$

Thus unlike in the case of a discrete set of sinusoids the unknown "coefficients" $\hat{f}(\omega')$ now span a continuum. In fact, according to (2.106), to find $\hat{f}(\omega')$ we must solve an integral equation.

2.2.2   Transition to an Infinite Observation Interval: The Fourier Transform

For any finite time interval T the solution of (2.106) for $\hat{f}(\omega')$ can be expressed in terms of spheroidal functions [23]. Here we confine our attention to the case of an infinite time interval, which is the conventional domain of the Fourier integral. In that case we can employ the limiting form of the Fourier kernel in (1.269) (with Ω replaced by $T/2$) so that the right side of (2.106) becomes

$$\lim_{T \to \infty} \int_{-\Omega}^{\Omega} \hat{f}(\omega') \frac{2 \sin\left[(\omega - \omega')T/2\right]}{(\omega - \omega')} d\omega' = 2\pi \hat{f}(\omega). \tag{2.107}$$

Hence as the expansion interval in the time domain is allowed to approach infinity the solution of (2.106) reads

$$\int_{-\infty}^{\infty} f(t) e^{-i\omega t} dt = F(\omega), \tag{2.108}$$

where we have set $F(\omega) = 2\pi \hat{f}(\omega)$ which shall be referred to as the Fourier Integral (or the Fourier transform) of $f(t)$. Substituting this in (2.104) and integrating with respect to ω we get

$$f^{\Omega}(t) = \int_{-\infty}^{\infty} f(t') \frac{\sin\left[(t - t')\Omega\right]}{\pi(t - t')} dt'. \tag{2.109}$$

The corresponding LMS error $\varepsilon_{\Omega \min}$ is

$$\begin{aligned} \varepsilon_{\Omega \min} &= \left(f - f^{\Omega}, f - f^{\Omega}\right) \\ &= (f, f) - \left(f, f^{\Omega}\right) \geq 0, \end{aligned} \tag{2.110}$$

where the inner products are taken over the infinite time domain and account has been taken of the projection theorem (1.75). Substituting for f^{Ω} from (2.104) the preceding is equivalent to

$$\begin{aligned} \varepsilon_{\Omega \min} &= \int_{-\infty}^{\infty} |f(t)|^2 dt - \int_{-\infty}^{\infty} f^*(t) dt \int_{-\Omega}^{\Omega} \hat{f}(\omega) e^{i\omega t} d\omega \\ &= \int_{-\infty}^{\infty} |f(t)|^2 dt - 2\pi \int_{-\Omega}^{\Omega} \left|\hat{f}(\omega)\right|^2 d\omega \\ &= \int_{-\infty}^{\infty} |f(t)|^2 dt - \frac{1}{2\pi} \int_{-\Omega}^{\Omega} |F(\omega)|^2 d\omega \geqslant 0, \end{aligned} \tag{2.111}$$

which is the Bessel inequality for the Fourier transform. As $\Omega \longrightarrow \infty$ the integrand in (2.109) approaches a delta function and in accordance with (1.285) we have

$$\lim_{\Omega \to \infty} f^{\Omega}(t) = \frac{1}{2} \left[f\left(t^{+}\right) + f\left(t^{-}\right) \right] \tag{2.112}$$

or, equivalently, using (2.104) with $F(\omega) = 2\pi \hat{f}(\omega)$

$$\lim_{\Omega \to \infty} \frac{1}{2\pi} \int_{-\Omega}^{\Omega} F(\omega) e^{i\omega t} d\omega = \frac{1}{2} \left[f\left(t^{+}\right) + f\left(t^{-}\right) \right]. \tag{2.113}$$

At the same time the MS error in (2.111) approaches zero and we obtain

$$\int_{-\infty}^{\infty} |f(t)|^2 dt = \frac{1}{2\pi} \int_{-\infty}^{\infty} |F(\omega)|^2 d\omega, \tag{2.114}$$

which is Parseval theorem for the Fourier transform. Equation (2.113) is usually written in the abbreviated form

$$f(t) = \frac{1}{2\pi} \int_{-\infty}^{\infty} F(\omega) e^{i\omega t} d\omega \tag{2.115}$$

and is referred to as the inverse Fourier transform or the Fourier transform inversion formula. It will be frequently convenient to designate both (2.115) and the direct transform (2.108) by the concise statement

$$f(t) \overset{\mathcal{F}}{\Longleftrightarrow} F(\omega). \tag{2.116}$$

In addition, we shall at times find it useful to express the direct and inverse transform pair as

$$\mathcal{F}\{f(t)\} = F(\omega), \tag{2.117}$$

which is just an abbreviation of the statement "the Fourier transform of $f(t)$ is $F(\omega)$." We shall adhere to the convention of designating the time domain signal by a lowercase letter and its Fourier transform by the corresponding uppercase letter.

2.2.3 Completeness Relationship and Relation to Fourier Series

Proceeding in a purely formal way we replace $F(\omega)$ in (2.115) by (2.108) and interchange the order of integration and obtain

$$f(t) = \int_{-\infty}^{\infty} f(t') \left\{ \frac{1}{2\pi} \int_{-\infty}^{\infty} e^{i\omega(t-t')} d\omega \right\} dt'. \tag{2.118}$$

The quantity in braces can now be identified as the delta function

$$\delta(t - t') = \frac{1}{2\pi} \int_{-\infty}^{\infty} e^{i\omega(t-t')} d\omega, \tag{2.119}$$

which is a slightly disguised version of (1.254). To see this we merely have to rewrite (2.119) as the limiting form

$$\lim_{\Omega \longrightarrow \infty} \frac{1}{2\pi} \int_{-\Omega}^{\Omega} e^{i\omega(t-t')} d\omega$$

and note that for any finite Ω the integration yields $\sin\left[\Omega\left(t-t'\right)\right]/\pi(t-t')$.

The representation (2.119) bears a formal resemblance to the completeness relationship for orthonormal discrete function sets, (1.302), and, more directly, to the completeness statement for Fourier series in (2.31). This resemblance can be highlighted by rewriting (2.119) to read

$$\delta\left(t-t'\right) = \int_{-\infty}^{\infty} \left(\frac{1}{\sqrt{2\pi}} e^{i\omega t}\right) \left(\frac{1}{\sqrt{2\pi}} e^{i\omega t'}\right)^{*} d\omega \qquad (2.120)$$

so that a comparison with (2.31) shows that the functions $\phi_\omega\left(t\right) \equiv 1/\sqrt{2\pi} exp\left(i\omega t\right)$ play an analogous role to the orthonormal functions $\phi_n\left(t\right) \equiv 1/\sqrt{T} exp\left(2\pi i n t/T\right)$ provided we view the continuous variable ω in (2.120) as proportional to a summation index. In fact a direct comparison of the variables between (2.31) and (2.120) gives the correspondence

$$\omega \quad \longleftrightarrow \quad \frac{2\pi n}{T} \qquad (2.121a)$$

$$d\omega \quad \longleftrightarrow \quad \frac{2\pi}{T}. \qquad (2.121b)$$

Thus as the observation period T of the signal increases, the quantity $2\pi/T$ may be thought of as approaching the differential $d\omega$ while the discrete spectral lines occurring at $2\pi n/T$ merge into a continuum corresponding to the frequency variable ω. Moreover the orthogonality over the finite interval $-T/2, T/2$, as in (1.213), becomes in the limit as $T \longrightarrow \infty$

$$\begin{aligned} \delta\left(\omega-\omega'\right) &= \frac{1}{2\pi} \int_{-\infty}^{\infty} e^{it\left(\omega-\omega'\right)} dt \\ &= \int_{-\infty}^{\infty} \left(\frac{1}{\sqrt{2\pi}} e^{it\omega}\right) \left(\frac{1}{\sqrt{2\pi}} e^{it\omega'}\right)^{*} dt \end{aligned} \qquad (2.122)$$

i.e., the identity matrix represented by the Kronecker symbol δ_{mn} goes over into a delta function, which is the proper identity transformation for the continuum.

A more direct but qualitative connection between the Fourier series and the Fourier transform can be established if we suppose that the function $f\left(t\right)$ is initially truncated to $|t| < T/2$ in which case its Fourier transform is

$$F\left(\omega\right) = \int_{-T/2}^{T/2} f\left(t\right) e^{-i\omega t} dt. \qquad (2.123)$$

The coefficients in the Fourier series that represents this function within the interval $-T/2, T/2$ can now be expressed as $\hat{f}_n = F\left(2\pi n/T\right)/T$ so that

$$f\left(t\right) = \sum_{n=-\infty}^{\infty} F\left(2\pi n/T\right) e^{i2\pi n t/T} \left(\frac{1}{T}\right). \qquad (2.124)$$

Thus in view of (2.121) we can regard the Fourier transform inversion formula (2.115) as a limiting form of (2.124) as $T \longrightarrow \infty$. Figure 2.22 shows the

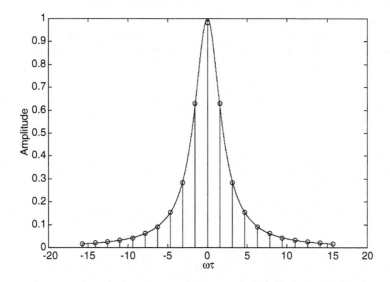

Figure 2.22: Continuous and discrete spectra

close correspondence between the discrete spectrum defined by Fourier series coefficients and the continuous spectrum represented by the Fourier transform. The time domain signal is the exponential $\exp -2 |t/\tau|$. For the discrete spectrum the time interval is truncated to $-T/2 \le t \le T/2$ (with $T/2\tau = 2$) and the Fourier series coefficients $\left|(T/\tau)\,\hat{f}_n\right|$ (stem diagram) plotted as a function of $2\pi n\tau/T$. Superposed for comparison is the continuous spectrum represented by $4/\left[4 + (\omega\tau)^2\right]$, the Fourier transform of $(1/\tau)\exp -2 |t/\tau|$.

2.2.4 Convergence and the Use of CPV Integrals

The convergence properties of the Fourier integral are governed by the delta function kernel (2.109). In many respects they are qualitatively quite similar to the convergence properties of Fourier series kernel (2.5). For example, as we shall show explicitly in 2.2.7, the convergence at points of discontinuity is again accompanied by the Gibbs oscillatory behavior. The one convergence issue that does not arise with Fourier series, but is unavoidable with the Fourier Integral, relates to the behavior of the functions at infinity, a problem we had already dealt with in Chap. 1 in order to arrive at the limit statement (1.269). There we found that it was sufficient to require that $f(t)$ satisfy (1.266) which, in particular, is satisfied by square integrable functions (Problem 1-18). Unfortunately this constraint does not apply to several idealized signals that have been found to be of great value in simplifying system analysis. To accommodate these signals, the convergence of the Fourier transform has to be examined on a case-by-case basis. In certain cases this requires a special definition of the limiting process underlying the improper integrals that define the Fourier transform and its inverse. In the following we provide a brief account of this limiting process.

An improper integral of the form $\int_{-\infty}^{\infty} f(t)\, dt$, unless stated to the contrary, implies the limit (1.278c)

$$\lim_{T_1 \to \infty} \lim_{T_2 \to \infty} \int_{-T_1}^{T_2} f(t)\, dt, \tag{2.125}$$

which means that integral converges when the upper and lower limits approach infinity independently. This definition turns out to be too restrictive in many situations of physical interest. An alternative and more encompassing definition is the following:

$$\lim_{T \to \infty} \int_{-T}^{T} f(t)\, dt. \tag{2.126}$$

Here we stipulate that upper and lower limits must approach infinity at the same rate. It is obvious that (2.126) implies (2.125). The converse is, however, not true. The class of functions for which the integral exists in the sense of (2.126) is much larger than under definition (2.125). In particular, all (piecewise differentiable) bounded odd functions are integrable in the sense of (2.126) and the integral yields zero. Under these circumstances (2.125) would generally diverge, unless of course the growth of the function at infinity is suitably restricted. When the limit is taken symmetrically in accordance with (2.126) the integral is said to be defined in terms of the Cauchy Principal Value (CPV). We have in fact already employed this definition implicitly on several occasions, in particular in (2.113). A somewhat different form of the CPV limit is also of interest in Fourier transform theory. This form arises whenever the integral is improper in virtue of one or more simple pole singularities within the integration interval. For example, the integral $\int_{-2}^{8} \frac{dt}{t-1}$ has a singularity at $t = 1$ where the integrand becomes infinite. The first inclination would be to consider this integral simply as divergent. On the other hand since the integrand changes sign as one moves through the singularity it is not unreasonable to seek a definition of a limiting process which would facilitate the mutual cancellation of the positive and negative infinite contributions. For example, suppose we define

$$I(\epsilon_1, \epsilon_2) = \int_{-2}^{1-\epsilon_1} \frac{dt}{t-1} + \int_{1+\epsilon_2}^{8} \frac{dt}{t-1},$$

where ϵ_1 and ϵ_2 are small positive numbers so that the integration is carried out up to and past the singularity. By direct calculation we find $I(\epsilon_1, \epsilon_2) = \ln(7\epsilon_1/3\epsilon_2)$. We see that if we let ϵ_1 and ϵ_2 approach zero independently the integral diverges. On the other by setting $\epsilon_1 = \epsilon_2 = \epsilon$ the result is always finite. Apparently when the singularity is approached symmetrically from both sides results in a cancellation of the positive and negative infinite contributions and yield a convergent integral. The formal expression for the limit is

$$\lim_{\epsilon \to 0} \left\{ \int_{-2}^{1-\epsilon} \frac{dt}{t-1} + \int_{1+\epsilon}^{8} \frac{dt}{t-1} \right\} = \ln(7/3).$$

This limiting procedure constitutes the CPV definition of the integral whenever the singularity falls within the integration interval. Frequently a special symbol is used to indicate a CPV evaluation. We shall indicate it by prefixing the letter P to the integration symbol. Thus $P \int_{-2}^{8} \frac{dt}{t-1} = \ln(7/3)$. When more than one singularity is involved the CPV limiting procedure must be applied to each. For example,

$$
\begin{aligned}
I &= P \int_{-5}^{9} \frac{dt}{(t-1)(t-2)} \\
&= \lim_{\epsilon \to 0} \left\{ \int_{-5}^{1-\epsilon} \frac{dt}{(t-1)(t-2)} + \int_{1+\epsilon}^{2-\varepsilon} \frac{dt}{(t-1)(t-2)} \right. \\
&\quad \left. + \int_{2+\varepsilon}^{9} \frac{dt}{(t-1)(t-2)} \right\} \\
&= \ln 3/4.
\end{aligned}
$$

The following example illustrates the CPV evaluation of an integral with infinite limits of integration:

$$
\begin{aligned}
I &= P \int_{-\infty}^{\infty} \frac{dt}{t-2} = \lim_{\substack{\epsilon \to 0 \\ T \to \infty}} \left\{ \int_{-T}^{2-\epsilon} \frac{dt}{t-2} + \int_{2+\epsilon}^{T} \frac{dt}{t-2} \right\} \\
&= \lim_{\substack{\epsilon \to 0 \\ T \to \infty}} \ln \frac{(2-\epsilon-2)(T-2)}{(-T-2)(2+\varepsilon-2)} = 0.
\end{aligned}
$$

Note that the symbol P in this case pertains to a CPV evaluation at $t = -\infty$ and $t = \infty$. A generic form of an integral that is frequently encountered is

$$I = P \int_{a}^{b} \frac{f(t)}{t-q} dt, \tag{2.127}$$

where $a < q < b$ and $f(t)$ is a bounded function within a, b and differentiable at $t = q$. We can represent this integral as a sum of an integral of a bounded function and a CPV integral which can be evaluated in closed form as follows:

$$
\begin{aligned}
I &= P \int_{a}^{b} \frac{f(t) - f(q) + f(q)}{t-q} dt \\
&= \int_{a}^{b} \frac{f(t) - f(q)}{t-q} dt + f(q) P \int_{a}^{b} \frac{dt}{t-q} \\
&= \int_{a}^{b} \frac{f(t) - f(q)}{t-q} dt + f(q) \ln \frac{b-q}{q-a}. \tag{2.128}
\end{aligned}
$$

Note that the integrand in the first integral in the last expression is finite at $t = q$ so that the integral can be evaluated, if necessary, numerically using standard techniques.

Let us now apply the CPV procedure to the evaluation of the Fourier transform of $f(t) = 1/t$. Even though a signal of this sort might appear quite artificial it will be shown to play a pivotal role in the theory of the Fourier transform. Writing the transform as a CPV integral we have

$$F(\omega) = P \int_{-\infty}^{\infty} \frac{e^{-i\omega t}}{t} dt = P \int_{-\infty}^{\infty} \frac{\cos \omega t}{t} dt - iP \int_{-\infty}^{\infty} \frac{\sin \omega t}{t} dt.$$

Since $P \int_{-\infty}^{\infty} \frac{\cos \omega t}{t} dt = 0$, and $\frac{\sin \omega t}{t}$ is free of singularities we have

$$F(\omega) = -i \int_{-\infty}^{\infty} \frac{\sin \omega t}{t} dt \tag{2.129}$$

Recalling that $\int_{-\infty}^{\infty} \frac{\sin x}{x} dx = \pi$ we obtain by setting $\omega t = x$ in (2.129)

$$\frac{1}{\pi} \int_{-\infty}^{\infty} \frac{\sin \omega t}{t} dt = sign(\omega) = \left\{ \begin{array}{l} 1 \; ; \; \omega > 0, \\ -1 \; ; \; \omega < 0. \end{array} \right. \tag{2.130}$$

Thus we arrive at the transform pair

$$\frac{1}{\pi t} \overset{\mathcal{F}}{\longleftrightarrow} -isign(\omega). \tag{2.131}$$

By using the same procedure for the inverse transform of $1/\omega$ we arrive at the pair

$$sign(t) \overset{\mathcal{F}}{\longleftrightarrow} \frac{2}{i\omega}. \tag{2.132}$$

Several idealized signals may be termed canonical in that they form the essential building blocks in the development of analytical techniques for evaluation of Fourier transforms and also play a fundamental role in the characterization of linear system. One such canonical signal is the sign function just considered. We consider several others in turn.

2.2.5 Canonical Signals and Their Transforms

The Delta Function

That Fourier transform of $\delta(t)$ equals 1 follows simply from the basic property of the delta function as an identity transformation. The consistency of this with the inversion formula follows from (2.119). Hence

$$\delta(t) \overset{\mathcal{F}}{\longleftrightarrow} 1. \tag{2.133}$$

In identical fashion we get

$$1 \overset{\mathcal{F}}{\longleftrightarrow} 2\pi \delta(\omega). \tag{2.134}$$

The Unit Step Function

From the identity $U(t) = \frac{1}{2}[1 + sign(t)]$ we get in conjunction with (2.132) and (2.134)

$$U(t) \overset{\mathcal{F}}{\Longleftrightarrow} \pi\delta(\omega) + \frac{1}{i\omega}. \tag{2.135}$$

The Rectangular Pulse Function

Using the definition for $p_T(t)$ in (1.6-40b) we obtain by direct integration the pair

$$p_T(t) \overset{\mathcal{F}}{\Longleftrightarrow} \frac{2\sin(\omega T)}{\omega}, \tag{2.136}$$

where we again find the familiar Fourier integral kernel. If, on the other hand, $p_\Omega(\omega)$ describes a rectangular frequency window, then a direct evaluation of the inverse transform yields

$$\frac{\sin(\Omega t)}{\pi t} \overset{\mathcal{F}}{\Longleftrightarrow} p_\Omega(\omega). \tag{2.137}$$

The transition of (2.137) to (2.133) as $\Omega \to \infty$ and of (2.136) to (2.134) as $T \to \infty$ should be evident.

Triangular Pulse Function

Another signal that we should like to add to our catalogue of canonical transforms is the triangular pulse $q_T(t)$ defined in (1.278d) for which we obtain the pair

$$q_T(t) \overset{\mathcal{F}}{\Longleftrightarrow} \frac{T\sin^2(\omega T/2)}{(\omega T/2)^2}. \tag{2.137*}$$

Exponential Functions

Since the Fourier transform is a representation of signals in terms of exponentials we would expect exponential functions to play a special role in Fourier analysis. In the following we distinguish three cases: a purely imaginary argument, a purely real argument with the function truncated to the positive time axis, and a real exponential that decays symmetrically for both negative and positive times. In the first case we get from the definition of the delta function (2.119) and real ω_0

$$e^{i\omega_0 t} \overset{\mathcal{F}}{\Longleftrightarrow} 2\pi\delta(\omega - \omega_0). \tag{2.138}$$

This result is in perfect consonance with the intuitive notion that a single tone, represented in the time domain by a unit amplitude sinusoidal oscillation of infinitely long duration, should correspond in the frequency domain to a single number, i.e., the frequency of oscillation, or, equivalently, by a spectrum consisting of a single spectral line. Here this spectrum is represented symbolically by a delta function at $\omega = \omega_0$. Such a single spectral line, just like the corresponding tone of infinite duration, are convenient abstractions never realizable in practice. A more realistic model should consider a tone of finite duration, say $-T < t < T$. We can do this either by truncating the limits of integration in the evaluation of the direct transform, or, equivalently, by specifying this truncation in terms of the pulse function $p_T(t)$. The resulting transform pair then reads

$$p_T(t)e^{iw_0t} \overset{\mathcal{F}}{\Longleftrightarrow} \frac{2\sin\left[(\omega - \omega_0)\,T\right]}{(\omega - \omega_0)}, \tag{2.139}$$

so that the form of the spectrum is the Fourier kernel (2.136) whose peak has been shifted to ω_0. One can show that slightly more than 90% of the energy is contained within the frequency band defined by the first two nulls on either side of the principal peak. It is therefore reasonable to take this bandwidth as the nominal spectral linewidth of the tone. Thus we see that a tone of duration $2T$ has a spectral width of $2\pi/T$ which is sometimes referred to as the Rayleigh resolution limit. This inverse relationship between the signal duration and spectral width is of fundamental importance in spectral analysis. Its generalization to a wider class of signals is embodied in the so-called uncertainty principle discussed in 2.5.1.

With $\alpha > 0$ and the exponential truncated to the nonnegative time axis we get

$$e^{-\alpha t}U\,(t) \overset{\mathcal{F}}{\Longleftrightarrow} \frac{1}{\alpha + i\omega}. \tag{2.140}$$

For the exponential $e^{-\alpha|t|}$ defined over the entire real line the transform pair reads

$$e^{-\alpha|t|} \overset{\mathcal{F}}{\Longleftrightarrow} \frac{2\alpha}{\alpha^2 + \omega^2}. \tag{2.141}$$

Formula (2.140) also holds when α is replaced by the complex number $p_0 = \alpha - i\omega_0$ where ω_0 is real. A further generalization follows if we differentiate the right side of (2.140) $n - 1$ times with respect to ω. The result is

$$\frac{t^{n-1}}{(n-1)!}e^{-p_0t}U\,(t) \overset{\mathcal{F}}{\Longleftrightarrow} \frac{1}{(p_0 + i\omega)^n}; \quad n \geq 1. \tag{2.142}$$

Using this formula in conjunction with the partial fraction expansion technique constitutes one of the basic tools in the evaluation of inverse Fourier transforms of rational functions.

Gaussian Function

A rather important idealized signal is the Gaussian function

$$f(t) = \frac{1}{\sqrt{2\pi\sigma_t^2}}e^{-\frac{t^2}{2\sigma_t^2}},$$

where we have adopted the normalization $\left(\sqrt{f}, \sqrt{f}\right) = 1$. We compute its FT as follows:

$$
\begin{aligned}
F\left(\omega\right) &= \frac{1}{\sqrt{2\pi\sigma_t^2}} \int_{-\infty}^{\infty} e^{-\frac{t^2}{2\sigma_t^2}} e^{-i\omega t} dt = \frac{1}{\sqrt{2\pi\sigma_t^2}} \int_{-\infty}^{\infty} e^{-\frac{1}{2\sigma_t^2}\left[t^2 + 2i\omega\sigma_t^2 t\right]} dt \\
&= \frac{e^{-\frac{1}{2}\sigma_t^2\omega^2}}{\sqrt{2\pi\sigma_t^2}} \int_{-\infty}^{\infty} e^{-\frac{1}{2\sigma_t^2}\left[t + i\omega\sigma_t^2\right]^2} dt = \frac{e^{-\frac{1}{2}\sigma_t^2\omega^2}}{\sqrt{2\pi\sigma_t^2}} \int_{-\infty+i\omega\sigma_t^2}^{\infty+i\omega\sigma_t^2} e^{-\frac{z^2}{2\sigma_t^2}} dz.
\end{aligned}
$$

The last integral may be interpreted as an integral in the complex z plane with the path of integration running along the straight line with endpoints $(-\infty + i\omega\sigma_t^2, \infty + i\omega\sigma_t^2)$. Since the integrand is analytic in the entire finite z plane we can shift this path to run along the axis of reals so that

$$
\int_{-\infty+i\omega\sigma_t^2}^{\infty+i\omega\sigma_t^2} e^{-\frac{z^2}{2\sigma_t^2}} dz = \int_{-\infty}^{\infty} e^{-\frac{z^2}{2\sigma_t^2}} dz = \sqrt{2\pi\sigma_t^2}.
$$

Thus we obtain the transform pair

$$
\frac{1}{\sqrt{2\pi\sigma_t^2}} e^{-\frac{t^2}{2\sigma_t^2}} \stackrel{\mathcal{F}}{\Longleftrightarrow} e^{-\frac{1}{2}\sigma_t^2\omega^2}. \tag{2.142*}
$$

Note that except for a scale factor the Gaussian function is its own FT. Here we see another illustration of the inverse relationship between the signal duration and bandwidth. If we take σ_t as the nominal duration of the pulse in the time domain, then a similar definition for the effective bandwidths of $F\left(\omega\right)$ yields $\sigma_\omega = 1/\sigma_t$.

2.2.6   Basic Properties of the FT

Linearity

The Fourier transform is a linear operator which means that for any set of functions $f_n\left(t\right)$ $n = 1, 2, \ldots N$ and corresponding transforms $F_n\left(\omega\right)$ we have

$$
\mathcal{F}\left\{\sum_{n=1}^{N} \alpha_n f_n\left(t\right)\right\} = \sum_{n=1}^{N} \alpha_n F_n\left(\omega\right),
$$

where the α_n are constants. This property is referred to as the superposition principle. We shall return to it in Chap. 3 when we discuss linear systems. This superposition principle carries over to a continuous index. Thus if

$$
f\left(\xi, t\right) \stackrel{\mathcal{F}}{\Longleftrightarrow} F\left(\xi, \omega\right)
$$

holds for a continuum of values of ξ, then

$$
\mathcal{F}\left\{\int f\left(\xi, t\right) d\xi\right\} = \int F\left(\xi, \omega\right) d\xi.
$$

Symmetries

For any Fourier transform pair

$$f(t) \overset{\mathcal{F}}{\Longleftrightarrow} F(\omega)$$

we also have, by a simple substitution of variables,

$$F(t) \overset{\mathcal{F}}{\Longleftrightarrow} 2\pi f(-\omega). \tag{2.143}$$

For example, using this variable replacement in (2.141), we obtain

$$\frac{\alpha}{\pi(\alpha^2 + t^2)} \overset{\mathcal{F}}{\Longleftrightarrow} e^{-\alpha|\omega|}. \tag{2.143*}$$

The Fourier transform of the complex conjugate of a function follows through the variable replacement

$$f^*(t) \overset{\mathcal{F}}{\Longleftrightarrow} F^*(-\omega). \tag{2.144}$$

Frequently we shall be interested in purely real signals. If $f(t)$ is real, the preceding requires

$$F^*(-\omega) = F(\omega). \tag{2.145}$$

If we decompose $F(\omega)$ into its real and imaginary parts

$$F(\omega) = R(\omega) + iX(\omega), \tag{2.146}$$

we note that (2.145) is equivalent to

$$R(\omega) \;\;=\;\; R(-\omega), \tag{2.147a}$$
$$X(\omega) \;\;=\;\; -X(-\omega), \tag{2.147b}$$

so that for a real signal the real part of the Fourier transforms is even function while the imaginary part an odd function of frequency. The even and odd symmetries carry over to the amplitude and phase of the transform. Thus writing

$$F(\omega) = A(\omega) e^{i\theta(\omega)}, \tag{2.148}$$

wherein

$$A(\omega) \;\;=\;\; |F(\omega)| = \sqrt{[R(\omega)]^2 + [X(\omega)]^2}, \tag{2.149a}$$
$$\theta(\omega) \;\;=\;\; \tan^{-1}\frac{X(\omega)}{R(\omega)}, \tag{2.149b}$$

we have in view of (2.147)

$$A(\omega) \;\;=\;\; A(-\omega) \tag{2.150a}$$
$$\theta(\omega) \;\;=\;\; -\theta(-\omega). \tag{2.150b}$$

As a result the inversion formula can be put into the form

$$
\begin{aligned}
f(t) &= \frac{1}{\pi} \int_0^\infty A(\omega) \cos\left[\omega t + \theta(\omega)\right] d\omega \\
&= \Re\left\{ \frac{1}{2\pi} \int_0^\infty 2F(\omega) e^{i\omega t} d\omega \right\}.
\end{aligned}
\tag{2.151}
$$

The last expression shows that a real physical signal can be represented as the real part of a fictitious complex signal whose spectrum equals twice the spectrum of the real signal for positive frequencies but is identically zero for negative frequencies. Such a complex signal is referred to as an analytic signal, a concept that finds extensive application in the study of modulation to be discussed in 2.3.

Time Shift and Frequency Shift

For any real T we have

$$
f(t-T) \overset{\mathcal{F}}{\Longleftrightarrow} F(\omega) e^{-i\omega T}
\tag{2.152}
$$

and similarly for any real ω_0

$$
f(t) e^{i\omega_0 t} \overset{\mathcal{F}}{\Longleftrightarrow} F(\omega - \omega_0).
\tag{2.153}
$$

The last formula is the quantification of the modulation of a high frequency CW carrier by a baseband signal comprised of low frequency components. For example, for the carrier of $A\cos(\omega_0 t + \theta_0)$ and a baseband signal $f(t)$ we get

$$
f(t) A\cos(\omega_0 t + \theta_0) \overset{\mathcal{F}}{\Longleftrightarrow} \frac{A}{2} e^{i\theta_0} F(\omega - \omega_0) + \frac{A}{2} e^{-i\theta_0} F(\omega + \omega_0).
\tag{2.154}
$$

If we suppose that $F(\omega)$ is negligible outside the band defined by $|\omega| < \Omega$, and also assume that $\omega_0 > 2\Omega$, the relationship among the spectra in (2.154) may be represented schematically as in Fig. 2.23

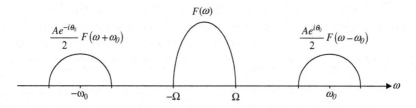

Figure 2.23: Modulation by a CW carrier

Differentiation

If $f(t)$ is everywhere differentiable, then a simple integration by parts gives

$$\int_{-\infty}^{\infty} f'(t) e^{-i\omega t} dt = f(t) e^{-i\omega t} \big|_{-\infty}^{\infty} + i\omega \int_{-\infty}^{\infty} f(t) e^{-i\omega t} dt$$

$$= i\omega F(\omega). \qquad (2.155)$$

Clearly if $f(t)$ is differentiable n times we obtain by repeated integration

$$f^{(n)}(t) \overset{\mathcal{F}}{\Longleftrightarrow} (i\omega)^n F(\omega). \qquad (2.156)$$

Actually this formula may still be used even if $f(t)$ is only piecewise differentiable and discontinuous with discontinuous first and even higher order derivatives at a countable set of points. We merely have to replace $f^{(n)}(t)$ with a generalized derivative defined in terms of singularity functions, an approach we have already employed for the first derivative in (1.280). For example, the Fourier transform of (1.280) is

$$f'(t) \overset{\mathcal{F}}{\Longleftrightarrow} i\omega F(\omega) = \mathcal{F}\{f'_s(t)\} + \sum_k \left[f(t_k^+) - f(t_k^-) \right] e^{-i\omega t_k}. \qquad (2.157)$$

In the special case of only one discontinuity at $t = 0$ and $f(0^-) = 0$ (2.157) becomes

$$f'_s(t) \overset{\mathcal{F}}{\Longleftrightarrow} i\omega F(\omega) - f(0^+). \qquad (2.158)$$

What about the Fourier transform of higher order derivatives? Clearly if the first derivative is continuous at $t = 0$, the Fourier transform of $f''_s(t)$ may be obtained by simply multiplying the right side of (2.158) by $i\omega$. However in case of a discontinuity in the first derivative the magnitude of the jump in the derivative must be subtracted. Again assuming $f'(0^-) = 0$ we have

$$f''_s(t) \overset{\mathcal{F}}{\Longleftrightarrow} i\omega \left[i\omega F(\omega) - f(0^+) \right] - f'(0^+). \qquad (2.159)$$

Higher order derivatives can be handled similarly.

Since an $n-th$ order derivative in the time domain transforms in the frequency domain to a multiplication by $(i\omega)^n$, the Fourier transform of any linear differential operator with constant coefficients is a polynomial in $i\omega$. This feature makes the Fourier transform a natural tool for the solution of linear differential equations with constant coefficients. For example, consider the following differential equation:

$$x''(t) + 2x'(t) + x(t) = 0. \qquad (2.160)$$

We seek a solution for $x(t)$ for $t \geq 0$ with initial conditions $x(0^+) = 2$ and $x'(0^+) = 6$. Then

$$x'(t) \overset{\mathcal{F}}{\Longleftrightarrow} i\omega X(\omega) - 2$$

$$x''(t) \overset{\mathcal{F}}{\Longleftrightarrow} -\omega^2 X(\omega) - i\omega 2 - 6.$$

The solution for $X(\omega)$ reads

$$X(\omega) = \frac{i2\omega + 10}{-\omega^2 + i2\omega + 1},$$

while the signal $x(t)$ is to be computed from

$$x(t) = \frac{1}{2\pi} \int_{-\infty}^{\infty} \frac{i2\omega + 10}{-\omega^2 + i2\omega + 1} e^{i\omega t} d\omega. \tag{2.161}$$

The integral can be evaluated by contour integration as will be shown in 2.4.4 (see also (A.96) in the Appendix).

Inner Product Invariance

We compute the inner product of two functions in the time domain and with the aid of the inversion formulas transform it into an inner product in the frequency domain as follows:

$$
\begin{aligned}
(f_1, f_2) &= \int_{-\infty}^{\infty} f_1^*(t) f_2(t) \, dt \\
&= \int_{-\infty}^{\infty} \left\{ \frac{1}{2\pi} \int_{-\infty}^{\infty} F_1^*(\omega) e^{-i\omega t} d\omega \frac{1}{2\pi} \int_{-\infty}^{\infty} F_2(\omega') e^{i\omega' t} d\omega' \right\} dt \\
&= \frac{1}{2\pi} \int_{-\infty}^{\infty} \int_{-\infty}^{\infty} F_1^*(\omega) F_2(\omega') \left\{ \frac{1}{2\pi} \int_{-\infty}^{\infty} e^{i(\omega' - \omega)t} dt \right\} d\omega' d\omega \\
&= \frac{1}{2\pi} \int_{-\infty}^{\infty} \int_{-\infty}^{\infty} F_1^*(\omega) F_2(\omega') \delta(\omega - \omega') \, d\omega' d\omega \\
&= \frac{1}{2\pi} \int_{-\infty}^{\infty} F_1^*(\omega) F_2(\omega) \, d\omega.
\end{aligned}
$$

The final result may be summarized to read

$$\int_{-\infty}^{\infty} f_1^*(t) f_2(t) \, dt = \frac{1}{2\pi} \int_{-\infty}^{\infty} F_1^*(\omega) F_2(\omega) \, d\omega, \tag{2.162}$$

which is recognized as a generalization of Parseval's formula.

Convolution

We have already encountered the convolution of two functions in connection with Fourier series, (2.49). Since in the present case the time domain encompasses the entire real line the appropriate definition is

$$h(t) = \int_{-\infty}^{\infty} f(\tau) g(t - \tau) \, d\tau.$$

We shall frequently employ the abbreviated notation

$$\int_{-\infty}^{\infty} f(\tau) g(t - \tau) \, d\tau = f * g. \tag{2.163}$$

Note that

$$\int_{-\infty}^{\infty} f(\tau) g(t-\tau) d\tau = \int_{-\infty}^{\infty} g(\tau) f(t-\tau) d\tau$$

as one can readily convince oneself through a change of the variable of integration. This can also be expressed as $f * g = g * f$, i.e., the convolution operation is commutative. In view of (2.152) $g(t-\tau) \overset{\mathcal{F}}{\Longleftrightarrow} G(\omega) e^{-i\omega\tau}$ so that

$$\int_{-\infty}^{\infty} f(\tau) g(t-\tau) d\tau \overset{\mathcal{F}}{\Longleftrightarrow} \int_{-\infty}^{\infty} f(\tau) G(\omega) e^{-i\omega\tau} d\tau = F(\omega) G(\omega). \quad (2.164)$$

In identical manner we establish that

$$f(t) g(t) \overset{\mathcal{F}}{\Longleftrightarrow} \frac{1}{2\pi} \int_{-\infty}^{\infty} F(\eta) G(\omega-\eta) d\eta = \frac{1}{2\pi} F * G. \quad (2.165)$$

Integration

When the Fourier transform is applied to integro-differential equations one sometimes needs to evaluate the transform of the integral of a function. For example with $g(t) = \int_{-\infty}^{t} f(\tau) d\tau$ we would like to determine $G(\omega)$ in terms of $F(\omega)$. We can do this by first recognizing that $\int_{-\infty}^{t} f(\tau) d\tau = \int_{-\infty}^{\infty} f(\tau) U(t-\tau) d\tau$. Using (2.164) and (2.135) we have

$$\int_{-\infty}^{\infty} f(\tau) U(t-\tau) d\tau \overset{\mathcal{F}}{\Longleftrightarrow} F(\omega) \left[\pi\delta(\omega) + \frac{1}{i\omega}\right]$$

with the final result

$$\int_{-\infty}^{t} f(\tau) d\tau \overset{\mathcal{F}}{\Longleftrightarrow} \pi F(0) \delta(\omega) + \frac{F(\omega)}{i\omega} = G(\omega). \quad (2.166)$$

Note that the integral implies $g'(t) = f(t)$ so that

$$i\omega G(\omega) = F(\omega). \quad (2.167)$$

This is certainly compatible with (2.166) since $\omega\delta(\omega) = 0$. However the solution of (2.167) for $G(\omega)$ by simply dividing both sides by $i\omega$ is in general not permissible since $G(\omega) \neq F(\omega)/i\omega$ unless $F(0) = 0$.

Causal Signals and the Hilbert Transform [16]

Let

$$f_e(t) = \frac{f(t) + f(-t)}{2}, \quad (2.168a)$$

$$f_o(t) = \frac{f(t) - f(-t)}{2}, \quad (2.168b)$$

so that $f(t) = f_e(t) + f_o(t)$ for any signal. Since $f_e(t) = f_e(-t)$ and $f_o(t) = -f_o(-t)$ (2.168a) and (2.168b) are referred to as the even and odd parts of $f(t)$, respectively. Now

$$
\begin{aligned}
\mathcal{F}\{f_e(t)\} &= \frac{1}{2}\int_{-\infty}^{\infty}[f(t) + f(-t)][\cos(\omega t) - i\sin(\omega t)]\,dt \\
&= \int_{-\infty}^{\infty} f(t)\cos(\omega t)\,dt \tag{2.169a}
\end{aligned}
$$

and

$$
\begin{aligned}
\mathcal{F}\{f_o(t)\} &= \frac{1}{2}\int_{-\infty}^{\infty}[f(t) - f(-t)][\cos(\omega t) - i\sin(\omega t)]\,dt \\
&= -i\int_{-\infty}^{\infty} f(t)\sin(\omega t)\,dt. \tag{2.169b}
\end{aligned}
$$

In view of the definition (2.146), for a real $f(t)$ (2.169a) and (2.169b) are equivalent to

$$
f_e(t) \overset{\mathcal{F}}{\Longleftrightarrow} R(\omega), \tag{2.170a}
$$
$$
f_o(t) \overset{\mathcal{F}}{\Longleftrightarrow} iX(\omega). \tag{2.170b}
$$

In the following we shall be concerned only with real signals.

As will be discussed in Chap. 3, signals that vanish for negative values of the argument play a special role in linear time-invariant systems. Such signals are said to be causal. Suppose $f(t)$ is a causal signal. Then according to (2.168)

$$
f(t) = \begin{cases} 2f_e(t) = 2f_o(t) & ; t > 0, \\ 0 & ; t < 0. \end{cases} \tag{2.171}
$$

Evidently the even and odd parts are not independent for

$$
\begin{aligned}
f_e(t) &= f_o(t) \;\; ; t > 0, \\
f_e(t) &= -f_o(t) \;\; ; t < 0, .
\end{aligned}
$$

which can be rephrased in more concise fashion with the aid of the *sign* function as follows:

$$
\begin{aligned}
f_o(t) &= \text{sign}(t)\, f_e(t) \tag{2.172a} \\
f_e(t) &= \text{sign}(t)\, f_o(t). \tag{2.172b}
\end{aligned}
$$

Taking account of (2.170), (2.132), and (2.165) Fourier Transformation of both sides of (2.172) results in the following pair of equations:

$$
X(\omega) = -\frac{1}{\pi} P \int_{-\infty}^{\infty} \frac{R(\eta)\,d\eta}{\omega - \eta}, \tag{2.173a}
$$
$$
R(\omega) = \frac{1}{\pi} P \int_{-\infty}^{\infty} \frac{X(\eta)\,d\eta}{\omega - \eta}. \tag{2.173b}
$$

These relations show explicitly that the real and imaginary parts of the Fourier transform of a causal signal may not be prescribed independently. For example if we know $R(\omega)$, then $X(\omega)$ can be determined uniquely by (2.173a). Since $P \int_{-\infty}^{\infty} \frac{d\eta}{\omega - \eta} = 0$, an $R(\omega)$ that is constant for all frequencies gives a null result for $X(\omega)$. Consequently, (2.173b) determines $R(\omega)$ from $X(\omega)$ only within a constant.

The integral transform $\frac{1}{\pi} P \int_{-\infty}^{\infty} \frac{R(\eta) d\eta}{\omega - \eta}$ is known as the Hilbert Transform which shall be denoted by $\mathcal{H}\{R(\omega)\}$. Using this notation we rewrite (2.173) as

$$X(\omega) = -\mathcal{H}\{R(\omega)\}, \tag{2.174a}$$
$$R(\omega) = \mathcal{H}\{X(\omega)\}. \tag{2.174b}$$

Since (2.174b) is the inverse of (2.174a) the inverse Hilbert transform is obtained by a change in sign. As an example, suppose $R(\omega) = p_{\Omega}(\omega)$. Carrying out the simple integration yields

$$X(\omega) = \frac{1}{\pi} \ln \left| \frac{\omega - \Omega}{\omega + \Omega} \right|, \tag{2.175}$$

which is plotted in Fig. 2.24. The Hilbert Transform in the time domain is defined similarly. Thus for a signal $f(t)$

$$\mathcal{H}\{f(t)\} = \frac{1}{\pi} P \int_{-\infty}^{\infty} \frac{f(\tau) d\tau}{t - \tau}. \tag{2.176}$$

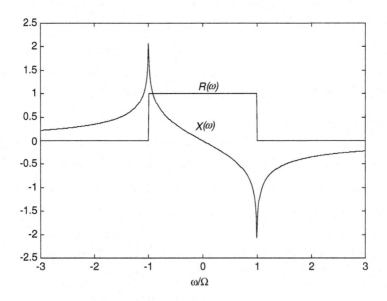

Figure 2.24: $R(\omega)$ and its Hilbert transform

Particularly simple results are obtained for Hilbert transforms of sinusoids. For example, with $f(t) = \cos(\omega t)$ (with ω a real constant) we have

$$
\begin{aligned}
\frac{1}{\pi} P \int_{-\infty}^{\infty} \frac{\cos(\omega\tau)\,d\tau}{t-\tau} &= \frac{1}{\pi} P \int_{-\infty}^{\infty} \frac{\cos[\omega(t-\tau)]\,d\tau}{\tau} \\
&= \cos(\omega t)\frac{1}{\pi} P \int_{-\infty}^{\infty} \frac{\cos(\omega\tau)\,d\tau}{\tau} \\
&\quad + \sin(\omega t)\frac{1}{\pi} P \int_{-\infty}^{\infty} \frac{\sin(\omega\tau)\,d\tau}{\tau}.
\end{aligned}
$$

We note that the first of the two preceding integrals involves an odd function and therefore vanishes, while in virtue of (2.132) the second integral yields $sign(\omega)$. Hence

$$
\mathcal{H}\{\cos(\omega t)\} = sign(\omega)\sin(\omega t). \tag{2.177}
$$

In identical fashion we obtain

$$
\mathcal{H}\{\sin(\omega t)\} = -sign(\omega)\cos(\omega t). \tag{2.178}
$$

We shall have occasion to employ the last two formulas in connection with analytic signal representations.

The Hilbert transform finds application in signal analysis, modulation theory, and spectral analysis. In practical situations the evaluation of the Hilbert transform must be carried out numerically for which purpose direct use of the defining integral is not particularly efficient. The preferred approach is to carry out the actual calculations in terms of the Fourier transform which can be computed efficiently using the FFT algorithm. To see how this may be arranged, let us suppose that $R(\omega)$ is given and we wish to find $X(\omega)$. By taking the inverse FT we first find $f_e(t)$, in accordance with (2.170a). In view of (2.171), if we now multiply the result by 2, truncate it to nonnegative t, and take the direct FT, we should obtain $F(\omega)$. Thus

$$
2f_e(t)U(t) \overset{\mathcal{F}}{\Longleftrightarrow} F(\omega) \tag{2.179}
$$

and $X(\omega)$ follows by taking the imaginary part of $F(\omega)$. In summary we have

$$
\mathcal{H}\{R\} = -\Im m\left\{ \int_0^\infty 2\mathcal{F}^{-1}\{R\}e^{-i\omega t}dt \right\} = -X(\omega), \tag{2.180}
$$

where

$$
\mathcal{F}^{-1}\{R\} \equiv \frac{1}{2\pi}\int_{-\infty}^{\infty} R(\omega')e^{i\omega' t}d\omega'.
$$

Initial and Final Value Theorems

Again assume that $f(t)$ is a causal signal and that it is piecewise differentiable for all $t > 0$. Then

$$
\begin{aligned}
\mathcal{F}\{f'(t)\} &= \int_{0+}^{\infty} f'(t)e^{-i\omega t}dt = f(t)e^{-i\omega t}\,|_{0+}^{\infty} + i\omega \int_{0+}^{\infty} f(t)e^{-i\omega t}dt \\
&= i\omega F(\omega) - f(0^+).
\end{aligned}
$$

Since by assumption $f'(t)$ exists for $t > 0$, or, equivalently, $f(t)$ is smooth, $\mathcal{F}\{f'(t)\}$ approaches zero as $\omega \to \infty$ (c.f. (2.158)). Under these conditions the last equation yields

$$\lim_{\omega \to \infty} i\omega F(\omega) = f(0^+),\tag{2.181}$$

a result known as the initial value theorem. Note that $f_e(0) \equiv f(0)$ but according to (2.171) $2 f_e(0) = f(0^+)$. Hence

$$f(0) = \frac{1}{2} f(0^+),\tag{2.182}$$

which is consistent with the fact that the FT converges to the arithmetic mean of the step discontinuity.

Consider now the limit

$$
\begin{aligned}
\lim_{\omega \to 0} \left[i\omega F(\omega) - f(0^+) \right] &= \lim_{\omega \to 0} \int_{0+}^{\infty} f'(t) e^{-i\omega t} dt \\
&= \int_{0+}^{\infty} f'(t) \lim_{\omega \to 0} \left[e^{-i\omega t} \right] dt \\
&= \lim_{t \to \infty} f(t) - f(0^+).
\end{aligned}
$$

Upon cancelling $f(0^+)$ we get

$$\lim_{\omega \to 0} \left[i\omega F(\omega) \right] = \lim_{t \to \infty} f(t),\tag{2.183}$$

which is known as the final value theorem.

Fourier Series and the Poisson Sum Formula

Given a function $f(t)$ within the finite interval $-T/2, T/2$ we can represent it either as a Fourier integral, (2.123), comprised of a continuous spectrum of

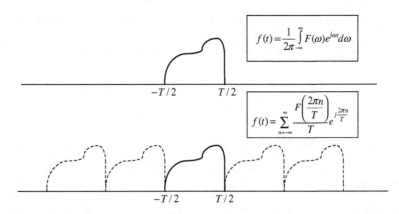

$$f(t) = \frac{1}{2\pi} \int_{-\infty}^{\infty} F(\omega) e^{j\omega t} d\omega$$

$$f(t) = \sum_{n=-\infty}^{\infty} \frac{F\left(\dfrac{2\pi n}{T}\right)}{T} e^{j\frac{2\pi n}{T}}$$

Figure 2.25: Fourier integral and Fourier series representations

sinusoids, or as a Fourier series, (2.124), comprised of discrete harmonically related sinusoids. In the former case the representation converges to zero outside the interval in question while in the latter case we obtain a periodic repetition (extension) of the given function, as illustrated in Fig. 2.25. The significant point to note here is that the Fourier series coefficients are given by the FT formula. Note also that the Fourier transform of $f(t)$ and its periodic extension (taken over the entire real-time axis) is a infinite series comprised of delta functions, i.e.,

$$\sum_{n=-\infty}^{\infty} \hat{f}_n e^{i2\pi nt/T} \overset{\mathcal{F}}{\Longleftrightarrow} 2\pi \sum_{n=-\infty}^{\infty} \hat{f}_n \delta\left(\omega - 2\pi n/T\right). \tag{2.184}$$

In the following we present a generalization of (2.124), known as the Poisson sum formula wherein the function $f(t)$ may assume nonzero values over the entire real line. We start by defining the function $g(t)$ through the sum

$$g(t) = \sum_{n=-\infty}^{\infty} f(t - nT). \tag{2.185}$$

It is easy to see that $g(t)$ is periodic with period T. We take the FT to obtain

$$\sum_{n=-\infty}^{\infty} f(t - nT) \overset{\mathcal{F}}{\Longleftrightarrow} \sum_{n=-\infty}^{\infty} F(\omega) e^{-i\omega nT}.$$

In view of (2.31) the sum of exponentials can be replaced by a sum comprised of delta functions. Thus

$$\sum_{n=-\infty}^{\infty} F(\omega) e^{-i\omega nT} = \sum_{\ell=-\infty}^{\infty} F(\omega) 2\pi\delta\left(\omega T - 2\pi\ell\right)$$

$$= \frac{2\pi}{T} \sum_{\ell=-\infty}^{\infty} F(2\pi\ell/T) \delta\left(\omega - 2\pi\ell/T\right).$$

Inverting the FT gives

$$\sum_{\ell=-\infty}^{\infty} F(2\pi\ell/T)/T e^{i2\pi\ell t/T} \overset{\mathcal{F}}{\Longleftrightarrow} \frac{2\pi}{T} \sum_{\ell=-\infty}^{\infty} F(2\pi\ell/T) \delta\left(\omega - 2\pi\ell/T\right).$$

Since the left side in the last expression must be identical to (2.185) we are justified in writing

$$\sum_{n=-\infty}^{\infty} f(t - nT) = \sum_{\ell=-\infty}^{\infty} F(2\pi\ell/T)/T e^{i2\pi\ell t/T}, \tag{2.186}$$

which is the desired Poisson sum formula.

As an example, suppose $f(t) = 1/(1 + t^2)$. Then $F(\omega) = \pi e^{-|\omega|}$ (see (2.143*)) and with $T = 2\pi$ we get

$$\sum_{n=-\infty}^{\infty} \frac{1}{1 + (t - 2\pi n)^2} = \frac{1}{2} \sum_{\ell=-\infty}^{\infty} e^{-(|\ell| - i\ell t)} = \frac{e^2 - 1}{2\left[e^2 - 2e\cos(t) + 1\right]}. \tag{2.187}$$

2.2.7 Convergence at Discontinuities

The convergence of the FT at a step discontinuity exhibits the Gibbs oscillatory behavior similar to Fourier series. Thus suppose $f(t)$ has step discontinuities at $t = t_k$, $k = 1, 2, \ldots$ and we represent it as in (1.282). Then with $f^\Omega(t)$ as in (2.109) we have

$$
\begin{aligned}
f^\Omega(t) &= f_s^\Omega(t) + \sum_k \left[f\left(t_k^+\right) - f\left(t_k^-\right)\right] \int_{t_k}^\infty \frac{\sin\left[(t - t')\,\Omega\right]}{\pi\,(t - t')} dt' \\
&= f_s^\Omega(t) + \sum_k \left[f\left(t_k^+\right) - f\left(t_k^-\right)\right] \frac{1}{\pi} \int_{-\infty}^{(t - t_k)\Omega} \frac{\sin x}{x} dx \\
&= f_s^\Omega(t) + \sum_k \left[f\left(t_k^+\right) - f\left(t_k^-\right)\right] \left[\frac{1}{2} + \frac{1}{\pi} \operatorname{Si}\left[(t - t_k)\,\Omega\right]\right]. \quad (2.188)
\end{aligned}
$$

As $\Omega \to \infty$ the $f_s^\Omega(t)$ tends uniformly to $f_s(t)$ whereas the convergence of each member in the sum is characterized by the oscillatory behavior of the sine

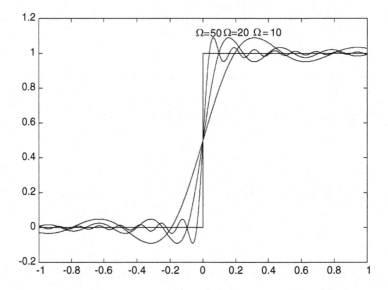

Figure 2.26: Convergence of the Fourier transform at a step discontinuity

integral function. This is illustrated in Fig. 2.26 which shows a unit step together with plots of $1/2 + (1/\pi)\operatorname{Si}(\Omega t)$ for $\Omega = 10, 20$, and 50.

2.2.8 Fejer Summation

In 2.1.5 it was shown that the Gibbs oscillations at step discontinuities arising in partial sums of Fourier series can be suppressed by employing the Fejer summation technique. An analogous procedure works for the Fourier Integral where instead of (2.32) we must resort to the following fundamental theorem from the theory of limits. Given a function $f(\Omega)$ integrable over any finite interval $0, \Omega$ we define, by analogy with (2.135), the average σ_Ω by

$$\sigma_\Omega = \frac{1}{\Omega} \int_0^\Omega f(\Omega) d\Omega. \tag{2.189}$$

It can be shown that if $\lim_{\Omega \to \infty} f(\Omega) = f$ exists then so does $\lim_{\Omega \to \infty} \sigma_\Omega = f$. Presently for the function $f(\Omega)$ we take the partial "sum" $f^\Omega(t)$ in (2.109) and denote the left side of (2.189) by $\sigma_\Omega(t)$. If we suppose that $\lim_{\Omega \to \infty} f^\Omega(t) = \frac{1}{2}[f(t^+) + f(t^-)]$, then by the above limit theorem we also have

$$\lim_{\Omega \to \infty} \sigma_\Omega(t) = \frac{1}{2}\left[f(t^+) + f(t^-)\right]. \tag{2.190}$$

By integrating the right side of (2.109) with respect to Ω and using (2.189) we obtain

$$\sigma_\Omega(t) = \int_{-\infty}^{\infty} f(t') \frac{\sin^2\left[(\Omega/2)(t - t')\right]}{\pi(\Omega/2)(t - t')^2} dt'. \tag{2.191}$$

Unlike the kernel (2.38) in the analogous formula for Fourier series in (2.37), the kernel

$$K_\Omega(t - t') = \frac{\sin^2\left[(\Omega/2)(t - t')\right]}{\pi(\Omega/2)(t - t')^2} \tag{2.192}$$

is not periodic. We leave it exercise to show that

$$\lim_{\Omega \to \infty} \frac{\sin^2\left[(\Omega/2)(t - t')\right]}{\pi(\Omega/2)(t - t')^2} = \delta(t - t'), \tag{2.193}$$

which may be taken as a direct verification of (2.190). A plot of the Fejer kernel together with the Fourier integral kernel is shown in Fig. 2.27, where the maximum of each kernel has been normalized to unity. Note that the Fejer kernel is always nonnegative with a wider main lobe than the Fourier kernel and exhibits significantly lower sidelobes. One can readily show that

$$\mathcal{F}\left\{\frac{\sin^2(\Omega/2)t}{\pi(\Omega/2)t^2}\right\} = \begin{cases} 1 - \frac{|\omega|}{\Omega}; & |\omega| < \Omega, \\ 0; & |\omega| > \Omega. \end{cases} \tag{2.194}$$

Since the right side of (2.191) is a convolution in the time domain, its FT yields a product of the respective transforms. Therefore using (2.194) we can rewrite (2.191) as an inverse FT as follows:

$$\sigma_\Omega(t) = \frac{1}{2\pi} \int_{-\Omega}^{\Omega} F(\omega)\left(1 - \frac{|\omega|}{\Omega}\right) e^{i\omega t} d\omega. \tag{2.195}$$

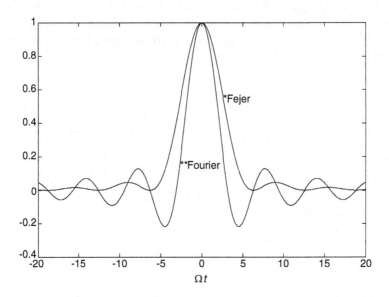

Figure 2.27: Fejer and Fourier integral kernels

We see that the Fejer "summation" (2.195) is equivalent to the multiplication of the signal transform $F(\omega)$ by the triangular spectral window:

$$W(\omega) = \left(1 - \frac{|\omega|}{\Omega}\right) p_\Omega(\omega) \tag{2.196}$$

quite analogous to the discrete spectral weighting of the Fourier series coefficients in (2.34). Figure 2.28 shows a rectangular pulse together with the Fejer and Fourier approximations using a spectral truncation of $\Omega = 40/T$. These results are seen to be very similar to those plotted in Fig. 2.9 for Fourier series. Just like for Fourier series, we can also introduce higher order Fejer approximations. For example, the second-order approximation $\sigma_\Omega^{(1)}(t)$ can be defined by

$$\sigma_\Omega^{(1)}(t) = \sigma_\Omega = \frac{1}{\Omega} \int_0^\Omega \sigma_a(t)\, da \tag{2.197}$$

again with the property

$$\lim_{\Omega \to \infty} \sigma_\Omega^{(1)}(t) = \frac{1}{2}\left[f(t^+) + f(t^-)\right]. \tag{2.198}$$

Substituting (2.191) with Ω replaced by the integration variable a into (2.197) one can show that

$$\sigma_\Omega^{(1)}(t) = \int_{-\infty}^\infty f(t')\, K_\Omega^{(1)}(t - t')\, dt', \tag{2.199}$$

where

$$K_\Omega^{(1)}(t) = \frac{1}{\pi t^2} \int_0^{\Omega t} \frac{1 - \cos x}{x}\, dx. \tag{2.200}$$

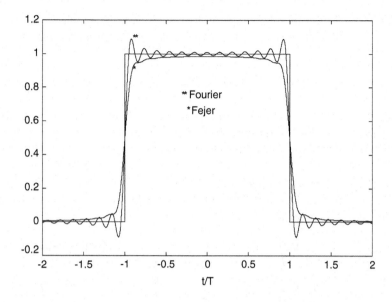

Figure 2.28: Comparison of Fejer and Fourier integral approximations

One can show directly that $\lim\limits_{\Omega \longrightarrow \infty} K_\Omega^{(1)}(t) = \delta(t)$, consistent with (2.198). A plot of $4\pi K_\Omega^{(1)}(t)/\Omega^2$ as a function of Ωt together with the (first-order) Fejer and Fourier kernels is shown in Fig. 2.29. Unlike the Fourier Integral and the (first-order) Fejer kernels, $K_\Omega^{(1)}(t)$ decreases monotonically on both sides of the

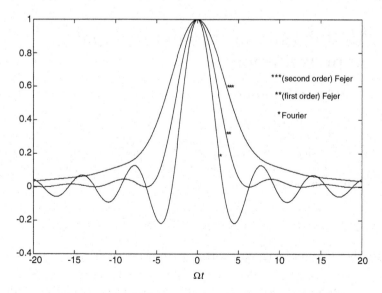

Figure 2.29: Comparison of Fourier and Fejer kernels

maximum, i.e., the functional form is free of sidelobes. At the same time its single lobe is wider than the main lobe of the other two kernels. It can be shown that for large Ωt

$$K_\Omega^{(1)}(t) \sim \frac{\ln(\Omega |t| \gamma)}{\pi (\Omega t)^2}, \tag{2.201}$$

where $\ln \gamma = 0.577215\ldots$ is the Euler constant. Because of the presence of the logarithmic term (2.201) represents a decay rate somewhere between that of the Fourier Integral kernel $(1/\Omega t)$ and that of the (first-order) Fejer kernel $(1/(\Omega t)^2)$.

The Fourier transform of $K_\Omega^{(1)}(t)$ furnishes the corresponding spectral window. An evaluation of the FT by directly transforming (2.200) is somewhat cumbersome. A simpler approach is the following:

$$\begin{aligned}
\mathcal{F}\{K_\Omega^{(1)}(t)\} &= \mathcal{F}\{\frac{1}{\Omega}\int_0^\Omega \frac{\sin^2[(a/2)t]}{\pi(a/2)t^2}da\} = \frac{1}{\Omega}\int_0^\Omega \mathcal{F}\left\{\frac{\sin^2[(a/2)t]}{\pi(a/2)t^2}\right\}da \\
&= \frac{1}{\Omega}\int_0^\Omega\left(1-\frac{|\omega|}{a}\right)p_a(\omega)\,da = \begin{cases} \frac{1}{\Omega}\int_{|\omega|}^\Omega\left(1-\frac{|\omega|}{a}\right)da \; ; \; |\omega| < \Omega, \\ 0 \; ; \; |\omega| > \Omega. \end{cases}
\end{aligned}$$

The last integral is easily evaluated with the final result

$$\mathcal{F}\{K_\Omega^{(1)}(t)\} \equiv W^{(1)}(\omega) = \left\{1 + \frac{|\omega|}{\Omega}\left(\ln\frac{|\omega|}{\Omega} - 1\right)\right\} p_\Omega(\omega). \tag{2.202}$$

A plot of this spectral window is shown in Fig. 2.30 which is seen to be quite similar to its discrete counterpart in Fig. 2.11.

2.3   Modulation and Analytic Signal Representation

2.3.1   Analytic Signals

Suppose $z(t)$ is a real signal with a Fourier transform $Z(\omega) = A(\omega)e^{i\theta(\omega)}$. According to (2.151) this signal can be expressed as a real part of the complex signal whose Fourier transform vanishes for negative frequencies and equals twice the transform of the given real signal for positive frequencies. Presently we denote this complex signal by $w(t)$ so that

$$w(t) = \frac{1}{2\pi}\int_0^\infty 2Z(\omega)e^{i\omega t}d\omega, \tag{2.203}$$

whence the real and imaginary parts are, respectively,

$$\Re\{w(t)\} = z(t) = \frac{1}{\pi}\int_0^\infty A(\omega)\cos[\omega t + \theta(\omega)]d\omega, \tag{2.204a}$$

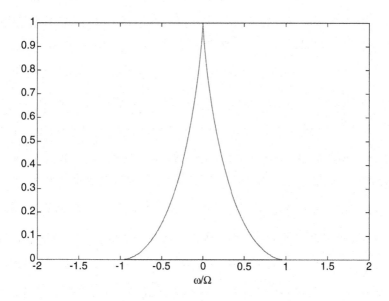

Figure 2.30: FT of second-order Fejer kernel

$$\Im m\{w(t)\} = \frac{1}{\pi}\int_0^\infty A(\omega)\sin[\omega t + \theta(\omega)]d\omega. \tag{2.204b}$$

We claim that $\Im m\{w(t)\}$, which we presently denote by $\hat{z}(t)$, is the Hilbert transform of $z(t)$, i.e.,

$$\hat{z}(t) = \frac{1}{\pi}P\int_{-\infty}^\infty \frac{z(\tau)}{t-\tau}d\tau. \tag{2.205}$$

Taking Hilbert transforms of both sides of (2.204a) and using trigonometric sum formulas together with (2.179) and (2.180) we obtain

$$
\begin{aligned}
\hat{z}(t) &= \mathcal{H}\{z(t)\} = \mathcal{H}\left\{\frac{1}{\pi}\int_0^\infty A(\omega)\cos[\omega t + \theta(\omega)]d\omega\right\}\\
&= \frac{1}{\pi}\int_0^\infty A(\omega)\mathcal{H}\{\cos[\omega t + \theta(\omega)]\}\,d\omega\\
&= \frac{1}{\pi}\int_0^\infty A(\omega)\left(\cos[\theta(\omega)]\mathcal{H}\{\cos(\omega t)\} - \sin[\theta(\omega)]\mathcal{H}\{\sin(\omega t)\}\right)d\omega\\
&= \frac{1}{\pi}\int_0^\infty A(\omega)\left(\cos[\theta(\omega)]\sin(\omega t) + \sin[\theta(\omega)]\cos(\omega t)\right)d\omega\\
&= \frac{1}{\pi}\int_0^\infty A(\omega)\sin[\omega t + \theta(\omega)]d\omega = \Im m\{w(t)\}
\end{aligned}
$$

as was to be demonstrated. As a by-product of this derivation we see that the evaluation of the Hilbert transform of any signal can always be carried out entirely in terms of the FT, as already remarked in connection with the frequency domain calculation in (2.180).

The complex function

$$w(t) = z(t) + i\hat{z}(t) \tag{2.206}$$

of a real variable t is referred to as *an analytic signal*.[1] By construction the Fourier transform of such a signal vanishes identically for negative frequencies. This can also be demonstrated directly by Fourier transforming both sides of (2.206). This entails recognition of (2.205) as a convolution of $z(t)$ with $1/\pi t$ and use of (2.164) and (2.131). As a result we get the transform pair

$$\hat{z}(t) \overset{\mathfrak{F}}{\Longleftrightarrow} -i\,sign(\omega)\,Z(\omega). \tag{2.207}$$

Using this in the FT of (2.206) yields $W(\omega) = Z(\omega) + i\left[-i\,sign(\omega)\,Z(\omega)\right]$ which is equivalent to

$$W(\omega) = \begin{cases} 2Z(\omega)\ ;\ \omega > 0 \\ 0\ ;\ \omega < 0. \end{cases} \tag{2.208}$$

In practical situations a signal will invariably have negligible energy above a certain frequency. It is frequently convenient to idealize this by assuming that the FT of the signal vanishes identically above a certain frequency. Such a signal is said to be bandlimited (or effectively bandlimited). For example if $z(t)$ is bandlimited to $|\omega| < \omega_{max}$ the magnitude of its FT may appear as shown in Fig. 2.31a. In conformance with (2.208) the magnitude of the Fourier spectrum of the corresponding analytic signal then appears as in Fig. 2.31b. It is common to refer to the spectrum in Fig. 2.31a as double sided and to that in Fig. 2.31b as the single sided. In practical applications the use of the latter is more common. The energy balance between the time and the frequency domains follows from Parseval theorem

$$\int_{-\infty}^{\infty} |w(t)|^2\, dt = \frac{2}{\pi} \int_0^{\omega_{max}} |Z(\omega)|^2\, d\omega.$$

Because of (2.207) the energy of an analytic signal is shared equally by the real signal and its Hilbert transform.

2.3.2 Instantaneous Frequency and the Method of Stationary Phase

The analytic signal furnishes a means of quantifying the amplitude, phase, and frequency of signals directly in the time domain. We recall that these concepts have their primitive origins in oscillatory phenomena described by

---

[1]The term "analytic" refers to the fact that a signal whose Fourier transform vanishes for real negative values of frequency, i.e., is represented by the integral (2.203), is an analytic function of t in the upper half of the complex t plane (i.e., $\Im m(t) > 0$). (See Appendix, pages 341–348).

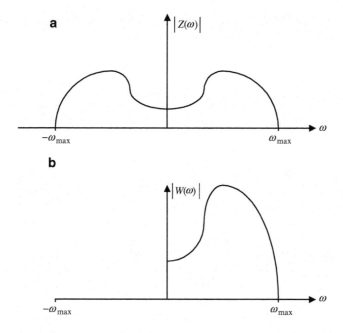

Figure 2.31: Spectrum of $z(t)$ and $z(t) + i\hat{z}(t)$

sinusoids. Thus we say that the signal $r\cos(\omega t + \psi_0)$ has amplitude r, frequency ω and a fixed phase reference ψ_0, where for purposes of analysis we sometimes find it more convenient to deal directly with a fictitious complex signal $r\exp[i(\omega t + \psi_0)]$ with the tacit understanding that physical processes are to be associated only with the real part of this signal. A generalization of this construct is an analytic signal. In addition to simplifying the algebra such complex notation also affords novel points of view. For example, the exponential of magnitude r and phase angle $\psi(t) = \omega t + \theta_0$ can be interpreted graphically as a phasor of length r rotating at the constant angular velocity $\omega = \frac{d}{dt}(\omega t + \theta_0)$. Classically for a general nonsinusoidal (real) signal $z(t)$ the concepts of frequency, amplitude, and phase are associated with each sinusoidal component comprising the signal Fourier spectrum, i.e., in this form these concepts appear to have meaning only when applied to each individual spectral component of the signal. On the other hand we can see intuitively that at least in special cases the concept of frequency must bear a close relationship to the rate of zero crossings of a real signal. For pure sinusoids this observation is trivial, e.g., the number of zero crossings of the signal $\cos(10t)$ per unit time is twice that of $\cos(5t)$. Suppose instead we take the signal $\cos(10t^2)$. Here the number of zero crossings varies linearly with time and the corresponding complex signal, as represented by the phasor $\exp[i(10t^2)]$, rotates at the rate $\frac{d}{dt}(10t^2) = 20t$ rps. Thus we conclude that the frequency of this signal varies linearly with time. The new concept here is that of *instantaneous frequency* which is clearly not identical with the frequency associated with each Fourier component of the signal (except of course in case of a pure sinusoid). We extend this definition to arbitrary real signals $z(t)$ through an analytic signal constructed in accordance with (2.203). We write it presently in the form

$$w(t) = r(t) e^{i\psi(t)}, \qquad (2.209)$$

where

$$r(t) = \sqrt{z^2(t) + \hat{z}^2(t)} \qquad (2.210)$$

is the (real, nonnegative) time-varying amplitude, or envelope, $\psi(t)$ the instantaneous phase, and

$$\tilde{\omega}(t) = \frac{d\psi}{dt} \qquad (2.211)$$

the instantaneous frequency. Note also that the interpretation of $\tilde{\omega}(t)$ as a zero crossing rate requires that it be nonnegative which is compatible with the analytic signal having only positive frequency components. To deduce the relationship between the instantaneous frequency and the signal Fourier spectrum let us formulate an estimate of the spectrum of $w(t)$:

$$W(\omega) = \int_{-\infty}^{\infty} r(t) e^{i[\psi(t) - \omega t]} dt. \qquad (2.212)$$

We can, of course, not "evaluate" this integral without knowing the specific signal. However for signals characterized by a large time-bandwidth product we can carry out an approximate evaluation utilizing the so-called principle of stationary phase. To illustrate the main ideas without getting sidetracked by peripheral generalities consider the real part of the exponential in (2.212), i.e., $\cos[q(t)]$ with $q(t) = \psi(t) - \omega t$. Figure 2.32 shows a plot of $\cos[q(t)]$ for

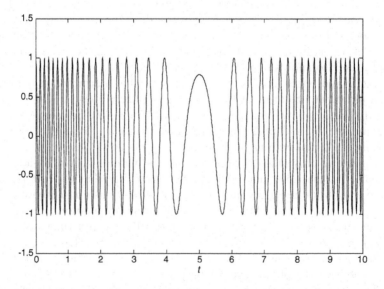

Figure 2.32: Plot of $\cos(5t^2 - 50t)$ (stationary point at $t = 5$)

the special choice $\psi(t) = 5t^2$ and $\omega = 50$. This function is seen to oscillate rapidly except in the neighborhood of $t = t_0 = 5 = \omega/10$ which point corresponds to $q'(5) = 0$. The value $t_0 = 5$ in the neighborhood of which the phase varies slowly is referred to as the stationary point of $q(t)$ (or a point of stationary phase). If we suppose that the function $r(t)$ is slowly varying relative to these oscillations, we would expect the contributions to an integral of the form $\int_{-\infty}^{\infty} r(t) \cos(5t^2 - 50t)\, dt$ from points not in the immediate vicinity of $t = 5$ to mutually cancel. Consequently the dominant contributions to the integral would arise only from the values of $r(t)$ and $\psi(t)$ in the immediate neighborhood of the point of stationary phase. We note in passing that in this example the product $t_0\omega = 250 >> 1$. It is not hard to show that the larger this dimensionless quantity (time bandwidth product) the narrower the time band within which the phase is stationary and therefore the more nearly localized the contribution to the overall integral. In the general case the stationary point is determined by

$$q'(t) = \psi'(t) - \omega = 0, \tag{2.213}$$

which coincides with the definition of the instantaneous frequency in (2.211). When we expand the argument of the exponential in a Taylor series about $t = t_0$ we obtain

$$q(t) = \psi(t_0) - \omega t_0 + \frac{1}{2}(t - t_0)^2 \psi''(t_0) + \ldots \tag{2.214}$$

Similarly we have for $r(t)$

$$r(t) = r(t_0) + (t - t_0)r'(t_0) + \ldots \tag{2.215}$$

In accordance with the localization principle just discussed we expect, given a sufficiently large ωt_0, that in the exponential function only the first two Taylor series terms need to be retained. Since $r(t)$ is assumed to be relatively slowly varying it may be replaced by $r(t_0)$. Therefore (2.212) may be approximated by

$$W(\omega) \sim r(t_0)\, e^{i[\psi(t_0) - \omega t_0]} \int_{-\infty}^{\infty} e^{i\frac{1}{2}(t - t_0)^2 \psi''(t_0)} dt. \tag{2.216}$$

When the preceding standard Gaussian integral is evaluated we obtain the final formula

$$W(\omega) \underset{\omega t_0 \sim \infty}{\sim} r(t_0)\, e^{i[\psi(t_0) - \omega t_0]} \sqrt{\frac{2\pi}{|\psi''(t_0)|}}\, e^{i\frac{\pi}{4} sign[\psi''(t_0)]}. \tag{2.217}$$

It should be noted that the variable t_0 is to be expressed in terms of ω by inverting (2.213), a procedure that in general is far from trivial. When this is done (2.217) provides an asymptotic approximation to the signal Fourier spectrum for large ωt_0.

To illustrate the relationship between the instantaneous frequency of a signal and its frequency content as defined by Fourier synthesis consider the signal

$$g(t) = \begin{cases} A\cos(at^2 + \beta t) \; ; \; 0 \le t \le T, \\ \qquad\qquad 0 \text{ elsewhere.} \end{cases}$$

whose instantaneous frequency increases linearly from $\omega_{min} = \beta$ $(\beta > 0)$ to $\omega_{max} = 2aT + \beta$ rps. Based on this observation it appears reasonable to define the nominal bandwidth of this signal by $B = aT/\pi$ Hz. The relationship between B and the bandwidth as defined by the signal Fourier spectrum is more readily clarified in terms of the dimensionless parameters $M = 2BT$ (the nominal time-bandwidth product) and $r = \omega_{min}/\omega_{max} < 1$. Using these parameters we put the signal in the form

$$g(t) = \begin{cases} A\cos\left[\left(\frac{\pi}{2}M\right)(t/T)^2 + \pi\frac{M}{1-r}\left(\frac{t}{T}\right)\right] \; ; \; 0 \le t \le T, \\ \qquad\qquad\qquad 0 \text{ elsewhere.} \end{cases} \qquad (2.218)$$

The FT of (2.218) can be expressed in terms of Fresnel integrals whose standard forms read

$$C(x) \;=\; \int_0^x \cos\left(\frac{\pi}{2}\xi^2\right) d\xi, \qquad (2.219a)$$

$$S(x) \;=\; \int_0^x \sin\left(\frac{\pi}{2}\xi^2\right) d\xi. \qquad (2.219b)$$

One then finds

$$G(\omega) \;=\; \frac{AT}{2\sqrt{M}}\left[e^{-i\frac{\pi}{2}M\left(\frac{r-f'}{1-r}\right)^2} [C\{\sqrt{M}\frac{1-f'}{1-r}\} + iS\{\sqrt{M}\frac{1-f'}{1-r}\} \right.$$
$$-C\{\sqrt{M}\frac{r-f'}{1-r}\} - iS\{\sqrt{M}\frac{r-f'}{1-r}\}]$$
$$+e^{i\frac{\pi}{2}M\left(\frac{r+f'}{1-r}\right)^2} [C\{\sqrt{M}\frac{1+f'}{1-r}\} - iS\{\sqrt{M}\frac{1+f'}{1-r}\}$$
$$\left. -C\{\sqrt{M}\frac{r+f'}{1-r}\} + iS\{\sqrt{M}\frac{r+f'}{1-r}\}]\right], \qquad (2.220)$$

where we have introduced the normalized frequency variable $f' = \omega(1-r)/2\pi B$. Using the asymptotic forms of the Fresnel integrals for large arguments, i.e., $C(\pm\infty) = \pm 1/2$ and $S(\pm\infty) = \pm 1/2$, we find that as the nominal time-bandwidth product $(M/2)$ approaches infinity, the rather cumbersome expression (2.220) assumes the simple asymptotic form

$$G(\omega) \underset{M\sim\infty}{\sim} \begin{cases} \frac{AT}{2}\sqrt{\frac{2}{M}}e^{-i\frac{\pi}{2}M\left(\frac{r-f'}{1-r}\right)^2} \; ; r < f' < 1, \\ \frac{AT}{2}\sqrt{\frac{2}{M}}e^{i\frac{\pi}{2}M\left(\frac{r+f'}{1-r}\right)^2} \; ; -1 < f' < -r, \\ \qquad 0 \; ; |f'| > 1 \text{ and } |f'| < r. \end{cases} \qquad (2.221)$$

From (2.221) we see that the FT of $g(t)$ approaches the constant $AT/2\sqrt{2/M}$ within the frequency band $r < |f'| < 1$ and vanishes outside this range, except at the band edges (i.e., $f' = \pm 1$ and $\pm r$) where it equals one-half this constant. Since $g(t)$ is of finite duration it is asymptotically simultaneously bandlimited and timelimited. Even though for any finite M the signal spectrum will not be bandlimited this asymptotic form is actually consistent with Parseval theorem. For applying Parseval formula to (2.221) we get

$$\frac{1}{2\pi} \int_{-\infty}^{\infty} |G(\omega)|^2 \, d\omega = \frac{A^2}{4} \frac{T}{B} 2B = (A^2/2) \, T \qquad (2.222)$$

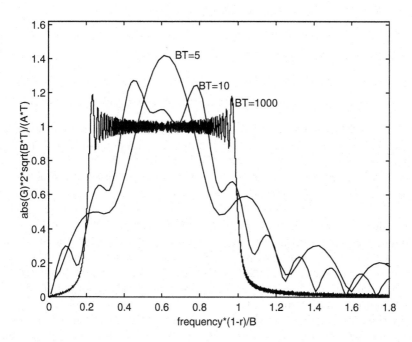

Figure 2.33: Magnitude of the FT of a linear FM pulse for different time-bandwidth products

On the other hand we recognize the last term as the asymptotic form (i.e., for large $\omega_0 T$) of the total energy of a sinusoid of fixed frequency, amplitude A, and duration T. Thus apparently if the time-bandwidth product is sufficiently large we may approximate the energy of a constant amplitude sinusoid with variable phase by the same simple formula $(A^2/2)T$. Indeed this result actually generalizes to signals of the form $A \cos[\phi(t)]$. How large must M be for (2.221) to afford a reasonable approximation to the signal spectrum? Actually quite large, as is illustrated by the plots in Fig. 2.33 for $BT = 5, 10$, and $1,000$, where the lower (nominal) band edge is defined by $r = 0.2$ and the magnitude of the asymptotic spectrum equals unity within $.2 < f' < 1$ and $1/2$ at $f' = .2$ and $f' = 1$.

2.3.3   Bandpass Representation

The construct of an analytic signal affords a convenient tool for describing the process of modulation of a low frequency (baseband) signal by a high frequency carrier as well as the demodulation of a transmitted bandpass signal down to baseband frequencies. We have already encountered a modulated signal in simplified form in connection with the frequency shifting properties of the FT in (2.154). Adopting now a more general viewpoint we take an arbitrary real signal $z(t)$ together with its Hilbert transform $\hat{z}(t)$ and a positive constant ω_0 to define two functions $x(t)$ and $y(t)$ as follows:

$$x(t) = z(t)\cos(\omega_0 t) + \hat{z}(t)\sin(\omega_0 t), \qquad (2.223a)$$

$$y(t) = -z(t)\sin(\omega_0 t) + \hat{z}(t)\cos(\omega_0 t), \qquad (2.223b)$$

which are easily inverted to yield

$$z(t) = x(t)\cos(\omega_0 t) - y(t)\sin(\omega_0 t) \qquad (2.224a)$$

$$\hat{z}(t) = x(t)\sin(\omega_0 t) + y(t)\cos(\omega_0 t). \qquad (2.224b)$$

Equations (2.223) and (2.224) constitute a fundamental set of relations that are useful in describing rather general modulation and demodulation processes. In fact the form of (2.224a) suggests an interpretation of $z(t)$ as a signal modulated by a carrier of frequency ω_0 a special case of which is represented by the left side of (2.154). Comparison with (2.224a) yields $x(t) = Af(t)\cos(\theta_0)$ and $y(t) = Af(t)\sin(\theta_0)$. We note that in this special case $x(t)$ and $y(t)$ are linearly dependent which need not be true in general.

Let us now suppose that the only datum at our disposal is the signal $z(t)$ and that the carrier frequency ω_0 is left unspecified. As far as the mathematical representation (2.223) and (2.224) is concerned it is, of course, perfectly valid and consistent for any choice of (real) ω_0. However, if $x(t)$ and $y(t)$ in (2.223) are to represent baseband signals at the receiver resulting from the demodulation of $z(t)$ by the injection of a local oscillator with frequency ω_0, then the bandwidth of $x(t)$ and $y(t)$ (centered at $\omega = 0$) should certainly be less than $2\omega_0$. A more precise interrelation between the constraints on signal bandwidth and carrier frequency ω_0 is readily deduced from the FT of $x(t)$ and $y(t)$. Denoting these, respectively, by $X(\omega)$ and $Y(\omega)$, we obtain using (2.223) and (2.207)

$$X(\omega) = U(\omega_0 - \omega)Z(\omega - \omega_0) + U(\omega + \omega_0)Z(\omega + \omega_0), \qquad (2.225a)$$

$$Y(\omega) = iU(\omega_0 - \omega)Z(\omega - \omega_0) - iU(\omega + \omega_0)Z(\omega + \omega_0). \qquad (2.225b)$$

On purely physical grounds we would expect $Z(\omega)$ to be practically zero above some finite frequency, say ω_{max}. If the bandwidth of $X(\omega)$ and $Y(\omega)$ is to be limited to $|\omega| \leq \omega_0$, then $\omega_{max} - \omega_0$ may not exceed ω_0. This follows directly from (2.225) or from the graphical superposition of the spectra shown in Fig. 2.34. In other words if $x(t)$ and $y(t)$ are to represent baseband signals, we must have

$$\omega_0 \geq \omega_{max}/2. \qquad (2.226)$$

When this constraint is satisfied the spectrum of $z(t)$ may in fact extend down to zero frequency (as, e.g., in Fig. 2.31a) so that theoretically the spectra of $x(t)$ and $y(t)$ are allowed to occupy the entire bandwidth $|\omega| \leq \omega_0$. However in practice there will generally also be a lower limit on the band occupancy of $Z(\omega)$, say ω_{min}. Thus the more common situation is that of a bandpass spectrum illustrated in Fig. 2.34 wherein the nonzero spectral energy of $z(t)$ occupies the band $\omega_{min} < \omega < \omega_{max}$ for positive frequencies and the band $-\omega_{max} < \omega < -\omega_{min}$ for negative frequencies.

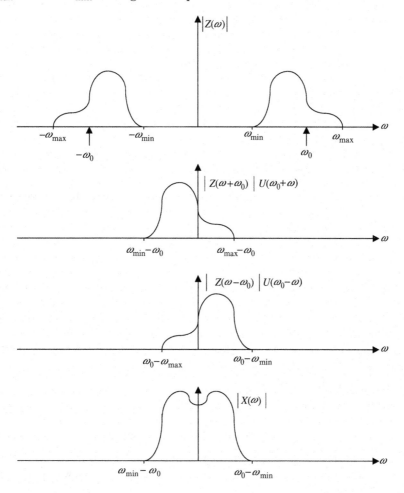

Figure 2.34: Bandpass and demodulated baseband spectra

In the case depicted $\omega_{min} < \omega_0 < \omega_{max}$ and $\omega_0 - \omega_{min} > \omega_{max} - \omega_0$. The synthesis of $X(\omega)$ from the two frequency-shifted sidebands follows from (2.225a) resulting in a total band occupancy of $2\,|\omega_0 - \omega_{min}|$. It is easy to see from (2.225b) that $Y(\omega)$ must occupy the same bandwidth. Observe that shifting ω_0 closer to ω_{min} until $\omega_{max} - \omega_0 > \omega_0 - \omega_{min}$ results in a total band occupancy of $2 \left| \omega_{\max} - \omega_0 \right|$ and that the smallest possible baseband bandwidth is obtained by positioning ω_0 midway between $\omega_{\max}$ and $\omega_{\min}$.

The two real baseband signals $x(t)$ and $y(t)$ are referred to as the inphase and quadrature signal components. It is convenient to combine them into the single complex baseband signal

$$b(t) = x(t) + iy(t). \tag{2.227}$$

The analytic signal $w(t) = z(t) + i\hat{z}(t)$ follows from a substitution of (2.224a) and (2.224b)

$$
\begin{aligned}
w(t) &= x(t) \cos(\omega_0 t) - y(t) \sin(\omega_0 t) \\
&\quad + i[x(t) \sin(\omega_0 t) + y(t) \cos(\omega_0 t)] \\
&= [x(t) + iy(t)] e^{i\omega_0 t} = b(t) e^{i\omega_0 t}.
\end{aligned} \tag{2.228}
$$

The FT of (2.228) reads

$$W(\omega) = X(\omega - \omega_0) + iY(\omega - \omega_0) = B(\omega - \omega_0) \tag{2.229}$$

or, solving for $B(\omega)$,

$$B(\omega) = W(\omega + \omega_0) = 2U(\omega + \omega_0) Z(\omega + \omega_0) = X(\omega) + iY(\omega). \tag{2.230}$$

In view of (2.224a) the real bandpass $z(t)$ signal is given by the real part of (2.228), i.e.,

$$z(t) = \Re \left\{ b(t) e^{i\omega_0 t} \right\}. \tag{2.228*}$$

Taking the FT we get

$$Z(\omega) = \frac{1}{2} \left[B(\omega - \omega_0) + B^*(-\omega - \omega_0) \right], \tag{2.228**}$$

which reconstructs the bandpass spectrum in terms of the baseband spectrum.

As the preceding formulation indicates, given a bandpass signal $z(t)$, the choice of ω_0 at the receiver effectively *defines* the inphase and quadrature components. Thus a different choice of (local oscillator) frequency, say ω_1, $\omega_1 \neq \omega_0$ leads to the representation

$$z(t) = x_1(t) \cos(\omega_1 t) - y_1(t) \sin(\omega_1 t), \tag{2.229a}$$

$$\hat{z}(t) = x_1(t) \sin(\omega_1 t) + y_1(t) \cos(\omega_1 t), \tag{2.229b}$$

wherein the $x_1(t)$ and $y_1(t)$ are the new inphase and quadrature components. The relationship between $x(t)$, $y(t)$ and $x_1(t)$ and $y_1(t)$ follows upon equating (2.229) to (2.224):

$$
\begin{bmatrix} \cos(\omega_0 t) & -\sin(\omega_0 t) \\ \sin(\omega_0 t) & \cos(\omega_0 t) \end{bmatrix} \begin{bmatrix} x(t) \\ y(t) \end{bmatrix} = \begin{bmatrix} \cos(\omega_1 t) & -\sin(\omega_1 t) \\ \sin(\omega_1 t) & \cos(\omega_1 t) \end{bmatrix} \begin{bmatrix} x_1(t) \\ y_1(t) \end{bmatrix},
$$

which yields

$$
\begin{bmatrix} x(t) \\ y(t) \end{bmatrix} = \begin{bmatrix} \cos[(\omega_0 - \omega_1)t] & \sin[(\omega_0 - \omega_1)t] \\ -\sin[(\omega_0 - \omega_1)t] & \cos[(\omega_0 - \omega_1)t] \end{bmatrix} \begin{bmatrix} x_1(t) \\ y_1(t) \end{bmatrix}. \tag{2.230}
$$

The linear transformations defined by the 2×2 matrices in (2.224), (2.229), and (2.230) are all orthogonal so that

$$z^2\left(t\right) + \hat{z}^2\left(t\right) = x^2\left(t\right) + y^2\left(t\right) = x_1^2\left(t\right) + y_1^2\left(t\right) = \left|b\left(t\right)\right|^2 = r^2\left(t\right). \qquad (2.231)$$

This demonstrates directly that the analytic signal and the complex baseband signal have the same envelope $r\left(t\right)$ which is in fact independent of the frequency of the reference carrier. We shall henceforth refer to $r(t)$ as the signal envelope. Unlike the signal envelope, the phase of the complex baseband signal does depend on the carrier reference. Setting

$$\theta\left(t\right) = \tan^{-1}\frac{x\left(t\right)}{y\left(t\right)}, \quad \theta_1\left(t\right) = \tan^{-1}\frac{x_1\left(t\right)}{y_1\left(t\right)}, \qquad (2.232)$$

we see that with a change in the reference carrier the analytic signal undergoes the transformation

$$\begin{aligned} w\left(t\right) &= r\left(t\right)e^{i\theta(t)}e^{i\omega_0 t} \\ &= r\left(t\right)e^{i\theta_1(t)}e^{i\omega_1 t} \end{aligned} \qquad (2.233)$$

or, equivalently, that the two phase angles transform in accordance with

$$\theta\left(t\right) + \omega_0 t = \theta_1\left(t\right) + \omega_1 t. \qquad (2.234)$$

It should be noted that in general the real and imaginary parts of a complex baseband signal need not be related by Hilbert transforms. In fact suppose $x\left(t\right)$ and $y\left(t\right)$ are two arbitrary real signals, bandlimited to $\left|\omega\right| < \omega_x$ and $\left|\omega\right| < \omega_y$, respectively. Then, as may be readily verified, for any ω_0 greater than $\omega_x/2$ and $\omega_y/2$ the Hilbert transform of the bandpass signal $z\left(t\right)$ defined by (2.224a) is given by (2.224b).

2.3.4 Bandpass Representation of Random Signals*

In the preceding discussion it was tacitly assumed that the signals are deterministic. The notion of an analytic signal is equally useful when dealing with stochastic signals. For example, if $z\left(t\right)$ is a real wide-sense stationary stochastic process, we can always append its Hilbert transform to form the complex stochastic process

$$w\left(t\right) = z\left(t\right) + i\hat{z}\left(t\right). \qquad (2.235)$$

By analogy with (2.206) we shall refer to it as an analytic stochastic process. As we shall show in the sequel its power spectrum vanishes for negative frequencies. First we note that in accordance with (2.207) the magnitude of the transfer function that transforms $z\left(t\right)$ into $\hat{z}\left(t\right)$ is unity. Hence the power spectrum as well as the autocorrelation function of $\hat{z}\left(t\right)$ are the same as that of $z\left(t\right)$, i.e.,

$$\langle z\left(t+\tau\right)z\left(t\right)\rangle \equiv R_{zz}\left(\tau\right) = \langle \hat{z}\left(t+\tau\right)\hat{z}\left(t\right)\rangle = R_{\hat{z}\hat{z}}\left(\tau\right). \qquad (2.236)$$

The cross-correlation between $z(t)$ and its Hilbert transform is then

$$
\begin{aligned}
R_{\hat{z}z}(\tau) &= \langle \hat{z}(t+\tau) z(t) \rangle = \frac{1}{\pi} \int_{-\infty}^{\infty} \frac{\langle z(\tau') z(t) \rangle}{t+\tau-\tau'} d\tau' \\
&= \frac{1}{\pi} \int_{-\infty}^{\infty} \frac{R_{zz}(\tau'-t)}{t+\tau-\tau'} d\tau' = \frac{1}{\pi} \int_{-\infty}^{\infty} \frac{R_{zz}(\xi)}{\tau-\xi} d\xi. \quad (2.237)
\end{aligned}
$$

The last expression states that the cross-correlation between a stationary stochastic process and its Hilbert transform is the Hilbert transform of the autocorrelation function of the process. In symbols

$$
R_{\hat{z}z}(\tau) = \hat{R}_{zz}(\tau). \quad (2.238)
$$

Recall that the Hilbert transform of an even function is an odd function and conversely. Thus, since the autocorrelation function of a real stochastic process is always even, $R_{\hat{z}z}(\tau)$ is odd. Therefore we have

$$
R_{z\hat{z}}(\tau) \equiv R_{\hat{z}z}(-\tau) = -R_{\hat{z}z}(\tau). \quad (2.239)
$$

The autocorrelation function of $w(t)$ then becomes

$$
\begin{aligned}
R_{ww}(\tau) &= \langle w(t+\tau) \overset{*}{w}(t) \rangle = 2\left[R_{zz}(\tau) + iR_{\hat{z}z}(\tau)\right] \\
&= 2\left[R_{zz}(\tau) + i\hat{R}_{zz}(\tau)\right]. \quad (2.240)
\end{aligned}
$$

With

$$
R_{zz}(\tau) \overset{\mathfrak{F}}{\longleftrightarrow} S_{zz}(\omega) \quad (2.241)
$$

we have in view of (2.238) and (2.207)

$$
R_{\hat{z}z}(\tau) = \hat{R}_{zz}(\tau) \overset{\mathfrak{F}}{\longleftrightarrow} -iS_{zz}(\omega)\, sign(\omega). \quad (2.242)
$$

Denoting the spectral density of $w(t)$ by $S_{ww}(\omega)$, (2.240) together with (2.241) and (2.242) gives

$$
S_{ww}(\omega) = \begin{cases} 4S_{zz}(\omega); & \omega > 0, \\ 0; & \omega < 0. \end{cases} \quad (2.243)
$$

so that the spectral density of the analytic complex process has only positive frequency content. The correlation functions of the baseband (inphase) $x(t)$ and (quadrature) process $y(t)$ follow from (2.223). By direct calculation we get

$$
\begin{aligned}
R_{xx}(\tau) &= \langle \{z(t+\tau)\cos[\omega_0(t+\tau)] + \hat{z}(t+\tau)\sin[\omega_0(t+\tau)]\} \\
&\quad \{z(t)\cos(\omega_0 t) + \hat{z}(t)\sin(\omega_0 t)\} \rangle \\
&= R_{zz}(\tau)\cos(\omega_0\tau) + R_{\hat{z}z}(\tau)\sin(\omega_0\tau) \quad (2.244a) \\
R_{yy}(\tau) &= \langle \{-z(t+\tau)\sin[\omega_0(t+\tau)] + \hat{z}(t+\tau)\cos[\omega_0(t+\tau)]\} \\
&\quad \{-z(t)\sin(\omega_0 t) + \hat{z}(t)\cos(\omega_0 t)\} \rangle \\
&= R_{zz}(\tau)\cos(\omega_0\tau) + R_{\hat{z}z}(\tau)\sin(\omega_0\tau) = R_{xx}(\tau) \quad (2.244b)
\end{aligned}
$$

$$
\begin{aligned}
R_{xy}(\tau) &= \langle \{z(t+\tau)\cos[\omega_0(t+\tau)] + \hat{z}(t+\tau)\sin[\omega_0(t+\tau)]\} \\
&\quad \{-z(t)\sin(\omega_0 t) + \hat{z}(t)\cos(\omega_0 t)\}\rangle \\
&= -R_{\hat{z}z}(\tau)\cos(\omega_0\tau) + R_{zz}(\tau)\sin(\omega_0\tau) \\
&= -\hat{R}_{zz}(\tau)\cos(\omega_0\tau) + R_{zz}(\tau)\sin(\omega_0\tau).
\end{aligned}
\tag{2.244c}
$$

Recall that for any two real stationary processes $R_{xy}(\tau) = R_{yx}(-\tau)$. Using this relation in (2.244c) we get

$$
R_{yx}(\tau) = -R_{xy}(\tau).
\tag{2.245}
$$

Also according to (2.244a) and (2.244b) the autocorrelation functions of the in-phase and quadrature components of the stochastic baseband signal are identical and consequently so are the corresponding power spectra. These are

$$
\begin{aligned}
S_{xx}(\omega) &= S_{yy}(\omega) = \\
&\frac{1}{2}[1 - sign(\omega - \omega_0)]S_{zz}(\omega - \omega_0) \\
&+ \frac{1}{2}[1 + sign(\omega + \omega_0)]S_{zz}(\omega + \omega_0).
\end{aligned}
\tag{2.246}
$$

From (2.244c) we note that $R_{xy}(0) \equiv 0$ but that in general $R_{xy}(\tau) \neq 0$ when $\tau \neq 0$. The FT of this quantity, i.e., the cross-spectrum, is

$$
\begin{aligned}
S_{xy}(\omega) &= \frac{i}{2}[sign(\omega - \omega_0) - 1]S_{zz}(\omega - \omega_0) \\
&+ \frac{i}{2}[sign(\omega + \omega_0) + 1]S_{zz}(\omega + \omega_0).
\end{aligned}
\tag{2.247}
$$

By constructing a mental picture of the relative spectral shifts dictated by (2.247) it is not hard to see that the cross spectrum vanishes identically (or, equivalently, $R_{xy}(\tau) \equiv 0$) when $S_{zz}(\omega)$, the spectrum of the band-pass process, is symmetric about ω_0.

Next we compute the autocorrelation function $R_{bb}(\tau)$ of the complex stochastic baseband process $b(t) = x(t) + iy(t)$. Taking account of $R_{xx}(\tau) = R_{yy}(\tau)$ and (2.245) we get

$$
R_{bb}(\tau) = \langle b(t+\tau)b^*(t)\rangle = 2[R_{xx}(\tau) + iR_{yx}(\tau)].
\tag{2.248}
$$

In view of (2.240) and (2.228) the autocorrelation function of the analytic bandpass stochastic process is

$$
R_{ww}(\tau) = 2\left[R_{zz}(\tau) + i\hat{R}_{zz}(\tau)\right] = R_{bb}(\tau)e^{i\omega_0\tau}.
\tag{2.249}
$$

The autocorrelation function of the real bandpass process can then be represented in terms of the autocorrelation function of the complex baseband process as follows:

$$
R_{zz}(\tau) = \frac{1}{2}\Re\left\{R_{bb}(\tau)e^{i\omega_0\tau}\right\}.
\tag{2.250}
$$

With the definition

$$R_{bb}(\tau) \stackrel{\mathfrak{F}}{\Longleftrightarrow} S_{bb}(\omega)$$

the FT of (2.250) reads

$$S_{zz}(\omega) = \frac{1}{4}\left[S_{bb}(\omega + \omega_0) + S_{bb}(-\omega - \omega_0)\right]. \qquad (2.251)$$

Many sources of noise can be modeled (at least on a limited timescale) as stationary stochastic processes. The spectral distribution of such noise is usually of interest only in a relatively narrow pass band centered about some frequency, say ω_0. The measurement of the power spectrum within a predetermined pass band can be accomplished by using a synchronous detector that separates the inphase and quadrature channels, as shown in Fig. 2.35.

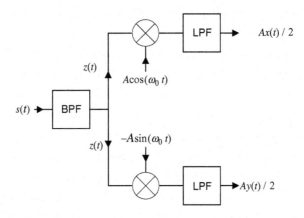

Figure 2.35: Synchronous detection

The signal $s(t)$ is first bandpass filtered to the bandwidth of interest and then split into two separate channels each of which is heterodyned with a local oscillator with a 90 degree relative phase shift. The inphase and quadrature components are obtained after lowpass filtering to remove the second harmonic contribution generated in each mixer. To determine the power spectral density of the bandpass signal requires measurement of the auto and cross spectra of $x(t)$ and $y(t)$. The power spectrum can then be computed with the aid of (2.246) and (2.247) which give

$$S_{zz}(\omega + \omega_0) = \frac{1}{2}\left[S_{xx}(\omega) - iS_{xy}(\omega)\right]. \qquad (2.252)$$

This procedure assumes that the process $s(t)$ is stationary so that $S_{xx}(\omega) = S_{yy}(\omega)$. Unequal powers in the two channels would be an indication of nonstationarity on the measurement timescale. A rather common form of nonstationarity is the presence of an additive deterministic signal within the bandpass process.

A common model is a rectangular bandpass power spectral density. Assuming that ω_0 is chosen symmetrically disposed with respect to the bandpass power spectrum, the power spectra corresponding to the analytic and baseband stochastic processes are shown in Fig. 2.36. In this case the baseband autocorrelation function is purely real and equal to

$$R_{bb}\left(\tau\right) = N_0\frac{\sin(2\pi B\tau)}{\pi\tau} = 2R_{xx}\left(\tau\right) = 2R_{yy}\left(\tau\right) \qquad (2.253)$$

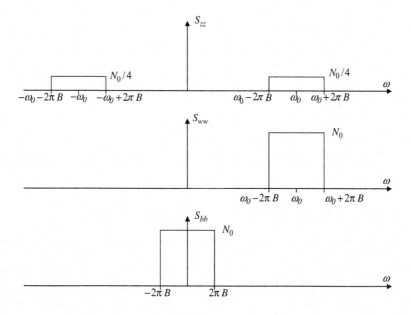

Figure 2.36: Bandpass-to-baseband transformation for a symmetric power spectrum

while $R_{xy}\left(\tau\right) \equiv 0$. What happens when the local oscillator frequency is set to $\omega = \omega_1 = \omega_0 - \Delta\omega$, i.e., off the passband center by $\Delta\omega$? In that case the baseband power spectral density will be displaced by $\Delta\omega$ and equal to

$$S_{bb}\left(\omega\right) = \begin{cases} N_0 ; & -2\pi B + \Delta\omega < \omega < 2\pi B + \Delta\omega, \\ 0 ; & otherwise. \end{cases} \qquad (2.254)$$

The baseband autocorrelation function is now the complex quantity

$$R_{bb}\left(\tau\right) = e^{i\tau\Delta\omega}N_0\frac{\sin(2\pi B\tau)}{\pi\tau}. \qquad (2.255)$$

In view of (2.248) the auto and crosscorrelation functions of the inphase and quadrature components are

$$R_{xx}\left(\tau\right) = R_{yy}\left(\tau\right) = \cos\left(\tau\Delta\omega\right)N_0\frac{\sin(2\pi B\tau)}{2\pi\tau}, \qquad (2.256a)$$

$$R_{yx}\left(\tau\right) = \sin\left(\tau\Delta\omega\right)N_0\frac{\sin(2\pi B\tau)}{2\pi\tau}. \qquad (2.256b)$$

The corresponding spectrum $S_{xx}(\omega) = S_{yy}(\omega)$ occupies the band $|\omega| \leq 2\pi B + \Delta\omega$. Unlike in the symmetric case, the power spectrum is no longer flat but exhibits two steps caused by the spectral shifts engendered by $\cos(\tau\Delta\omega)$, as shown in Fig. 2.37.

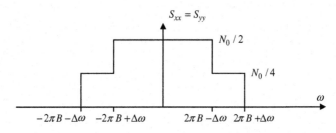

Figure 2.37: Baseband I&Q power spectra for assymmetric local oscillator frequency positioning

2.4   Fourier Transforms and Analytic Function Theory

2.4.1   Analyticity of the FT of Causal Signals

Even though both the direct and the inverse FT have been initially defined strictly for functions of a real variables one can always formally replace t and (or) ω by complex numbers and, as long as the resulting integrals converge, define the signal $f(t)$ and (or) the frequency spectrum $F(\omega)$ as functions of a complex variable. Those unfamiliar with complex variable theory should consult the Appendix, and in particular A.4.

Let us examine the analytic properties of the FT in the complex domain of a causal signal. To this end we replace ω by the complex variable $z = \omega + i\delta$ and write

$$F(z) = \int_0^\infty f(t) e^{-izt} dt, \qquad (2.257)$$

wherein $F(\omega)$ is $F(z)$ evaluated on the axis of reals. Furthermore let us assume that

$$\int_0^\infty |f(t)| \, dt < \infty. \qquad (2.258)$$

To put the last statement into the context of a physical requirement let us suppose that the signal $f(t)$ is the impulse response of a linear time-invariant system. In that case, as will be shown in 3.1.4, absolute integrability in the sense of (2.258) is a requirement for system stability. Using (2.257) we obtain in view of (2.258) for all $\operatorname{Im} z = \delta \leq 0$ the bound

$$\left| \int_0^\infty f(t) e^{-izt} dt \right| \leq \int_0^\infty |f(t)| e^{\delta t} dt < \infty. \qquad (2.259)$$

From this follows (see Appendix) that $F(z)$ is an analytic function of the complex variable z in the closed lower half of the complex z plane, i.e., $\text{Im}\, z \leq 0$. Moreover, for $\text{Im}\, z \leq 0$,

$$\lim_{|z| \to \infty} F(z) \to 0 \tag{2.260}$$

as we see directly from (2.259) by letting δ approach $-\infty$. In other words, the FT of the impulse response of a causal linear time-invariant system is an analytic function of the complex frequency variable z in the closed lower half plane. This feature is of fundamental importance in the design and analysis of frequency selective devices (filters).

2.4.2 Hilbert Transforms and Analytic Functions

A direct consequence of the analyticity of $F(z)$ is that the real and imaginary parts of $F(\omega)$ may not be specified independently. In fact we have already established in 2.2.6 that for a causal signal they are linearly related through the Hilbert transform. The properties of analytic functions afford an alternative derivation. For this purpose consider the contour integral

$$I_R(\omega_0) = \oint_{\Gamma_R} \frac{F(z)}{z - \omega_0} dz, \tag{2.261}$$

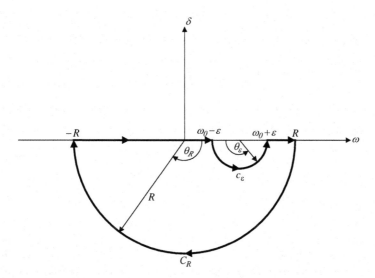

Figure 2.38: Integration contour Γ_R for the derivation of the Hilbert transforms

wherein ω_0 is real, taken in the clockwise direction along the closed path Γ_R as shown in Fig. 2.38. We note that Γ_R is comprised of the two linear segments $(-R, \omega_0 - \varepsilon)$, $(\omega_0 + \varepsilon, R)$ along the axis of reals, the semicircular contour c_ε of radius ε with the circle centered at $\omega = \omega_0$, and the semicircular contour C_R

of radius R in the lower half plane with the circle centered at $\omega = 0$. Since the integrand in (2.261) is analytic within Γ_R, we have $I_R(\omega_0) \equiv 0$, so that integrating along each of the path-segments indicated in Fig. 2.38 and adding the results in the limit as $\varepsilon \to 0$ and $R \to \infty$, we obtain

$$
0 = \lim_{\varepsilon \to 0,\, R \to \infty} \left(\int_{-R}^{\omega_0 - \varepsilon} \frac{F(\omega)}{\omega - \omega_0} d\omega + \int_{\omega_0 + \varepsilon}^{R} \frac{F(\omega)}{\omega - \omega_0} d\omega \right)
$$
$$
+ \lim_{\varepsilon \to 0} \int_{c_\varepsilon} \frac{F(z)}{z - \omega_0} dz + \lim_{R \to \infty} \int_{C_R} \frac{F(z)}{z - \omega_0} dz. \tag{2.262}
$$

On C_R we set $z = Re^{i\theta_R}$ so that $dz = iRe^{i\theta_R} d\theta_R$ and we have

$$
\left| \int_{C_R} \frac{F(z)}{z - \omega_0} dz \right| = \left| \int_{0}^{-\pi} \frac{F\left(Re^{i\theta_R}\right)}{Re^{i\theta_R} - \omega_0} R d\theta_R \right|
$$

so that in view of (2.260) in the limit of large R the last integral in (2.262) tends to zero. On c_ε we set $z - \omega_0 = \varepsilon e^{i\theta_\varepsilon}$ and substituting into the third integral in (2.262) evaluate it as follows:

$$
\lim_{\varepsilon \to 0} \int_{c_\varepsilon} \frac{F(z)}{z - \omega_0} dz = \lim_{\varepsilon \to 0} \int_{-\pi}^{0} F\left(\omega_0 + \varepsilon e^{i\theta_\varepsilon}\right) i d\theta = i\pi F(\omega_0).
$$

Now the limiting form of the first two integrals in (2.262) are recognized as the definition a CPV integral so that collecting our results we have

$$
0 = P \int_{-\infty}^{\infty} \frac{F(\omega)}{\omega - \omega_0} d\omega + i\pi F(\omega_0). \tag{2.263}
$$

By writing $F(\omega) = R(\omega) + iX(\omega)$ and similarly for $F(\omega_0)$, substituting in (2.263), and setting the real and the imaginary parts to zero we obtain

$$
X(\omega_0) = \frac{1}{\pi} P \int_{-\infty}^{\infty} \frac{R(\omega)}{\omega - \omega_0} d\omega, \tag{2.264a}
$$
$$
R(\omega_0) = -\frac{1}{\pi} P \int_{-\infty}^{\infty} \frac{R(\omega)}{\omega - \omega_0} d\omega, \tag{2.264b}
$$

which, apart from a different labeling of the variables, are the Hilbert Transforms in (2.173a) and (2.173b). Because the real and imaginary parts of the FT evaluated on the real frequency axis are not independent it should be possible to determine the analytic function $F(z)$ either from $R(\omega)$ of from $X(\omega)$. To obtain such formulas let z_0 be a point in the lower half plane (i.e., Im $z_0 < 0$) and apply the Cauchy integral formula

$$
F(z_0) = -\frac{1}{2\pi i} \oint_{\hat{\Gamma}_R} \frac{F(z)}{z - z_0} dz \tag{2.265}
$$

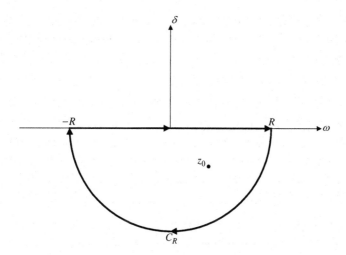

Figure 2.39: Integration contour for the evaluation of Eq. (2.265)

taken in the counterclockwise direction over the contour $\hat{\Gamma}_R$ as shown in Fig. 2.39 and comprised of the line segment $(-R, R)$ and the semicircular contour C_R of radius R. Again because of (2.260) the contribution over C_R vanishes as R is allowed to approach infinity so that (2.265) may be replaced by

$$
\begin{aligned}
F(z_0) &= -\frac{1}{2\pi i} \int_{-\infty}^{\infty} \frac{F(\omega)}{\omega - z_0} d\omega \\
&= -\frac{1}{2\pi i} \int_{-\infty}^{\infty} \frac{R(\omega)}{\omega - z_0} d\omega - \frac{1}{2\pi} \int_{-\infty}^{\infty} \frac{X(\omega)}{\omega - z_0} d\omega. \quad (2.266)
\end{aligned}
$$

In the last integral we now substitute for $X(\omega)$ its Hilbert Transform from (2.264a) to obtain

$$
\begin{aligned}
-\frac{1}{2\pi} \int_{-\infty}^{\infty} \frac{X(\omega)}{\omega - z_0} d\omega &= -\frac{1}{2\pi^2} \int_{-\infty}^{\infty} d\omega P \int_{-\infty}^{\infty} \frac{R(\eta)}{(\omega - z_0)(\eta - \omega)} d\eta \\
&= \frac{1}{2\pi^2} \int_{-\infty}^{\infty} R(\eta) d\eta P \int_{-\infty}^{\infty} \frac{d\omega}{(\omega - z_0)(\omega - \eta)}. \\
&\quad (2.267)
\end{aligned}
$$

The last CPV integral over ω is evaluated using the calculus of residues as follows:

$$
P \int_{-\infty}^{\infty} \frac{d\omega}{(\omega - z_0)(\omega - \eta)} = \oint_{\Gamma_R} \frac{dz}{(z - z_0)(z - \eta)} - i\pi \frac{1}{\eta - z_0}, \quad (2.268)
$$

where Γ_R is the closed contour in Fig. 2.38 and where the location of the simple pole at ω_0 is now designated by η. The contour integral in (2.268) is performed in the clockwise direction and the term $-i\pi/(\eta - z_0)$ is the negative of the contribution from the integration over the semicircular contour c_ε. The only contribution to the contour integral arises from the simple pole at $z = z_0$ which equals $-i2\pi/(z_0 - \eta)$ resulting in a net contribution in (2.268) of $i\pi/(\eta - z_0)$. Substituting this into (2.267) and then into (2.266) gives the final result

$$F(z) = \frac{i}{\pi} \int_{-\infty}^{\infty} \frac{R(\eta)}{\eta - z} d\eta, \tag{2.269}$$

where we have replaced the dummy variable ω by η and z_0 by $z \equiv \omega + i\delta$. Unlike (2.257), the integral (2.269) defines the analytic function $F(z)$ only in the open lower half plane, i.e., for $\operatorname{Im} z < 0$. On the other hand, one would expect that in the limit as $\delta \to 0$, $F(z) \to F(\omega)$. Let us show that this limit is actually approached by the real part. Thus using (2.269) we get with $z = \omega + i\delta$

$$\operatorname{Re} F(z) = R(\omega, \delta) = \int_{-\infty}^{\infty} R(\eta) \frac{-\delta}{\pi \left[(\eta - \omega)^2 + \delta^2 \right]} d\eta. \tag{2.270}$$

The factor multiplying $R(\eta)$ in integrand will be recognized as the delta function kernel in (1.250) so that $\lim R(\omega, \delta)$ as $-\delta \to 0$ is in fact $R(\omega)$.

2.4.3   Relationships Between Amplitude and Phase

We again suppose that $F(\omega)$ is the FT of a causal signal. Presently we write it in terms of its amplitude and phase

$$F(\omega) = A(\omega)e^{i\theta(\omega)} \tag{2.271}$$

and set

$$A(\omega) = e^{\alpha(\omega)}. \tag{2.272}$$

Taking logarithms we have

$$\ln F(\omega) = \alpha(\omega) + i\theta(\omega). \tag{2.273}$$

Based on the results of the preceding subsection it appears that if $\ln F(\omega)$ can be represented as an analytic function in the lower half plane one should be able to employ Hilbert Transforms to relate the phase to the log amplitude of the signal FT. From the nature of the logarithmic function we see that this is not possible for an arbitrary FT of a causal signal but only for signals whose FT, when continued analytically into the complex z-domain via formula (2.257) or (2.269), has no zeros in the lower half of the z-plane. Such transforms are said to be of the minimum-phaseshift type. If $f(t)$ is real so that $A(\omega)$ and $\theta(\omega)$ is, respectively, an even and an odd function of ω, we can express $\theta(\omega)$ in terms of $\alpha(\omega)$ using contour integration, provided the FT decays at infinity in accordance with

$$|F(\omega)| \underset{\omega \to \infty}{\sim} O\left(|\omega|^{-k}\right) \text{ for some } k > 0. \tag{2.274}$$

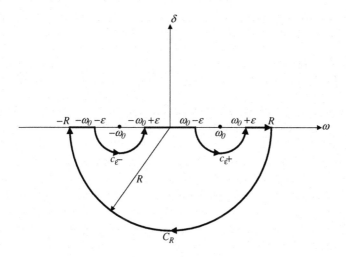

Figure 2.40: Integration contour for relating amplitude to phase

For this purpose we consider the integral

$$I_R = \oint_{\Gamma_R} \frac{\ln F(z)}{\omega_0^2 - z^2} dz \qquad (2.275)$$

taken in the clockwise direction over the closed contour Γ_R comprised of the three linear segments $(-R, -\omega_0 - \varepsilon)$, $(-\omega_0 + \varepsilon, \omega_0 - \varepsilon)$, $(\omega_0 + \varepsilon, R)$, the two semicircular arcs $c_{\varepsilon-}$ and $c_{\varepsilon+}$ each with radius ε, and the semicircular arc C_R with radius R, as shown in Fig. 2.40. By assumption $F(z)$ is free of zeros within the closed contour so that $I_R \equiv 0$. In the limit as $R \to \infty$ and $\varepsilon \to 0$ the integral over the line segments approaches a CPV integral while the integrals c_{ε}^- and c_{ε}^+ each approach $i\pi$ times the residue at the respective poles. The net result can then be written as follows:

$$0 = P \int_{-\infty}^{\infty} \frac{\ln F(\omega)}{\omega_0^2 - \omega^2} d\omega + i\pi \frac{\ln F(-\omega_0)}{2\omega_0} + i\pi \frac{\ln F(\omega_0)}{-2\omega_0}$$
$$+ \lim_{R \to \infty} \oint_{C_R} \frac{\ln F(z)}{\omega_0^2 - z^2} dz. \qquad (2.276)$$

In view of (2.274) for sufficiently large R the last integral may be bounded as follows:

$$\left| \oint_{C_R} \frac{\ln F(z)}{\omega_0^2 - z^2} dz \right| \leq constant \times \int_0^{\pi} \frac{k \ln R}{|\omega_0^2 - R^2 e^{i2\theta}|} R d\theta. \qquad (2.277)$$

Since $\ln R < R$ for $R > 1$, the last integral approaches zero as $R \to \infty$ so that the contribution from C_R in (2.276) vanishes. Substituting from (2.273) into the first three terms on the right of (2.276) and taking account of the fact that $\alpha(\omega)$ is even while $\theta(\omega)$ is odd, one obtains

$$0 = P \int_{-\infty}^{\infty} \frac{\alpha(\omega) + i\theta(\omega)}{\omega_0^2 - \omega^2} d\omega + i\pi \frac{\alpha(\omega_0) - i\theta(\omega_0)}{2\omega_0} + i\pi \frac{\alpha(\omega_0) + i\theta(\omega_0)}{-2\omega_0}$$

Observe that the terms on the right involving $\alpha(\omega_0)$ cancel while the integration involving $\theta(\omega)$ vanishes identically. As a result we can solve for $\theta(\omega_0)$ with the result

$$\theta(\omega_0) = \frac{2\omega_0}{\pi} P \int_0^{\infty} \frac{\alpha(\omega)}{\omega^2 - \omega_0^2} d\omega. \tag{2.278}$$

Proceeding similarly with the aid of the contour integral

$$I_R = \oint_{\Gamma_R} \frac{\ln F(z)}{z(\omega_0^2 - z^2)} dz \tag{2.279}$$

one obtains the formula

$$\alpha(\omega_0) = \alpha(0) - \frac{2\omega_0^2}{\pi} P \int_0^{\infty} \frac{\theta(\omega)}{\omega(\omega^2 - \omega_0^2)} d\omega. \tag{2.280}$$

It is worth noting that the assumed rate of decay at infinity in (2.274) is crucial to the vanishing of the contribution over the semicircular contour C_R in Fig. 2.40 and hence the validity of (2.278). Indeed if the decay of the FT is too rapid the contribution from C_R will not vanish and can in fact diverge as, e.g., for $A(\omega) = \exp(-\omega^2)$. Note that in this case (2.278) also diverges. This means that for an arbitrary $A(\omega)$ one cannot find a $\theta(\omega)$ such that $A(\omega)\exp{-i\theta(\omega)}$ has a causal inverse, i.e., an $f(t)$ that vanishes for negative t. What properties must $A(\omega)$ possess for this to be possible? An answer can be given if $A(\omega)$ is square integrable over $(-\infty, \infty)$. In that case the necessary and sufficient condition for a $\theta(\omega)$ to exist is the convergence of the integral

$$\int_{-\infty}^{\infty} \frac{|\ln A(\omega)|}{1 + \omega^2} d\omega < \infty,$$

which is termed the Paley–Wiener condition [15]. Note that it precludes $A(\omega)$ from being identically zero over any finite segment of the frequency axis.

2.4.4   Evaluation of Inverse FT Using Complex Variable Theory

The theory of functions of a complex variable provides a convenient tool for the evaluation of inverse Fourier transforms. The evaluation is particularly straightforward when the FT is a rational function. For example, let us evaluate

$$f(t) = \frac{1}{2\pi} \int_{-\infty}^{\infty} \frac{e^{i\omega t} d\omega}{\omega^2 + i\omega + 2}. \tag{2.281}$$

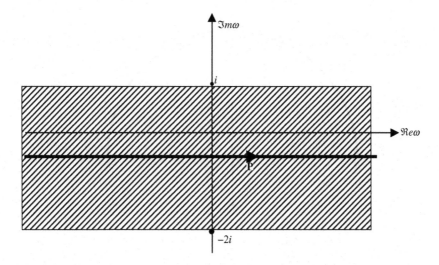

Figure 2.41: Deformation of integration path within the strip of analyticity

The only singularities of $F(\omega) = 1/(\omega^2 + i\omega + 2)$ in the complex ω plane are poles corresponding to the two simple zeros of $\omega^2 + i\omega + 2 = (\omega - i)(\omega + 2i) = 0$, namely $\omega_1 = i$ and $\omega_2 = -2i$. Therefore the integration path in (2.281) may be deformed away from the real axis into any path P lying within the strip of analyticity bounded by $-2 < \text{Im}\,\omega < 1$, as depicted in Fig. 2.41. The exponential multiplying $F(\omega)$ decays for $t > 0$ in the upper half plane ($\text{Im}\,\omega > 0$) and for $t < 0$ in the lower half plane ($\text{Im}\,\omega < 0$). For $t > 0$ we form the contour integral

$$I_R = \oint e^{i\omega t} F(\omega) \frac{d\omega}{2\pi} \tag{2.282}$$

taken in the counterclockwise direction over the closed path formed by the linear segment $(-R, R)$ along P and the circular contour C_{R+} lying in the upper half plane, as shown in Fig. 2.42. The residue evaluation at the simple pole at $\omega = i$ gives $I_R = e^{-t}/3$. As R is allowed to approach infinity the integral over the linear segment becomes just $f(t)$. Therefore

$$e^{-t}/3 = f(t) + \lim_{R \to \infty} \oint_{C_{R+}} e^{i\omega t} F(\omega) \frac{d\omega}{2\pi}.$$

Since $F(\omega) \to 0$ as $\omega \to \infty$, and the exponential decays on C_{R+} Jordan lemma (see Appendix A) applies so that in the limit the integral over C_{R+} vanishes and we obtain $f(t) = e^{-t}/3$; $t > 0$. When $t < 0$ the contour integral (2.282) is evaluated in the clockwise direction over the closed path in Fig. 2.43 with a circular path C_{R-} in the lower half plane. The residue evaluation at the simple pole at $\omega = -2i$ now gives $I_R = e^{2t}/3$ so that

$$e^{2t}/3 = f(t) + \lim_{R \to \infty} \oint_{C_{R-}} e^{i\omega t} F(\omega) \frac{d\omega}{2\pi}.$$

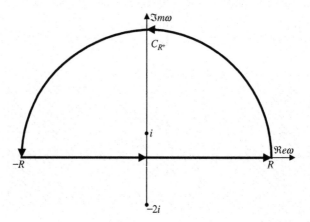

Figure 2.42: Integration contour for $t > 0$

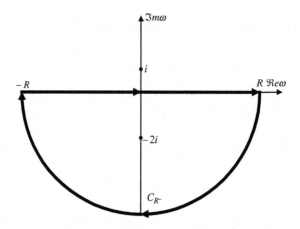

Figure 2.43: Integration contour for $t < 0$

Since now the exponential decays in the lower half plane, Jordan's lemma again guarantees that the limit of the integral over C_{R-} vanishes. Thus the final result reads

$$f(t) = \begin{cases} e^{-t}/3 \;;\; t \geq 0, \\ e^{2t}/3 \;;\; t \leq 0. \end{cases} \qquad (2.283)$$

This procedure is readily generalized to arbitrary rational functions. Thus suppose $F(\omega) = N(\omega)/D(\omega)$ with $N(\omega)$ and $D(\omega)$ polynomials in ω. We shall assume that[2] *degree* $N(\omega) <$ *degree* $D(\omega)$ so that $F(\omega)$ vanishes at infinity,

---

[2] If N and D are of the same degree, then the FT contains a delta function which can be identified by long division to obtain $N/D =$ constant $+\hat{N}/D$, with degree $\hat{N} <$ degree D. The inverse FT then equals constant $\times \delta(t) + \mathfrak{F}^{-1}\left(\hat{N}/D\right)$.

as required by the Jordan lemma. If $D(\omega)$ has no real zeros, then proceeding as in the preceding example we find that the inverse FT is given by the residue sums

$$f(t) = \begin{cases} i\sum_{k;\,\text{Im}\,\omega_k>0} res\left[\frac{N(\omega)}{D(\omega)}e^{i\omega t}\right]_{\omega=\omega_k} & ;t\geq 0, \\ -i\sum_{k;\,\text{Im}\,\omega_k<0} res\left[\frac{N(\omega)}{D(\omega)}e^{i\omega t}\right]_{\omega=\omega_k} & ;t\leq 0. \end{cases} \tag{2.284}$$

For example, suppose $F(\omega) = i/(\omega + 2i)^2(\omega - i)$ which function has a double pole at $\omega = -2i$ and a simple pole at $\omega = i$. For $t \geq 0$ the contribution comes from the simple pole in the upper half plane and we get

$$f(t) = i\frac{ie^{-t}}{(i+2i)^2} = \frac{e^{-t}}{9} \ ; \ t \geq 0.$$

For $t \leq 0$ the double pole in the lower half plane contributes. Hence

$$\begin{aligned} f(t) &= -i\,i\frac{d}{d\omega}\frac{e^{i\omega t}}{(\omega - i)}\,|_{\omega=-2i} = \frac{(\omega - i)\,it e^{i\omega t} - e^{i\omega t}}{(\omega - i)^2}\,|_{\omega=-2i} \\ &= \frac{1-3t}{9}e^{-2t} \ ; \ t \leq 0. \end{aligned}$$

The case of $D(\omega)$ having real roots requires special consideration. First, if the order of any one of the zeros is greater than 1, the inverse FT does not exist.[3] On the other hand, as will be shown in the sequel, if the zeros are simple the inverse FT can computed by suitably modifying the residue formulas (2.284). Before discussing the general case we illustrate the procedure by a specific example. For this purpose consider the time function given by the inversion formula

$$f(t) = \frac{1}{2\pi}P\int_{-\infty}^{\infty}\frac{e^{i\omega t}}{(\omega^2 - 4)(\omega^2 + 1)}d\omega, \tag{2.285}$$

where $F(\omega) = 1/(\omega^2 - 4)(\omega^2 + 1)$ has two simple zeros at $\omega = \pm i$ and two at $\omega = \pm 2$ with the latter forcing a CPV interpretation of the integral. Before complementing (2.285) with a suitable contour integral it may be instructive to make the CPV form of (2.285) explicit. Thus

$$f(t) = \lim_{\varepsilon\to 0, R\to\infty} I_{R,\varepsilon} \tag{2.286}$$

with

$$I_{R,\varepsilon} = \frac{1}{2\pi}\left\{\int_{-R}^{-2-\varepsilon} + \int_{-2+\varepsilon}^{2-\varepsilon} + \int_{2+\varepsilon}^{R}\right\}\frac{e^{i\omega t}}{(\omega^2 - 4)(\omega^2 + 1)}d\omega. \tag{2.287}$$

---

[3]The corresponding time functions are unbounded at infinity and are best handled using Laplace transforms.

To evaluate (2.286) by residues we define a contour integral

$$\hat{I}_{R,\varepsilon} = \oint_{\Gamma} e^{i\omega t} F\left(\omega\right) \frac{d\omega}{2\pi} \tag{2.288}$$

over a closed path Γ that includes $I_{R,\varepsilon}$ as a partial contribution. For $t > 0$ the contour Γ is closed with the semicircle of radius R and includes the two semicircles $c_{\varepsilon+}$ and $c_{\varepsilon}-$ of radius ε centered, respectively, at $\omega = 2$ and $\omega = -2$, as shown in Fig. 2.44.

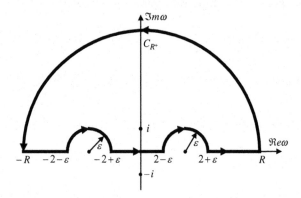

Figure 2.44: Integration contour for CPV integral

Writing (2.287) out in terms of its individual contributors we have

$$\hat{I}_{R,\varepsilon} = I_{R,\varepsilon} + \int_{c_{\varepsilon-}} e^{i\omega t} F\left(\omega\right) \frac{d\omega}{2\pi} + \int_{c_{\varepsilon+}} e^{i\omega t} F\left(\omega\right) \frac{d\omega}{2\pi} + \int_{C_{R+}} e^{i\omega t} F\left(\omega\right) \frac{d\omega}{2\pi}. \tag{2.289}$$

Taking account of the residue contribution at $\omega = i$ we get for the integral over the closed path

$$\hat{I}_{R,\varepsilon} = i \frac{e^{i\omega t}}{(\omega^2 - 4)(2\omega)} |_{\omega=i} = -\frac{e^{-t}}{10}.$$

As $\varepsilon \to 0$ the integrals over $c_{\varepsilon-}$ and $c_{\varepsilon}-$ each contribute $-2\pi i$ times one-half the residue at the respective simple pole (see Appendix A) and a $R \to \infty$ the integral over C_{R+} vanishes by the Jordan lemma. Thus taking the limits and summing all the contributions in (2.289) we get

$$-\frac{e^{-t}}{10} = f(t) - i\left[\frac{1}{2}\frac{e^{i\omega t}}{(2\omega)(\omega^2 + 1)}|_{\omega=-2} + \frac{1}{2}\frac{e^{i\omega t}}{(2\omega)(\omega^2 + 1)}|_{\omega=2}\right]$$

$$= f(t) + \frac{1}{20}\sin(2t)$$

and solving for $f(t)$,

$$f(t) = -\frac{1}{20}\sin(2t) - \frac{e^{-t}}{10} \quad ; \ t > 0. \tag{2.290}$$

For $t < 0$ we close the integration path $(-R, -2 - \varepsilon) + c_{\varepsilon-} + (-2 + \varepsilon, 2 - \varepsilon) + (2 + \varepsilon, R)$ in Fig. 2.44 with a semicircular path in the lower half plane and carry out the integration in the clockwise direction. Now Γ encloses in addition to the pole at the pole at $\omega = -i$, the two poles at $\omega = \pm 2$. Hence

$$
\begin{aligned}
\hat{I}_{R,\varepsilon} &= -i \frac{e^{i\omega t}}{(\omega^2 - 4)(2\omega)}\Big|_{\omega=-i} - i \left[\frac{e^{i\omega t}}{(2\omega)(\omega^2 + 1)}\Big|_{\omega=-2} + \frac{e^{i\omega t}}{(2\omega)(\omega^2 + 1)}\Big|_{\omega=2} \right] \\
&= -\frac{e^{-t}}{10} + \frac{1}{10}\sin(2t).
\end{aligned}
$$

Summing the contributions as in (2.289) and taking limits we get

$$
-\frac{e^t}{10} + \frac{1}{10}\sin(2t) = f(t) - \frac{1}{10}\sin(2t).
$$

Solving for $f(t)$ and combining with (2.290) we have for the final result

$$
f(t) = -\frac{1}{20}\sin(2t)\ sign(t) - \frac{e^{-|t|}}{10}. \tag{2.291}
$$

Note that we could also have used an integration contour with the semicircles $c_{\varepsilon-}$ and $c_{\varepsilon+}$ in the lower half plane. In that case we would have picked up the residue at $\omega = \pm 2$ for $t > 0$.

Based on the preceding example it is not hard to guess how to generalize (2.284) when $D(\omega)$ has simple zeros for real ω. Clearly for every real zero at $\omega = \omega_k$ we have to add the contribution $sign(t)\ (i/2)\ res\left[\frac{N(\omega)}{D(\omega)} e^{i\omega t} \right]\Big|_{\omega=\omega_k}$. Hence we need to replace (2.284) by

$$
\begin{aligned}
f(t) = {}&(i/2)\ sign(t) \sum_{k;\operatorname{Im}\omega_k = 0} res \left[\frac{N(\omega)}{D(\omega)} e^{i\omega t} \right]\Big|_{\omega=\omega_k} \\
&+ \begin{cases} i\sum_{k;\operatorname{Im}\omega_k > 0} res \left[\frac{N(\omega)}{D(\omega)} e^{i\omega t} \right]_{\omega=\omega_k} & ; t \geq 0, \\ -i\sum_{k;\operatorname{Im}\omega_k < 0} res \left[\frac{N(\omega)}{D(\omega)} e^{i\omega t} \right]_{\omega=\omega_k} & ; t \leq 0. \end{cases}
\end{aligned} \tag{2.292}
$$

For example, for $F(\omega) = i\omega/(\omega_0^2 - \omega^2)$, the preceding formula yields $f(t) = \frac{1}{2} sign(t) \cos\omega_0 t$ and setting $\omega_0 = 0$ we find that the FT of $sign(t)$ is $2/i\omega$, in agreement with our previous result.

2.5 Time-Frequency Analysis

2.5.1 The Uncertainty Principle

A common feature shared by simple idealized signals such as rectangular, triangular, or Gaussian pulses is the inverse scaling relationship between signal duration and its bandwidth. Qualitatively a relationship of this sort actually holds for a large class of signals but its quantitative formulation ultimately depends on the nature of the signal as well as on the definition of signal duration and bandwidth. A useful definition which also plays a prominent role not only in signal analysis but also in other areas where Fourier transforms are part of the basic theoretical framework is the so-called *rms* signal duration σ_t, defined by

$$\sigma_t^2 = \frac{1}{E} \int_{-\infty}^{\infty} (t- <t>)^2 \left| f(t) \right|^2 dt, \tag{2.293}$$

where

$$<t> = \frac{1}{E} \int_{-\infty}^{\infty} t \left| f(t) \right|^2 dt \tag{2.294}$$

and

$$E = \int_{-\infty}^{\infty} \left| f(t) \right|^2 dt \tag{2.295}$$

are the signal energies. We can accept this as a plausible measure of signal duration if we recall that σ_t^2 corresponds algebraically to the variance of a random variable with probability density $\left| f(t) \right|^2 / E$ wherein the statistical mean has been replaced by $<t>$. This quantity we may term "the average time of signal occurrence".[4] Although definition (2.295) holds formally for any signal (provided, of course, that the integral converges), it is most meaningful, just like the corresponding concept of statistical average in probability theory, when the magnitude of the signal is unimodal. For example, using these parameters a real Gaussian pulse takes the form

$$f(t) = \frac{\sqrt{E}}{\left(2\pi\sigma_t^2\right)^{1/4}} \exp -\frac{(t- <t>)^2}{4\sigma_t^2}. \tag{2.296}$$

To get an idea how the signal spectrum $F(\omega)$ affect the *rms* signal duration we first change the variables of integration in (2.293) from t to $t' = t- <t>$ and write it in the following alternative form:

$$\sigma_t^2 = \frac{1}{E} \int_{-\infty}^{\infty} t'^2 \left| f(t'+ <t>) \right|^2 dt'. \tag{2.297}$$

Using the identities $\mathfrak{F}\{-itf(t)\} = dF(\omega)/d\omega$ and $\mathfrak{F}\{f(t+ <t>)\} = F(\omega) \exp i\omega <t>$ we apply Parseval's theorem to (2.297) to obtain

$$\begin{aligned}
\sigma_t^2 &= \frac{1}{2\pi E} \int_{-\infty}^{\infty} \left| \frac{d\left[F(\omega) \exp i\omega <t>\right]}{d\omega} \right|^2 d\omega \\
&= \frac{1}{2\pi E} \int_{-\infty}^{\infty} \left| \frac{dF(\omega)}{d\omega} + i <t> F(\omega) \right|^2 d\omega. \tag{2.298}
\end{aligned}$$

This shows that the *rms* signal duration is a measure of the integrated fluctuations of the amplitude and phase of the signal spectrum. We can also express

[4]For a fuller discussion of this viewpoint see Chap. 3 in Leon Cohen, "Time-Frequency Analysis," Prentice Hall PTR, Englewood Cliffs, New Jersey (1995).

the average time of signal occurrence $< t >$ in terms of the signal spectrum by first rewriting the integrand in (2.294) as the product $t f(t) f(t)^*$ and using $\mathfrak{F} \{t f(t)\} = i dF(\omega)/d\omega$ together with Parseval's theorem. This yields

$$< t >= \frac{1}{2\pi E} \int_{-\infty}^{\infty} i \frac{dF(\omega)}{d\omega} F^*(\omega) \, d\omega.$$

With $F(\omega) = A(\omega) e^{i\theta(\omega)}$ the preceding becomes

$$< t >= \frac{1}{2\pi E} \int_{-\infty}^{\infty} [-\theta'(\omega)] |F(\omega)|^2 \, d\omega, \qquad (2.299)$$

where $\theta'(\omega) = d\theta(\omega)/d\omega$. In 2.6.1 we shall identify the quantity $-\theta'(\omega)$ as the signal group delay. Equation then (2.299) states that the group delay, when averaged with the "density" function $|F(\omega)|^2/2\pi E$, is identical to the average time of signal occurrence.

We now apply the preceding definitions of spread and average location in the frequency domain. Thus the *rms* bandwidth σ_ω will be defined by

$$\sigma_\omega^2 = \frac{1}{2\pi E} \int_{-\infty}^{\infty} (\omega - < \omega >)^2 |F(\omega)|^2 \, d\omega, \qquad (2.300)$$

where

$$< \omega >= \frac{1}{2\pi E} \int_{-\infty}^{\infty} \omega |F(\omega)|^2 \, d\omega. \qquad (2.301)$$

We can view $< \omega >$ as the center of mass of the amplitude of the frequency spectrum. Clearly for real signals $< \omega > \equiv 0$. By analogy with (2.297) we change the variable of integration in (2.300) to $\omega' = \omega - < \omega >$ and rewrite it as follows:

$$\sigma_\omega^2 = \frac{1}{2\pi E} \int_{-\infty}^{\infty} \omega'^2 |F(\omega' + < \omega >)|^2 \, d\omega \qquad (2.302)$$

$\mathfrak{F} \{df(t)/dt\} = i\omega F(\omega)$ and Parseval's theorem obtain the dual to (2.295), viz.,

$$\sigma_\omega^2 = \frac{1}{E} \int_{-\infty}^{\infty} \left| \frac{d[f(t) \exp -i < \omega > t]}{dt} \right|^2 dt$$

$$= \frac{1}{E} \int_{-\infty}^{\infty} \left| \frac{df(t)}{dt} - i < \omega > f(t) \right|^2 dt. \qquad (2.303)$$

Thus the *rms* bandwidth increases in proportion to the norm of the rate of change of the signal. In other words, the more rapid the variation of the signal in a given time interval the greater the frequency band occupancy. This is certainly compatible with the intuitive notion of frequency as a measure of the number of zero crossings per unit time as exemplified, for instance, by signals of the form $\cos[\varphi(t)]$.

Again using $\mathfrak{F} \{df(t)/dt\} = i\omega F(\omega)$ and Parseval's theorem we transform (2.301) into

$$< \omega >= \frac{1}{E} \int_{-\infty}^{\infty} -i \frac{df(t)}{dt} f^*(t) \, dt.$$

If $f(t) = r(t) \exp i\psi(t)$ is an analytic signal, the preceding yields

$$<\omega> = \frac{1}{E} \int_{-\infty}^{\infty} \psi'(t) |f(t)|^2 \, dt, \tag{2.304}$$

where $\psi'(t) = d\psi(t)/dt$ is the instantaneous frequency. This equation provides another interpretation of $<\omega>$, viz., as the average instantaneous frequency with respect to the density $|f(t)|^2/E$, a result which may be considered a sort of dual to (2.299).

The *rms* signal duration and *rms* bandwidth obey a fundamental inequality, known as the uncertainty relationship, which we now proceed to derive. For this purpose let us apply the Schwarz inequality to the following two functions: $(t - <t>) f(t)$ and $df(t)/dt - i <\omega> f(t)$. Thus

$$\int_{-\infty}^{\infty} (t - <t>)^2 |f(t)|^2 \, dt \int_{-\infty}^{\infty} \left| \frac{df(t)}{dt} - i <\omega> f(t) \right|^2 \, dt$$

$$\geq \left| \int_{-\infty}^{\infty} (t - <t>) f^*(t) \left[\frac{df(t)}{dt} - i <\omega> f(t) \right] dt \right|^2. \tag{2.305}$$

Substituting for the first two integrals in (2.305) the σ_t^2 and σ_ω^2 from (2.297) and (2.303), respectively, the preceding becomes

$$\sigma_t^2 \sigma_\omega^2 E^2 \geq \left| \int_{-\infty}^{\infty} (t - <t>) f^*(t) \left[\frac{df(t)}{dt} - i <\omega> f(t) \right] dt \right|^2$$

$$= \left| \int_{-\infty}^{\infty} (t - <t>) f^*(t) \frac{df(t)}{dt} \, dt \right|^2, \tag{2.306}$$

where in view of (2.294) we have set $\int_{-\infty}^{\infty} (t - <t>) |f(t)|^2 \, dt = 0$. We now integrate the last integral by parts as follows:

$$\int_{-\infty}^{\infty} (t - <t>) f^*(t) \frac{df(t)}{dt} \, dt$$

$$= (t - <t>) |f(t)|^2 \Big|_{-\infty}^{\infty} - \int_{-\infty}^{\infty} f(t) \frac{d\left[(t - <t>) f^*(t) \right]}{dt} \, dt$$

$$= (t - <t>) |f(t)|^2 \Big|_{-\infty}^{\infty} - E - \int_{-\infty}^{\infty} (t - <t>) f(t) \frac{df^*(t)}{dt} \, dt. \tag{2.307}$$

Because $f(t)$ has finite energy it must decay at infinity faster than $1/\sqrt{t}$ so that $(t - <t>) |f(t)|^2 \Big|_{-\infty}^{\infty} = 0$. Therefore after transposing the last term in (2.307) to the left of the equality sign we can rewrite (2.307) as follows:

$$\text{Re} \left\{ \int_{-\infty}^{\infty} (t - <t>) f^*(t) \frac{df(t)}{dt} \, dt \right\} = -E/2. \tag{2.308}$$

Since the magnitude of a complex number is always grater or equal to the magnitude of its real part the right side of (2.306) equals at least $E^2/4$. Cancelling of E^2 and taking the square root of both sides result in

$$\sigma_t \sigma_\omega \geq \frac{1}{2}, \qquad (2.309)$$

which is the promised uncertainty relation. Basically it states that simultaneous localization of a signal in time and frequency is not achievable to within arbitrary precision: the shorter the duration of the signal the greater its spectral occupancy and conversely. We note that except for a constant factor on the right (viz., Planck's constant), (2.309) is identical to the Heisenberg uncertainty principle in quantum mechanics where t and ω stand for any two canonically conjugate variables (e.g., particle position and particle momentum). When does (2.309) hold with equality? The answer comes from the Schwarz inequality (2.305) wherein equality can be achieved if and only if $(t- <t>)\, f(t)$ and $\frac{df(t)}{dt} - i<\omega> f(t)$ are proportional. Calling this proportionality constant $-\alpha$ results in the differential equation

$$\frac{df(t)}{dt} - i<\omega> f(t) + \alpha (t- <t>)\, f(t) = 0. \qquad (2.310)$$

This is easily solved for $f(t)$ with the result

$$f(t) = A \exp\left\{ -\frac{\alpha}{2}(t- <t>)^2 + \frac{\alpha}{2} <t>^2 + i<\omega> t \right\}, \qquad (2.311)$$

where A is a proportionality constant. Thus the optimum signal from the standpoint of simultaneous localization in time and frequency has the form of a Gaussian function. Taking account of the normalization (2.295) we obtain after a simple calculation

$$\alpha = 1/2\sigma_t^2, \quad A = \sqrt{E/2\pi\sigma_t^2} \exp\left\{ -<t>^2/2\sigma_t^2 \right\}. \qquad (2.312)$$

2.5.2 The Short-Time Fourier Transform

Classical Fourier analysis draws a sharp distinction between the time and frequency domain representations of a signal. Recall that the FT of a signal of duration T can be computed only *after* the signal has been observed in its entirety. The computed spectrum furnishes the relative amplitude concentrations within the frequency band and the relative phases but information as to the times at which the particular frequency components have been added to the spectrum is not provided. Asking for such information is of course not always sensible particularly in cases of simple and essentially single scale signals such as isolated pulses. On the other hand for signals of long duration possessing complex structures such as speech, music, or time series of environmental parameters the association of particular spectral features with the times of their generation not only is meaningful but in fact also constitutes an essential step in data analysis. A possible approach to the frequency/time localization problem is to multiply $f(t)$, the signal to be analyzed, by a sliding window function $g(t - \tau)$ and take the FT of the product. Thus

$$S(\omega, \tau) = \int_{-\infty}^{\infty} f(t)g(t - \tau) e^{-i\omega t} dt \qquad (2.313)$$

whence in accordance with the FT inversion formula

$$f(t)g(t - \tau) = \frac{1}{2\pi} \int_{-\infty}^{\infty} S(\omega, \tau) e^{i\omega t} d\omega. \qquad (2.314)$$

We can obtain an explicit formula for determining $f(t)$ from $S(\omega, \tau)$ by requiring that the window function satisfies

$$\int_{-\infty}^{\infty} |g(t - \tau)|^2 d\tau = 1 \qquad (2.315)$$

for all t. For if we now multiply both sides of (2.314) by $g^*(t - \tau)$ and integrate with respect to τ we obtain

$$f(t) = \frac{1}{2\pi} \int_{-\infty}^{\infty} \int_{-\infty}^{\infty} S(\omega, \tau) g^*(t - \tau) e^{i\omega t} d\omega d\tau. \qquad (2.316)$$

The two-dimensional function $S(\omega, \tau)$ is referred to as the short-time Fourier transform[5] (STFT) of $f(t)$ and (2.316) the corresponding inversion formula. The STFT can be represented graphically in various ways. The most common is the spectrogram, which is a two-dimensional plot of the magnitude of $S(\omega, \tau)$ in the $\tau\omega$ plane. Such representations are commonly used as an aid in the analysis of speech and other complex signals.

Clearly the characteristics of the STFT will depend not only on the signal but also on the choice of the window. In as much as the entire motivation for the construction of the STFT arises from a desire to provide simultaneous localization in frequency and time it is natural to choose for the window function the Gaussian function since, as shown in the preceding, it affords the optimum localization properties. This choice was originally made by Gabor [6] and the STFT with a Gaussian window is referred to as the Gabor transform. Here we adopt the following parameterization:

$$g(t) = \frac{2^{1/4}}{\sqrt{s}} e^{-\frac{\pi t^2}{s^2}}. \qquad (2.317)$$

Reference to (2.311) and (2.312) shows that $\sigma_t = s/(2\sqrt{\pi})$. Using (2.142*) we have for the FT

$$G(\omega) = 2^{1/4}\sqrt{s} e^{-s^2\omega^2/4\pi} \qquad (2.318)$$

from which we obtain $\sigma_\omega = \sqrt{\pi}/s$ so that $\sigma_t \sigma_\omega = 1/2$, as expected.

As an example, let us compute the Gabor transform of $\exp(\alpha t^2/2)$. We obtain

$$S(\omega, \tau)/\sqrt{s} = 2^{1/4} \sqrt{\frac{\pi}{i\alpha s^2/2 - \pi}} \exp{-\frac{\left(\frac{2\pi\tau}{s} - i\omega s\right)^2}{4(i\alpha s^2/2 - \pi)} - \frac{\pi\tau^2}{s^2}}. \qquad (2.319)$$

---

[5] Also referred to as the sliding-window Fourier transform

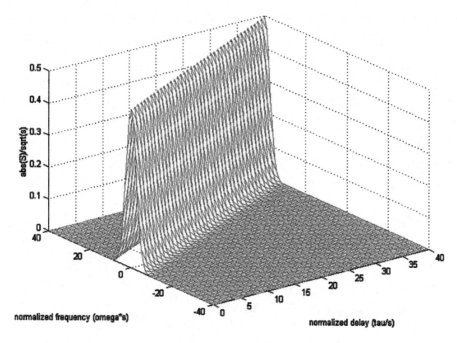

Figure 2.45: Magnitude of Gabor Transform of $\exp\left\{i\frac{1}{2}\alpha t^2\right\}$

A relief map of the magnitude of $S(\omega, \tau)/\sqrt{s}$ (spectrogram) as a function of the nondimensional variables τ/s (delay) and ωs (frequency) is shown in Fig. 2.45.

In this plot the dimensionless parameter $(1/2)\alpha s^2$ equals $1/2$. The map shows a single ridge corresponding to a straight line $\omega = \alpha\tau$ corresponding to the instantaneous frequency at time τ. As expected only positive frequency components are picked up by the transform. On the other hand, if instead we transform the real signal $\cos\left\{\frac{1}{2}\alpha t^2\right\}$, we get a plot as in Fig. 2.46. Since the cosine contains exponentials of both signs the relief map shows a second ridge running along the line $\omega = -\alpha\tau$ corresponding to negative instantaneous frequencies.

As a final example consider the signal plotted in Fig. 2.47. Even though this signal looks very much like a slightly corrupted sinusoid, it is actually comprised of a substantial band of frequencies with a rich spectral structure. This can be seen from Fig. 2.48 which shows a plot of the squared magnitude of the FT. From this spectral plot we can estimate the total signal energy and the relative contributions of the constitutive spectral components that make up the total signal but not their positions in the time domain. This information can be inferred from the Gabor spectrogram whose contour map is represented in Fig. 2.49. This spectrogram shows us that the spectral energy of the signal is in fact confined to a narrow sinuous band in the time-frequency plane. The width of this band is governed by the resolution properties of the sliding Gaussian window (5 sec. widths in this example) and its centroid traces out approximately the locus of the instantaneous frequency in the time-frequency plane.

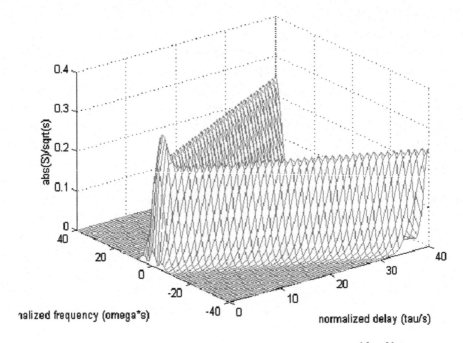

Figure 2.46: Magnitude of Gabor Transform of $\cos\left\{\frac{1}{2}\alpha t^2\right\}$

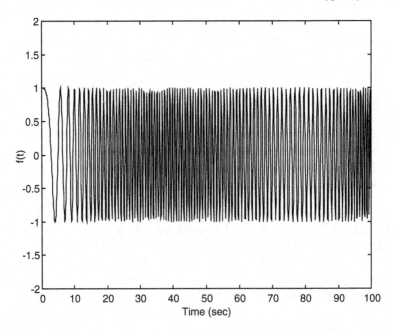

Figure 2.47: Constant amplitude signal comprising multiple frequencies

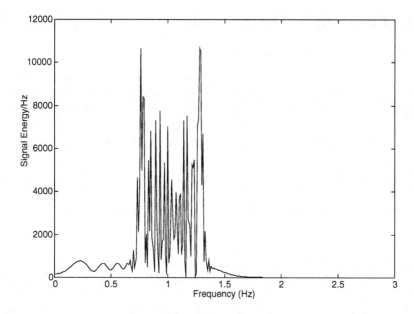

Figure 2.48: Squared magnitude of the FT of the signal in Fig. 2.47

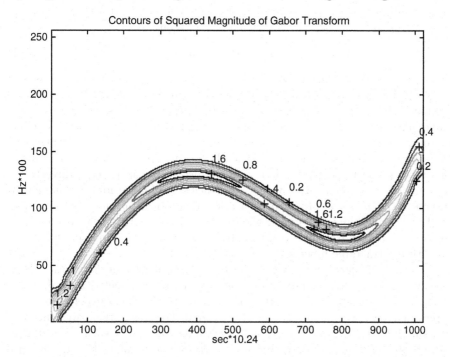

Figure 2.49: Contour map of the Gabor Transform of the signal in Fig. 2.48

2.6 Frequency Dispersion

2.6.1 Phase and Group Delay

In many physical transmission media the dominant effect on the transmitted signal is the distortion caused by unequal time delays experienced by different frequency components. In the frequency domain one can characterize such a transmission medium by the transfer function $e^{-i\psi(\omega)}$ where $\psi(\omega)$ is real. The FT $F(\omega)$ of the input signal $f(t)$ is then transformed into the FT $Y(\omega)$ of the output signal $y(t)$ in accordance with

$$Y(\omega) = e^{-i\psi(\omega)} F(\omega). \tag{2.320}$$

The time domain representation of the output then reads

$$y(t) = \frac{1}{2\pi} \int_{-\infty}^{\infty} e^{i\omega t} e^{-i\psi(\omega)} F(\omega) \, d\omega \tag{2.321}$$

so that by Parseval's theorem the total energy of the output signal is identical to that of the input signal. However its spectral components are in general delayed by different amounts so that in the time domain the output appears as a distorted version of the input. The exceptional case arises whenever the transfer phase $\psi(\omega)$ is proportional to frequency for then with $\psi(\omega) = \omega T$ the output is merely a time delayed version of the input:

$$y(t) = f(t - T). \tag{2.322}$$

Such distortionless transmission is attainable in certain special situations, the most notable of which is EM propagation through empty space. It may also be approached over limited frequency bands in certain transmission lines (coaxial cable, microstrip lines). In most practical transmission media however one has to count on some degree of phase nonlinearity with frequency, particularly as the signal bandwidth is increased. Clearly for any specific signal and transfer phase the quantitative evaluation of signal distortion can proceed directly via a numerical evaluation of (2.321). Nevertheless, guidance for such numerical investigations must be provided by a priori theoretical insights. For example, at the very minimum one should like to define and quantify measures of signal distortion. Fortunately this can usually be accomplished using simplified and analytically tractable models.

   Let us first attempt to define the delay experienced by a typical signal. Because each spectral component of the signal will be affected by a different amount, it is sensible to first attempt to quantify the delay experienced by a typical narrow spectral constituent of the signal. For this purpose we conceptually subdivided the signal spectrum $F(\omega)$ into narrow bands, each of width $\Delta\omega$, as indicated in Fig. 2.50 (also shown is a representative plot of $\psi(\omega)$, usually referred to as the medium dispersion curve). The contribution to the output signal from such a typical band (shown shaded in the figure) is

$$y_n(t) = \Re e \left\{ z_n(t) \right\}, \tag{2.323}$$

where

$$z_n(t) = \frac{1}{\pi} \int_{\omega_n-\Delta\omega/2}^{\omega_n+\Delta\omega/2} e^{i\omega t} e^{-i\psi(\omega)} F(\omega)\, d\omega \qquad (2.324)$$

$z_n(t)$ is the corresponding analytic signal (assuming real $f(t)$ and $\psi(\omega) = -\psi(-\omega)$) and the integration is carried out over the shaded band in Fig. 2.50.

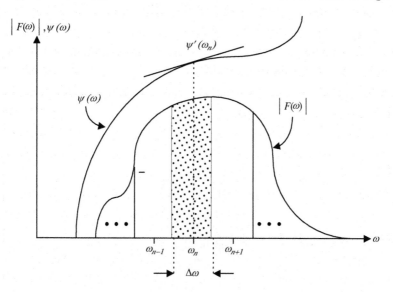

Figure 2.50: Group delay of signal component occupying a narrow frequency band

Clearly the complete signal $y(t)$ can be represented correctly by simply summing over the totality of such non-overlapping frequency bands, i.e.,

$$y(t) = \sum_n y_n(t). \qquad (2.325)$$

For sufficiently small $\Delta\omega/\omega_n$ the phase function within each band may be approximated by

$$\psi(\omega) \sim \psi(\omega_n) + (\omega - \omega_n)\psi'(\omega_n), \qquad (2.325^*)$$

where $\psi'(\omega_n)$ is the slope of the dispersion curve at the center of the band in Fig. 2.50. If we also approximate the signal spectrum $F(\omega)$ by its value at the band center, (2.324) can be replaced by

$$z_n(t) \sim \frac{1}{\pi} F(\omega_n) \int_{\omega_n-\Delta\omega/2}^{\omega_n+\Delta\omega/2} e^{i\omega t} e^{-i[\psi(\omega_n)+(\omega-\omega_n)\psi'(\omega_n)]}\, d\omega.$$

After changing the integration variable to $\eta = \omega - \omega_n$ this becomes

$$z_n(t) \sim \frac{1}{\pi} F(\omega_n)\, e^{i(\omega_n t - \psi(\omega_n))} \int_{-\Delta\omega/2}^{\Delta\omega/2} e^{i\eta(t-\psi'(\omega_n))}\, d\eta$$

$$= 2iF(\omega_n) e^{i(\omega_n t - \psi(\omega_n))} \frac{\sin\left[\Delta\omega/2\left(t - \psi'(\omega_n)\right)\right]}{\pi\left(t - \psi'(\omega_n)\right)} \qquad (2.326)$$

and upon setting $F(\omega_n) = A(\omega_n) e^{i\theta(\omega_n)}$ the real signal (2.323) assumes the form

$$y_n(t) \sim A(\omega_n) \sin\left[\omega_n t + \theta(\omega_n) + \pi - \psi(\omega_n)\right] \frac{\sin\left[\Delta\omega/2\left(t - \psi'(\omega_n)\right)\right]}{\pi\left(t - \psi'(\omega_n)\right)} \qquad (2.327)$$

a representative plot of which is shown in Fig. 2.51. Equation (2.327) has the form of a sinusoidal carrier at frequency ω_n that has been phase shifted by

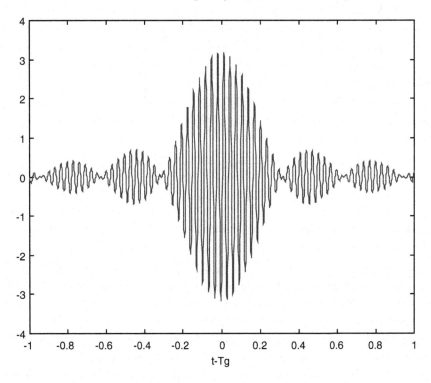

Figure 2.51: Plot of (2.327) for $\Delta\omega/2\theta'(\omega_n) = 10$, $\omega_n = 200rps$ and $A(\omega_n) = 1$

$\psi(\omega_n)$ radians. Note that the carrier is being modulated by an envelope in form of a sinc function delayed in time by $\psi'(\omega_n)$. Evidently this envelope is the time domain representation of the spectral components contained within the band $\Delta\omega$ all of which are undergoing the same time delay as a "group." Accordingly $\psi'(\omega_n)$ is referred to as the group delay (T_g) while the time (epoch) delay of the carrier $\theta(\omega_n)/\omega_n$ is referred to as the phase delay ($T\varphi$). One may employ these concepts to form a semi-quantitative picture of signal distortion by assigning to each narrow band signal constituent in the sum in (2.325) its own phase and group delay. Evidently if the dispersion curve changes significantly over the signal bandwidth no single numerical measure of distortion is possible. Thus it is not surprising that the concept of group delay is primarily of value for signals having sufficiently narrow bandwidth. How narrow must $\Delta\omega$ be chosen for the representation (2.327) to hold? Clearly in addition to $\Delta\omega/\omega_n << 1$ the next term in the Taylor expansion in (2.325*) must be negligible by comparison with $(\omega - \omega_n)\,\psi'(\omega_n)$. Since $|\omega - \omega_n| \leq \Delta\omega/2$ this additional constraint translates into

$$\Delta\omega \ll \left|\frac{4\psi'(\omega_n)}{\psi''(\omega_n)}\right| \qquad (2.328)$$

which evidently breaks down when $\psi'(\omega_n) = 0$.

2.6.2   Phase and Group Velocity

Phase and group delay are closely linked to phase and group velocities associated with wave motion. To establish the relationship we start with the definition of an the elementary wave

$$f(t,x) = f(t - x/v), \qquad (2.329)$$

where t is time x, represents space, and v a constant. Considered as a function of

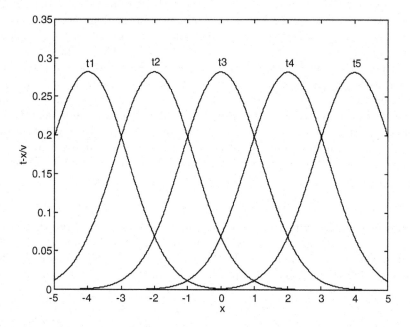

Figure 2.52: Self-preserving spatial pattern at successive instants of time ($t1 < t2 < t3 < t4 < t5$)

x which is sampled at discrete instances of time we can display it as in Fig. 2.52. Such a spatial display may be regarded as a sequence of snapshots of the function $f(\xi)$ which executes a continuous motion in the direction of the positive x-axis.

Clearly the speed of this translation may be defined unambiguously by the condition that the functional argument $t - x/v$ be maintained constant in time for a continuum of x. The derivative of the argument is then zero so that

$$\frac{dx}{dt} = v. \tag{2.330}$$

We take (2.330) as the *definition of* the velocity of the wave. Note that this definition is based entirely on the requirement that the functional form $f(\xi)$ be preserved exactly. This characterizes what is usually designated as dispersionless propagation. It is an idealization just as is distortionless transmission mentioned in the preceding subsection. Evidently as long as x is fixed the two concepts are identical as we see by setting $x/v = T$ in (2.322). In general the preservation of the waveform is approached only by narrow band signals. Hence we can again examine initially the propagation of a single sinusoid and appeal to Fourier synthesis to formulate the general case. For a time-harmonic signal the elementary wave function (2.329) reads

$$e^{i\omega(t - x/v(\omega))} = e^{i\omega t}e^{-i\beta(\omega)x}, \tag{2.331}$$

wherein we now allow the speed of propagation $v_\varphi(\omega)$ to depend on frequency. Note, however, that even though mathematically the functional forms (2.329) and (2.331) are identical, (2.331) represents an infinitely long periodic pattern so that we cannot really speak of the velocity of the translation of an identifiable space limited pattern (as, e.g., displayed in Fig. 2.52). Thus if we want to associate $v_\varphi(\omega)$ with the motion of some identifiable portion of the spatial pattern, we have only a phase reference at our disposal. Quite aptly then $v_\varphi(\omega)$ is referred to as the phase velocity. The quantity $\beta(\omega) = \omega/v(\omega)$ in (2.331) represents the propagation constant and may be taken as a fundamental characteristic of the propagation medium. The time domain representation of a general signal with spectrum $F(\omega)$ that has propagated through a distance x is obtained by multiplying (2.331) by $F(\omega)$ and taking the inverse FT. Thus

$$y(t, x) = \frac{1}{2\pi} \int_{-\infty}^{\infty} e^{i\omega t}e^{-i\beta(\omega)x}F(\omega)\,d\omega, \tag{2.332}$$

which is just (2.321) with the phase shift relabeled as $\beta(\omega)x$. Note that in the special case $\beta(\omega) = \omega/v$ and with v a constant (2.332) reduces to (2.329), i.e., the propagation is dispersionless. In the general case we proceed as in (2.326). After replacing $\psi(\omega)$ with $\beta(\omega)x$ in (2.327) we obtain

$$y_n(t, x) \sim A(\omega_n)\sin[\omega_n t + \theta(\omega_n) + \pi - \beta(\omega_n)x]\frac{\sin[\Delta\omega/2(t - \beta'(\omega_n)x)]}{\pi(t - \beta'(\omega_n)x)}. \tag{2.333}$$

Unlike (2.327), (2.333) depends on both space and time. It does not, however, have the same simple interpretation as the wavefunction defined in (2.329) because the speed of propagation of the carrier phase and the envelope differ. Thus while the carrier phase moves with the phase velocity

$$v_{\varphi n} = \omega_n/\beta(\omega_n) \tag{2.334}$$

the envelope[6] moves with velocity

$$v_{gn} = 1/\beta'(\omega_n) = \frac{d\omega}{d\beta}\big|_{\beta=\beta(\omega_n)}. \tag{2.335}$$

The latter is referred to as the group velocity and is the speed of propagation of the energy contained within the frequency band $\Delta\omega$ in Fig. 2.50. By contrast, the phase velocity has generally no connection with energy transport but represents merely the translation of a phase reference point.

2.6.3 Effects of Frequency Dispersion on Pulse Shape

Thus far we have not explicitly addressed quantitative measures of signal distortion. For this purpose consider a pulse with a (baseband) spectrum $P(\omega)$ most of whose energy is confined to the nominal frequency band $(-\Omega, \Omega)$. The pulse, after having been modulated by a carrier of frequency ω_0, propagates through a medium of length L characterized by the propagation constant $\beta(\omega)$. The output signal occupies the frequency band $\omega_0 - \Omega < \omega < \omega_0 + \Omega$ with and has the time domain representation

$$
\begin{aligned}
y(t, \omega_0) &= \Re\left\{2\int_{\omega_0-\Omega}^{\omega_0+\Omega}(1/2)P(\omega-\omega_0)e^{-i\beta(\omega)L}e^{i\omega t}\frac{d\omega}{2\pi}\right\} \\
&= \Re\left\{e^{i\omega_0 t}\int_{-\Omega}^{\Omega}P(\eta)e^{-i\beta(\eta+\omega_0)L}e^{i\eta t}\frac{d\eta}{2\pi}\right\},
\end{aligned}
\tag{2.336}
$$

where we have assumed that the pulse is a real function. From the last expression we identify the complex baseband output signal as

$$s(t) = \int_{-\Omega}^{\Omega}P(\eta)e^{-i\beta(\eta+\omega_0)L}e^{i\eta t}\frac{d\eta}{2\pi}. \tag{2.337}$$

Irrespective of the nature of the pulse spectrum the frequencies at the band center $\omega = \omega_0$ will be delayed by the group delay $\beta'(\omega_0)L$. In order to focus on pulse distortion (e.g., pulse broadening) it will be convenient to subtract this delay. We do this by initially adding and subtracting $\eta\beta'(\omega_0)L$ from the phase of the integrand in (2.337) as follows:

$$s(t) = \int_{-\Omega}^{\Omega}P(\eta)e^{-i[\beta(\eta+\omega_0)-\beta'(\omega_0)\eta]L}e^{i\eta[t-\beta'(\omega_0)L]}\frac{d\eta}{2\pi}. \tag{2.338}$$

Observe that this integral defines the time delayed version of $\hat{s}(t)$ defined by

$$s(t) = \hat{s}[t - \beta'(\omega_0)L] \tag{2.339}$$

---

[6]When the emphasis is on wave propagation rather than signal analysis, it is customary to represent the wavefunction (2.332) as a superposition of propagation constants β, in terms of the so-called wavenumber spectrum. In that case the envelope in (2.333) (usually referred to as a wavepacket) assumes the form

$$\frac{\sin[\Delta\beta/2(v_{gn}t-x)]}{v_{gn}\pi(v_{gn}t-x)},$$

where $\Delta\beta$ is the range of propagation constants corresponding to the frequency band $\Delta\omega$.

or, explicitly, by

$$\hat{s}(t) = \int_{-\Omega}^{\Omega} P(\eta) e^{-i[\beta(\eta+\omega_0)-\beta'(\omega_0)\eta]L} e^{i\eta t} \frac{d\eta}{2\pi}. \qquad (2.340)$$

We shall obtain an approximation to this integral under the following two assumptions:

$$\omega_0 \gg \Omega, \qquad (2.341a)$$

$$\Omega^2 \beta''(\omega_0)L \gg 1. \qquad (2.341b)$$

The first of these is the conventional narrow band approximation while the second implies a long propagation path.[7] Thus in view of (2.341a) we may approximate $\beta(\eta + \omega_0)$ by

$$\beta(\eta + \omega_0) \sim \beta(\omega_0) + \beta'(\omega_0)\eta + \frac{1}{2}\beta''(\omega_0)\eta^2. \qquad (2.342)$$

Substituting this into (2.340) leads to the following series of algebraic steps:

$$
\begin{aligned}
\hat{s}(t) \quad &\sim \quad e^{-i\beta(\omega_0)L} \int_{-\Omega}^{\Omega} P(\eta) e^{-i\frac{L}{2}\beta''(\omega_0)\eta^2} e^{i\eta t} \frac{d\eta}{2\pi} \\
&= \quad e^{-i\beta(\omega_0)L} \int_{-\Omega}^{\Omega} P(\eta) e^{-i\frac{L}{2}\beta''(\omega_0)\Omega^2 \left[\left(\frac{\eta}{\Omega}\right)^2 - 2\left(\frac{\eta}{\Omega}\right)\left(\frac{t}{\Omega\beta''(\omega_0)L}\right)\right]} \frac{d\eta}{2\pi} \\
&= \quad \Omega e^{-i\beta(\omega_0)L} \int_{-1}^{1} P(\nu\Omega) e^{-i\frac{L}{2}\beta''(\omega_0)\Omega^2 \left[\nu^2 - 2\nu\left(\frac{t}{\Omega\beta''(\omega_0)L}\right)\right]} \frac{d\nu}{2\pi} \\
&= \quad \Omega e^{-i\beta(\omega_0)L} e^{-i\frac{t^2}{2L\beta''(\omega_0)}} \int_{-1}^{1} P(\nu\Omega) e^{-i\frac{L}{2}\beta''(\omega_0)\Omega^2 \left[\nu - \frac{t}{\Omega\beta''(\omega_0)L}\right]^2} \frac{d\nu}{2\pi} \\
&= \quad \Omega e^{-i\beta(\omega_0)L} e^{-i\frac{t^2}{2L\beta''(\omega_0)}} \\
&\qquad \int_{-1-t/\Omega\beta''(\omega_0)L}^{1-t/\Omega\beta''(\omega_0)L} P\left[x\Omega + t/\beta''(\omega_0)L\right] e^{-i\frac{L}{2}\beta''(\omega_0)\Omega^2 x^2} \frac{dx}{2\pi}. \qquad (2.343)
\end{aligned}
$$

Since we are interested primarily in assessing pulse distortion the range of the time variable of interest is on the order of $t \sim 1/\Omega$ we have in view of (2.341b)

$$t/\Omega\beta''(\omega_0)L \ll 1. \qquad (2.344)$$

Consequently the limits in the last integral in (2.343) may be replaced by $-1, 1$. Again in view of (2.341b) we may evaluate this integral by appealing to the principle of stationary phase. Evidently the point of stationary phase is at $x = 0$ which leads to the asymptotic result

$$\hat{s}(t) \sim \frac{1}{\sqrt{2\pi |\beta''(\omega_0)|L}} e^{-i\pi/4\,\text{sign}[\beta''(\omega_0)]} e^{-i\beta(\omega_0)L} e^{-i\frac{t^2}{2L\beta''(\omega_0)}} P\left(\frac{t}{\beta''(\omega_0)L}\right). \qquad (2.345)$$

---

[7]Note (2.341b) necessarily excludes the special case $\beta''(\omega_0) = 0$.

In many applications (e.g., intensity modulation in fiber optic communication systems) only the pulse envelope is of interest. In that case (2.345) assumes the compact form

$$|\hat{s}(t)|^2 \sim \frac{1}{2\pi |\beta''(\omega_0)| L} \left| P\left(\frac{t}{\beta''(\omega_0) L}\right) \right|^2 \tag{2.346}$$

Parseval's theorem tells us that the energies of the input and output signals must be identical. Is this still the case for the approximation (2.346)? Indeed it is as we verify by a direct calculation:

$$\int_{-\infty}^{\infty} |\hat{s}(t)|^2 \, dt = (1/2\pi |\beta''(\omega_0)| L) \int_{-\infty}^{\infty} |P(t/\beta''(\omega_0) L)|^2 \, dt$$

$$= \frac{1}{2\pi} \int_{-\infty}^{\infty} |P(\omega)|^2 \, d\omega \equiv \frac{1}{2\pi} \int_{-\Omega}^{\Omega} |P(\omega)|^2 \, d\omega.$$

Equation (2.346) states that the envelope of a pulse propagating over a sufficiently long path assumes the shape of its Fourier transform wherein the timescale is determined only by the path length and the second derivative of the propagation constant at the band center. For example, for a pulse of unit amplitude and duration T we obtain

$$|\hat{s}(t)|^2 \sim 4 \frac{\sin^2\left(\frac{tT}{2\beta''(\omega_0)L}\right)}{\left(\frac{t}{\beta''(\omega_0)L}\right)^2}$$

giving a peak-to-first null pulsewidth of

$$T_L = \left| \frac{2\pi \beta''(\omega_0) L}{T} \right|. \tag{2.347}$$

In optical communications pulse broadening is usually described by the group index $N(\omega)$ defined as the ratio of the speed of light in free space to the group velocity in the medium:

$$N(\omega) = \frac{c}{v_g(\omega)} = c\beta'(\omega). \tag{2.348}$$

Expressed in terms of the group index the pulse width in (2.347) reads

$$T_L = \left| \frac{2\pi L}{cT} \frac{d}{d\omega} N(\omega) \big|_{\omega=\omega_0} \right|. \tag{2.349}$$

In view of (2.341b) these results break down whenever $\beta''(\omega_0) = 0$, i.e., at the inflection points (if they exist) of the dispersion curve. To include the case of inflection points requires the retention of the third derivative in Taylor expansion (2.342), i.e.,

$$\beta(\eta + \omega_0) \sim \beta(\omega_0) + \beta'(\omega_0)\eta + \frac{1}{2}\beta''(\omega_0)\eta^2 + \frac{1}{6}\beta'''(\omega_0)\eta^3 \tag{2.350}$$

so that

$$\hat{s}(t) \sim e^{-i\beta(\omega_0)L} \int_{-\Omega}^{\Omega} P(\eta) e^{-i\frac{L}{2}\beta''(\omega_0)\eta^2 - i\frac{L}{6}\beta'''(\omega_0)\eta^3} e^{i\eta t} \frac{d\eta}{2\pi}. \tag{2.351}$$

We shall not evaluate (2.351) for general pulse shapes but confine our attention to a Gaussian pulse. In that case we may replace the limits in (2.351) by $\pm\infty$ and require only that (2.341a) hold but not necessarily (2.341b). Using the parameterization in (2.296) we have

$$p(t) = \frac{2^{1/4}}{\sqrt{T}} e^{-\frac{\pi t^2}{T^2}}, \tag{2.352}$$

where we have relabeled the nominal pulse width s by T. The corresponding FT then reads

$$P(\omega) = 2^{1/4}\sqrt{T} e^{-T^2\omega^2/4\pi} \tag{2.353}$$

so that (2.351) assumes the form

$$
\begin{aligned}
\hat{s}(t) \quad &\sim \quad 2^{1/4}\sqrt{T} e^{-i\beta(\omega_0)L} \int_{-\infty}^{\infty} e^{-T^2\eta^2/4\pi} e^{-i\frac{L}{2}\beta''(\omega_0)\eta^2 - i\frac{L}{6}\beta'''(\omega_0)\eta^3} e^{i\eta t} \frac{d\eta}{2\pi} \\
&= \quad 2^{1/4}\sqrt{T} e^{-i\beta(\omega_0)L} \int_{-\infty}^{\infty} e^{-i\frac{L\beta'''(\omega_0)}{6}[\eta^3 + B\eta^2 - C\eta]} \frac{d\eta}{2\pi}, \tag{2.354}
\end{aligned}
$$

where

$$B = \frac{3\beta''(\omega_0)}{\beta'''(\omega_0)} - i\frac{3T^2}{2\pi L\beta'''(\omega_0)}, \tag{2.355a}$$

$$C = \frac{6t}{L\beta'''(\omega_0)}. \tag{2.355b}$$

Changing the variable of integration to z via $\eta = z - B/3$ eliminates the quadratic term in the polynomial in the exponential (2.354) resulting in

$$\eta^3 + B\eta^2 - C\eta = z^3 - z\left(B^2/3 + C\right) + (2/27)B^3 + BC/3.$$

Because of the analyticity of the integrand the integration limits in (2.354) may kept at $\pm\infty$. A subsequent change of the integration from z to $w = \left[L\beta'''(\omega_0)/2\right]^{1/3} z$ transforms (2.354) into

$$
\begin{aligned}
\hat{s}(t) \quad \sim \quad & 2^{1/4}\sqrt{T} e^{-i\beta(\omega_0)L} e^{-i\frac{\beta'''(\omega_0)L}{6}[(2/27)B^3 + BC/3]} \\
& \left\{\frac{\beta'''(\omega_0)L}{2}\right\}^{-1/3} Ai\left\{-\left[\frac{\beta'''(\omega_0)L}{2}\right]^{2/.3}(B^2/9 + C/3)\right\}, \tag{2.356}
\end{aligned}
$$

where $Ai(x)$ is the Airy function defined by the integral

$$Ai(x) = \frac{1}{2\pi} \int_{-\infty}^{\infty} e^{-i(w^3/3 + xw)} dw. \tag{2.357}$$

The interpretation of (2.356) will be facilitated if we introduce the following dimensionless parameters:

$$q = \frac{\beta''(\omega_0) T}{\beta'''(\omega_0)}, \tag{2.358a}$$

$$p = \frac{\beta'''(\omega_0) L}{T^3}, \tag{2.358b}$$

$$\chi = 2qp = \frac{2\beta''(\omega_0) L}{T^2}. \tag{2.358c}$$

Introducing these into (2.357) we obtain

$$\hat{s}(t) \sim 2^{1/4}(1/\sqrt{T}) e^{-i\beta(\omega_0)L} e^{-i\left\{ \frac{\chi q^2}{6}\left(1 - \frac{i}{\pi\chi}\right)\left[\left(1 - \frac{i}{\pi\chi}\right)^2 + \frac{6}{\chi q}\left(\frac{t}{T}\right)\right]\right\}}$$
$$(p/2)^{-1/3} Ai\left\{ -q^2 \left(\frac{p}{2}\right)^{2/3}\left[\left(1 - \frac{i}{\pi\chi}\right)^2 + 4\frac{t}{q\chi T}\right]\right\}. \tag{2.359}$$

Let us first examine this expression for the case in which the third derivative term in (2.350) can be neglected. Clearly this is tantamount to dropping the cubic term in (2.351). The integral then represents the FT of a Gaussian function and can be evaluated exactly. On the other hand, from the definition of q in (2.358a) we note that $\beta'''(\omega_0) \to 0$ and $\beta''(\omega_0) \neq 0$ correspond to $q \to \infty$. Hence we should be able to obtain the same result by evaluating (2.359) in the limit as $q \to \infty$. We do this with the aid of the first-order asymptotic form of the Airy function for large argument the necessary formula for which is given in [1]. It reads

$$Ai(-z) \sim \pi^{-1/2} z^{-1/4} \sin(\zeta + \frac{\pi}{4}), \tag{2.360}$$

where

$$\zeta = \frac{2}{3} z^{3/2} \; ; \; |\arg(z)| < \pi. \tag{2.361}$$

Thus we obtain for[8] $|q| \sim \infty$

$$(p/2)^{-1/3} Ai\left\{ -q^2 \left(\frac{p}{2}\right)^{2/.3}\left[\left(1 - \frac{i}{\pi\chi}\right)^2 + 4\frac{t}{q\chi T}\right]\right\}$$
$$\sim -i\left[\pi\chi(1 - \frac{i}{\pi\chi})\right]^{-1/2}$$
$$\left(\begin{array}{c} \exp\left\{ i(p/3)q^3\left[\left(1 - \frac{i}{\pi\chi}\right)^2 + 4\frac{t}{q\chi T}\right]^{3/2} + i\frac{\pi}{4}\right\} \\ -\exp\left\{ -i(p/3)q^3\left[\left(1 - \frac{i}{\pi\chi}\right)^2 + 4\frac{t}{q\chi T}\right]^{3/2} - i\frac{\pi}{4}\right\} \end{array}\right), \tag{2.362}$$

---

[8] q is real but may be of either sign.

where in the algebraic term corresponding to $z^{-1/4}$ in (2.360) we have dropped the term $o(1/q)$. Next we expand the argument of the first exponentials term in (2.362) as follows:

$$i\left(\chi q^2/6\right)\left[\left(1-\frac{i}{\pi\chi}\right)^2+4\frac{t}{q\chi T}\right]^{3/2}$$

$$=i\left(\chi q^2/6\right)\left(1-\frac{i}{\pi\chi}\right)^3\left[1+4\frac{t}{q\chi T\left(1-\frac{i}{\pi\chi}\right)^2}\right]^{3/2}$$

$$=i\left(\chi q^2/6\right)\left(1-\frac{i}{\pi\chi}\right)^3$$

$$\left[1+6\frac{t}{q\chi T\left(1-\frac{i}{\pi\chi}\right)^2}+6\frac{t^2}{(q\chi T)^2\left(1-\frac{i}{\pi\chi}\right)^4}+o(1/q^3)\right]$$

$$=i\left(\chi q^2/6\right)\left(1-\frac{i}{\pi\chi}\right)^3+iq\frac{t}{T}\left(1-\frac{i}{\pi\chi}\right)$$

$$+i\frac{t^2}{\chi T^2}\left(1-\frac{i}{\pi\chi}\right)^{-1}+o(1/q). \tag{2.363}$$

In identical fashion we can expand the argument of the second exponential which would differ from (2.363) only by a minus sign. It is not hard to show that for sufficiently large $|q|$ is real part will be negative provided

$$\chi^2>\frac{1}{3\pi^2}. \tag{2.364}$$

In that case the second exponential in (2.362) asymptotes to zero and may be ignored. Neglecting the terms $o(1/q)$ in (2.363) we now substitute (2.362) into (2.359) and note that the first two terms in the last line of (2.363) cancel against the exponential in (2.359). The final result then reads

$$\hat{s}(t)\sim2^{1/4}(1/\sqrt{T})e^{-i\beta(\omega_0)L}$$

$$\left\{-i\left[\pi\chi(1-\frac{i}{\pi\chi})\right]^{-1/2}\right\}\exp\left\{i\frac{t^2}{\chi T^2}\left(1-\frac{i}{\pi\chi}\right)^{-1}+i\frac{\pi}{4}\right\}$$

$$=2^{1/4}(1/\sqrt{T})e^{-i\beta(\omega_0)L}(1+i\pi\chi)^{-1/2}\exp-\frac{\pi t^2}{T^2}(1+i\pi\chi)^{-1}. \tag{2.365}$$

For the squared magnitude of the pulse envelope we get

$$|\hat{s}(t)|^2\sim\frac{\sqrt{2}}{T}\left(1+\pi^2\chi^2\right)^{-1/2}\exp-\frac{\pi t^2}{(T^2/2)\left(1+\pi^2\chi^2\right)}. \tag{2.366}$$

The nominal duration of this Gaussian signal may be defined by $(T/2\sqrt{\pi})$ $\sqrt{1+\pi^2\chi^2}$ so that χ plays the role of a pulse-stretching parameter. When $\chi\gg1$ (2.366) reduces to

$$|\hat{s}(t)|^2 \sim \frac{\sqrt{2}}{\pi\chi T}\exp-\frac{2t^2}{\pi T^2\chi^2} = \frac{T}{\pi\sqrt{2}\beta''(\omega_0)L}\exp-\frac{T^2t^2}{2\pi\beta''(\omega_0)L}. \quad (2.367)$$

The same result also follows more directly from the asymptotic form (2.346) as is readily verified by the substitution of the FT of the Gaussian pulse (2.353) into (2.346). Note that with $\chi = 0$ in (2.366) we recover the squared magnitude of the original (input) Gaussian pulse (2.352). Clearly this substitution violates our original assumption $|q| \sim \infty$ under which (2.366) was derived for in accordance with (2.358) $\chi = 0$ implies $q = 0$. On the other hand if $\beta'''(\omega_0)$ is taken to be *identically zero* (2.366) is a valid representation of the pulse envelope for *all values of* χ. This turns out to be the usual assumption in the analysis of pulse dispersion effects in optical fibers. In that case formula (2.366) can be obtained directly from (2.351) by simply completing the square in the exponential and integrating the resulting Gaussian function. When $\beta'''(\omega_0) \neq 0$ with q arbitrary numerical calculations of the output pulse can be carried out using (2.359). For this purpose it is more convenient to eliminate χ in favor of the parameters p and q. This alternative form reads

$$\hat{s}(t) \sim 2^{1/4}(1/\sqrt{T})e^{-i\beta(\omega_0)L}e^{-i\left\{\frac{p}{3}\left(q-\frac{i}{2\pi p}\right)\left[\left(q-\frac{i}{2\pi p}\right)^2+\frac{3}{p}\left(\frac{t}{T}\right)\right]\right\}}$$

$$(|p|/2)^{-1/3}Ai\left\{-\left(\frac{|p|}{2}\right)^{2/3}\left[\left(q-\frac{i}{2\pi p}\right)^2+2\frac{t}{pT}\right]\right\}. \quad (2.368)$$

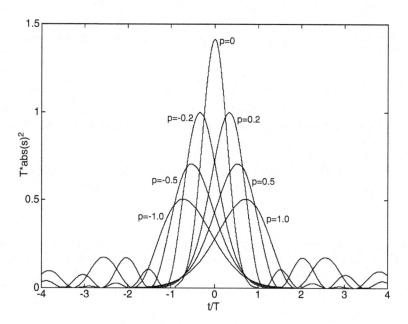

Figure 2.53: Distortion of Gaussian pulse envelope by cubic phase nonlinearities in the propagation constant

To assess the influence of the third derivative of the phase on the pulse envelope we set $q = 0$ and obtain the series of plots for several values of p as shown in Fig. 2.53. The center pulse labeled $p = 0$ corresponds to the undistorted Gaussian pulse ($\chi = 0$ in (2.366)). As p increases away from zero the pulse envelope broadens with a progressive increase in time delay. For sufficiently large p the envelope will tend toward multimodal quasi-oscillatory behavior the onset of which is already noticeable for p as low as 0.2. For negative p the pulse shapes are seen to be a mirror images with respect to $t = 0$ of those for positive p so that pulse broadening is accompanied by a time advance.

2.6.4 Another Look at the Propagation of a Gaussian Pulse When $\beta'''(\omega_0) = 0$

As was pointed out above in the absence of cubic (and higher order) nonlinearities (2.366) is an exact representation of the pulse envelope. In fact we can also get the complete waveform in the time domain with the aid of (2.336), (2.339), and (2.365). Thus

$$y(t, \omega_0) = 2^{1/4}/\sqrt{T}\,\Re\left\{ \frac{e^{i\omega_0\left[\tilde{t} - \frac{\pi^2\chi\tilde{t}^2}{(T^2)(1+\pi^2\chi^2)}\right]} e^{-i\left[\beta(\omega_0)L + (1/2)\tan^{-1}(\pi\chi)\right]}}{\left(1 + \pi^2\chi^2\right)^{-1/4}\exp-\frac{\pi t^2}{(T^2)(1+\pi^2\chi^2)}} \right\},$$

(2.369)

where

$$\tilde{t} = t - \beta'(\omega_0)\,L. \tag{2.370}$$

Note that the instantaneous frequency of this complex waveform varies linearly with time, i.e.,

$$\omega(t) = \omega_0 - \frac{2\pi^2\chi\left(t - \beta'(\omega_0)L\right)}{(T^2)\left(1 + \pi^2\chi^2\right)}. \tag{2.371}$$

In fiber optics such a pulse is referred to as a chirped pulse. This "chirping," (or linear FM modulation) is just a manifestation of the fact that the pulse distortion is due entirely to the quadratic nonlinearity in the phase rather than in the amplitude of the effective transfer function. On the other hand, chirping can occur also due to intrinsic characteristics of the transmitter generating the input pulse. We can capture this effect using the analytic form

$$p(t) = Ae^{-\frac{t^2}{2T_0^2}(1+i\kappa)}, \tag{2.372}$$

where A is a constant, κ the so-called chirp factor, and $2T_0$ the nominal pulse width.[9] Evidently when this pulse gets upconverted to the carrier frequency ω_0 its instantaneous frequency becomes

---

[9]Note that $T_0 = (1/\sqrt{2\pi})\,T$ where T represents the definition of pulse width in (2.352). Also $A = 2^{1/4}/\sqrt{T}$.

$$w(t) = w_0 \left[1 - \frac{\kappa}{\omega_0 T_0} \left(\frac{t}{T_0} \right) \right] \tag{2.373}$$

so that over the nominal pulse interval $-T_0 \leq t \leq T_0$ the fractional change in the instantaneous frequency is $2\kappa/\omega_0 T_0$. Presently we view this chirping as the intrinsic drift in the carrier frequency during the formation of the pulse. How does this intrinsic chirp affect pulse shape when this pulse has propagated over a transmission medium with transfer function $exp - \beta(\omega) L$? If we neglect the effects of the third and higher order derivatives of the propagation constant the answer is straightforward. We first compute the FT of (2.372) as follows:

$$
\begin{aligned}
P(\omega) &= A \int_{-\infty}^{\infty} e^{-\frac{t^2}{2T_0^2}(1+i\kappa)} e^{-i\omega t} dt = A \int_{-\infty}^{\infty} e^{-\frac{(1+i\kappa)}{2T_0^2} \left[\left(t - \frac{i\omega T_0^2}{1+i\kappa} \right)^2 + \frac{\omega^2 T_0^4}{(1+i\kappa)^2} \right]} dt \\
&= A e^{-\frac{\omega^2 T_0^2}{2(1+i\kappa)}} \int_{-\infty}^{\infty} e^{-\frac{(1+i\kappa)}{2T_0^2} \left(t - \frac{i\omega T_0^2}{1+i\kappa} \right)^2} dt \\
&= A T_0 \sqrt{\frac{2\pi}{1+i\kappa}} e^{-\frac{\omega^2 T_0^2}{2(1+i\kappa)}}, \tag{2.374}
\end{aligned}
$$

where the last result follows from the formula for the Gaussian error function with (complex) variance parameter $T_0^2/\sqrt{1+i\kappa}$. Next we substitute (2.374) in (2.340) with $\Omega = \infty$ together with the approximation (2.342) to obtain

$$\hat{s}(t) = e^{-i\beta(\omega_0)L} \int_{-\infty}^{\infty} P(\eta) e^{-i\frac{1}{2}\beta''(\omega_0)\eta^2 L} e^{i\eta t} \frac{d\eta}{2\pi} \tag{2.375}$$

Simplifying,

$$s(t) = A T_0 \sqrt{\frac{2\pi}{1+i\kappa}} e^{-i\beta(\omega_0)L} \int_{-\infty}^{\infty} e^{-\left[\frac{T_0^2}{2(1+i\kappa)} + i\frac{L\beta''(\omega_0)}{2} \right] \eta^2} e^{i\eta t} \frac{d\eta}{2\pi}. \tag{2.376}$$

Setting $Q = T_0^2 / [2(1+i\kappa)] + iL\beta''(\omega_0)/2$ we complete the square in the exponential as follows:

$$e^{-Q\eta^2 + i\eta t} = e^{-Q\left[\left(\eta - \frac{it}{2Q} \right)^2 + \frac{t^2}{4Q^2} \right]} = e^{-\frac{t^2}{4Q}} e^{-Q\left(\eta - \frac{it}{2Q} \right)^2}. \tag{2.377}$$

From this we note that the complex variance parameter is $1/(2Q)$ so that (2.376) integrates to

$$
\begin{aligned}
\hat{s}(t) &= \frac{A T_0}{2\pi} \sqrt{\frac{2\pi}{1+i\kappa}} e^{-i\beta(\omega_0)L} e^{-\frac{t^2}{4Q}} \sqrt{\frac{\pi}{Q}} \\
&= \frac{A}{\sqrt{1+i\kappa}} e^{-i\beta(\omega_0)L} \\
&\quad \frac{T_0}{\sqrt{T_0^2 + i\beta''(\omega_0)L(1+i\kappa)}} \exp -\frac{t^2(1+i\kappa)}{2\left[T_0^2 + i\beta''(\omega_0)L(1+i\kappa) \right]}. \\
&\tag{2.378}
\end{aligned}
$$

Expression for the pulse width and chirp is obtained by separating the argument of the last exponential into real and imaginary parts as follows:

$$\exp - \frac{t^2 \left(1 + i\kappa\right)}{2 \left[T_0^2 + i\beta'' \left(\omega_0\right) L \left(1 + i\kappa\right)\right]}$$
$$= \exp - \frac{T_0^2 t^2}{2 \left\{\left[T_0^2 - \beta'' \left(\omega_0\right) L\kappa\right]^2 + \left[\beta'' \left(\omega_0\right) L\right]^2\right\}} \exp -i\psi, \quad (2.379)$$

where

$$\psi = \frac{\kappa t^2 \left[T_0^2 - \beta'' \left(\omega_0\right) L(1 + \kappa)\right]}{2 \left\{\left[T_0^2 - \beta'' \left(\omega_0\right) L\kappa\right]^2 + \left[\beta'' \left(\omega_0\right) L\right]^2\right\}}. \quad (2.380)$$

Defining the magnitude of (2.379) as $\exp -t^2/\left(2T_L^2\right)$ we get for the pulse length T_L

$$T_L = T_0 \sqrt{\left(1 - \frac{\beta'' \left(\omega_0\right) L\kappa}{T_0^2}\right)^2 + \left(\frac{\beta'' \left(\omega_0\right) L}{T_0^2}\right)^2}. \quad (2.381)$$

When the input pulse is unchirped $\kappa = 0$, and we get

$$T_L = \sqrt{T_0^2 + \left(\frac{\beta'' \left(\omega_0\right) L}{T_0}\right)^2}. \quad (2.382)$$

We see from (2.381) that when $\kappa \neq 0$, T_L may be smaller or larger than the right side of (2.382) depending on the sign of κ. and the magnitude of L. Note, however, that for sufficiently large L, (2.381) is always larger than (2.382) regardless of the sign of κ. The quantity

$$L_D = T_0^2/\beta''(\omega_0) \quad (2.383)$$

is known as the dispersion length. Using this in (2.381) we have

$$T_L = T_0 \sqrt{\left(1 - \frac{L}{L_D}\kappa\right)^2 + \left(\frac{L}{L_D}\right)^2}. \quad (2.384)$$

The significance of L_D is that with $\kappa = 0$ for $L \ll L_D$ the effect of dispersion may be neglected.

2.6.5 Effects of Finite Transmitter Spectral Line Width*

In the preceding it was assumed that the carrier modulating the pulse is monochromatic, i.e., an ideal single frequency sinusoid with constant phase. In practice this will not be the case. Instead the carrier will have a fluctuating amplitude and phase which we may represent as

$$\widetilde{a(t)} \cos(\omega_0 t + \widetilde{\phi(t)}), \quad (2.385)$$

where $\widetilde{a(t)}$ and $\widetilde{\phi(t)}$ are random functions of time and ω_0 is the nominal carrier frequency which itself has to be quantified as a statistical average. In the following we assume that only the phase is fluctuating and that the carrier amplitude is fixed. Reverting to complex notation we then assume that the pulse $p(t)$ upon modulation is of the form

$$p(t)e^{i\omega_0 t}e^{i\widetilde{\phi(t)}}. \tag{2.386}$$

If we denote the FT of $e^{i\widetilde{\phi(t)}}$ by the random function $\widetilde{X(\omega)}$, we get for the FT of (2.386)

$$\int_{-\infty}^{\infty} p(t)e^{i\omega_0 t}e^{i\widetilde{\phi(t)}}e^{-i\omega t}dt = \frac{1}{2\pi}\int_{-\infty}^{\infty} P\left(\omega-\xi-\omega_0\right)\widetilde{X(\xi)}d\xi. \tag{2.387}$$

To get the response that results after this random waveform has propagated over a transmission medium with transfer function $\exp -i\beta(\omega)L$ we have to replace $P\left(\omega-\omega_0\right)$ in (2.336) by the right side of (2.387). Thus we obtain

$$
\begin{aligned}
y(t) &= \Re\left\{2\int_{\omega_0-\Omega}^{\omega_0+\Omega}(1/2)\left\{\frac{1}{2\pi}\int_{-\infty}^{\infty} P\left(\omega-\xi-\omega_0\right)\widetilde{X(\xi)}d\xi\right\}e^{-i\beta(\omega)L}e^{i\omega t}\frac{d\omega}{2\pi}\right\}\\
&= \Re\left\{e^{i\omega_0 t}\int_{-\Omega}^{\Omega}\left\{\frac{1}{2\pi}\int_{-\infty}^{\infty} P\left(\eta-\xi\right)\widetilde{X(\xi)}d\xi\right\}e^{-i\beta(\eta+\omega_0)L}e^{i\eta t}\frac{d\eta}{2\pi}\right\}\\
&= \Re\left\{e^{i\omega_0 t}\widetilde{s}(t-\beta'(\omega_0)L\right\},
\end{aligned}
$$

where

$$\widetilde{s}(t) = \int_{-\Omega}^{\Omega}\left\{\frac{1}{2\pi}\int_{-\infty}^{\infty} P\left(\eta-\xi\right)\widetilde{X(\xi)}d\xi\right\}e^{-i\left[\beta(\eta+\omega_0)-\beta'(\omega_0)\eta\right]L}e^{i\eta t}\frac{d\eta}{2\pi} \tag{2.388}$$

is the complex random envelope of the pulse. It is reasonable to characterize this envelope by its statistical average which we denote by

$$|ENV|^2 \equiv \langle|s(t)|^2\rangle. \tag{2.389}$$

In evaluating (2.389) we shall assume that $e^{i\widetilde{\phi(t)}}$ is a WSS process so that its spectral components are uncorrelated, i.e.,

$$\langle\widetilde{X}(\xi)\widetilde{X}^*(\xi')\rangle = 2\pi F\left(\xi\right)\delta\left(\xi-\xi'\right), \tag{2.390}$$

where $F\left(\xi\right)$ is the spectral power density of $e^{i\widetilde{\phi(t)}}$. If we approximate the propagation constant in (2.388) by the quadratic form (2.342), and substitute (2.388) into (2.389) we obtain with the aid of (2.390)

$$|ENV|^2 = \int_{-\infty}^{\infty} F\left(\xi\right)d\xi\frac{1}{(2\pi)^3}\left|\int_{-\Omega}^{\Omega} P(\eta-\xi)e^{-i\beta''(\omega_0)\eta^2 L/2}e^{i\eta t}d\eta\right|^2. \tag{2.391}$$

Assuming a Gaussian pulse with the FT as in (2.374) the inner integral in (2.391) can be expressed in the following form:

$$\frac{1}{(2\pi)^3} \left| \int_{-\Omega}^{\Omega} P(\eta - \xi) e^{-i\beta''(\omega_0)\eta^2 L/2} e^{i\eta t} d\eta \right|^2 = \frac{A^2 T_0^2 \pi}{(2\pi)^2 \sqrt{1 + \kappa^2} |Q|} f(\xi), \quad (2.392)$$

where $Q = T_0^2 / [2(1 + i\kappa)] + iL\beta''(\omega_0)/2$,

$$f(\xi) = e^{2\Re\{Qb^2\}} e^{-\frac{\xi^2 T_0^2}{2} \left[\frac{1}{1+i\kappa} + \frac{1}{1-i\kappa} \right]} \quad (2.393)$$

and

$$b = \frac{\xi T_0^2}{2Q(1 + i\kappa)} + \frac{it}{2Q}. \quad (2.394)$$

To complete the calculation of the average pulse envelope we need the functional form of the power spectral density of the phase fluctuations. The form depends on the physical process responsible for these fluctuations. For example for high quality solid state laser sources the spectral line width is Lorenzian, i.e., of the form

$$F(\omega - \omega_0) = \frac{2/W}{1 + \left(\frac{\omega - \omega_0}{W} \right)^2}. \quad (2.395)$$

Unfortunately for this functional form the integration in (2.391) has to be carried out numerically. On the other hand, an analytical expression is obtainable if we assume the Gaussian form

$$F(\omega - \omega_0) = \frac{1}{\sqrt{2\pi W^2}} \exp - (\omega - \omega_0)/2W^2. \quad (2.396)$$

After some algebra we get

$$|ENV|^2 = \frac{A^2 T_0^2 \pi}{(2\pi)^2 \sqrt{1 + \kappa^2} |Q|} \frac{\sqrt{\left(T_0^2 - \beta''(\omega_0) L\kappa\right)^2 + \left(\beta''(\omega_0) L\right)^2}}{\left[\left(T_0^2 - \beta''(\omega_0) L\kappa\right)^2 + (1 + 2W^2 T_0^2)\left(\beta''(\omega_0) L\right)^2 \right]}$$

$$\exp - \frac{t^2 T_0^2}{\left(T_0^2 - \beta''(\omega_0) L\kappa\right)^2 + (1 + 2W^2 T_0^2)\left(\beta''(\omega_0) L\right)^2}. \quad (2.397)$$

Note that the preceding is the squared envelope so that to get the effective pulse length of the envelope itself an additional factor of 2 needs to be inserted (see (2.381)). We then get

$$T_L = T_0 \sqrt{\left(1 - \frac{\beta''(\omega_0) L\kappa}{T_0^2}\right)^2 + (1 + 2W^2 T_0^2)\left(\frac{\beta''(\omega_0) L}{T_0^2}\right)^2}. \quad (2.398)$$

It should be noted that this expression is not valid when $\beta''(\omega_0) = 0$ as then the cubic phase term dominates. In that case the pulse is no longer Gaussian. The pulse width can then be defined as an r.m.s. duration. The result reads

$$T_L = T_0 \sqrt{\left(1 - \frac{\beta''(\omega_0) L\kappa}{T_0^2}\right)^2 + (1 + 2W^2 T_0^2)\left(\frac{\beta''(\omega_0) L}{T_0^2}\right)^2 + C}, \quad (2.399)$$

where

$$C = (1/4)\left(1 + \kappa^2 + 2W^2 T_0^2\right)\left(\frac{\beta'''(\omega_0)L}{T_0^3}\right)^2. \qquad (2.400)$$

2.7 Fourier Cosine and Sine Transforms

In Chap. 2.2 we took as the starting point in our development of the FT theory the LMS approximation of a function defined in $(-T/2, T/2)$ in terms of sinusoids with frequencies spanning the interval $(-\Omega, \Omega)$. The formal solution can then be phrased in terms of the integral equation (2.106) for the unknown coefficients (functions). For arbitrary finite intervals no simple analytical solutions of the normal equation appears possible. On the other hand, when both the expansion interval in the time domain and the range of admissible frequencies are allowed to approach infinity the normal equations admit a simple solution which we have identified with FT. As we shall see in the following a suitable set of normal equations can also be solved analytically when the expansion intervals in the time domain and in the frequency domain are chosen as semi-infinite.

We suppose that $f(t)$ is defined over $(0, T)$ and seek its LMS approximation in terms of $\cos(\omega t)$ with ω in the interval $(0, \Omega)$:

$$f(t) \sim \int_0^\Omega \cos(\omega t)\hat{f}_c(\omega)\,d\omega = f_c^\Omega(t), \qquad (2.401)$$

where $\hat{f}_c(\omega)$ if the expansion (coefficient) function. In accordance with (1.100) the normal equation reads

$$\int_0^T \cos(\omega t)f(t)\,dt = \int_0^\Omega \hat{f}_c(\omega')\,d\omega' \int_0^T \cos(\omega t)\cos(\omega' t)dt. \qquad (2.402)$$

Using the identity $\cos(\omega t)\cos(\omega' t) = (1/2)\{\cos[t(\omega - \omega')] + \cos[t(\omega + \omega')]\}$ we carry out the integration with respect to t to obtain

$$\int_0^T \cos(\omega t)f(t)\,dt = \frac{\pi}{2}\int_0^\Omega \hat{f}_c(\omega')\,d\omega'\{\frac{\sin[(\omega - \omega')T]}{\pi(\omega - \omega')} + \frac{\sin[(\omega + \omega')T]}{\pi(\omega + \omega')}\}. \qquad (2.403)$$

For arbitrary T this integral equation does not admit of simple analytical solutions. An exceptional case obtains when T is allowed to approach infinity for then the two Fourier Integral kernels approach delta functions. Because $\Omega > 0$ only the first of these contributes. Assuming that $\hat{f}_c(\omega')$ is a smooth function and we obtain in the limit

$$F_c(\omega) = \int_0^\infty \cos(\omega t)f(t)\,dt, \qquad (2.404)$$

where we have defined

$$F_c(\omega) = \frac{\pi}{2}\hat{f}_c(\omega). \qquad (2.405)$$

Inserting (2.404) into (2.401) the LMS approximation to $f(t)$ reads

$$
\begin{aligned}
f_c^\Omega(t) &= \frac{2}{\pi} \int_0^\Omega \cos(\omega t) \int_0^\infty \cos(\omega t') f(t')\, dt'\, d\omega \\
&= \int_0^\infty f(t')\, dt' \frac{2}{\pi} \int_0^\Omega \cos(\omega t) \cos(\omega t')\, d\omega \\
&= \int_0^\infty f(t')\, dt' (1/\pi) \int_0^\Omega \{\cos[\omega(t-t')] + \cos[\omega(t+t')]\}\, d\omega \\
&= \int_0^\infty f(t')\, dt' \left\{ \frac{\sin[(t-t')\Omega]}{\pi(t-t')} + \frac{\sin[(t+t')\Omega]}{\pi(t+t')} \right\}. \qquad (2.406)
\end{aligned}
$$

Using the orthogonality principle the corresponding LMS error $\varepsilon_{\Omega\min}$ is

$$
\varepsilon_{\Omega\min} = \int_0^\infty |f(t)|^2\, dt - \int_0^\infty f^*(t) f_c^\Omega(t)\, dt
$$

and using (2.401) and (2.404)

$$
\begin{aligned}
\varepsilon_{\Omega\min} &= \int_0^\infty |f(t)|^2\, dt - \int_0^\infty f^*(t) \int_0^\Omega \cos(\omega t) \hat{f}_c(\omega)\, d\omega dt \\
&= \int_0^\infty |f(t)|^2\, dt - \frac{2}{\pi} \int_0^\Omega |F_c(\omega)|^2\, d\omega \geq 0. \qquad (2.407)
\end{aligned}
$$

As $\Omega \to \infty$ the two Fourier kernels yield the limiting form

$$
\lim_{\Omega\to\infty} f_c^\Omega(t) = \frac{f(t^+) + f(t^-)}{2}. \qquad (2.408)
$$

We may then write in lieu of (2.401)

$$
\frac{f(t^+) + f(t^-)}{2} = \frac{2}{\pi} \int_0^\infty \cos(\omega t) F_c(\omega)\, d\omega. \qquad (2.409)
$$

At the same time $\lim_{\Omega\to\infty} \varepsilon_{\Omega\min} = 0$ so that (2.407) gives the identity

$$
\int_0^\infty |f(t)|^2\, dt = \frac{2}{\pi} \int_0^\infty |F_c(\omega)|^2\, d\omega. \qquad (2.410)
$$

When $f(t)$ is a smooth function (2.409) may be replaced by

$$
f(t) = \frac{2}{\pi} \int_0^\infty \cos(\omega t) F_c(\omega)\, d\omega. \qquad (2.411)
$$

The quantity $F_c(\omega)$ defined by (2.404) is the Fourier Cosine Transform (FCT) and (2.411) the corresponding inversion formula. Evidently (2.410) is the corresponding Parseval formula. As in the case of the FT we can use the compact notation

$$
f(t) \overset{F_c}{\Longleftrightarrow} F_c(\omega). \qquad (2.412)
$$

Replacing $F_c(\omega)$ in (2.411) by (2.404) yields the identity

$$\delta(t - t') = \int_0^\infty \sqrt{\frac{2}{\pi}} \cos(\omega t) \sqrt{\frac{2}{\pi}} \cos(\omega t')d\omega, \qquad (2.413)$$

which may be taken as the completeness relationship for the FCT.

Note that the derivative of $f_c^\Omega(t)$ at $t = 0$ vanishes identically. This means that pointwise convergence for the FCT is only possible for functions that possess a zero derivative at $t = 0$. This is, of course, also implied by the fact that the completeness relationship (2.413) is comprised entirely of cosine functions.

What is the relationship between the FT and the FCT? Since the FCT involves the cosine kernel one would expect that the FCT can be expressed in terms of the FT of an even function. This is actually the case. Thus suppose $f(t)$ is even then

$$F(\omega) = \int_0^\infty 2f(t)\cos(\omega t)dt \qquad (2.414)$$

so that $F(\omega)$ is also even. Therefore the inversion formula becomes

$$f(t) = \frac{1}{\pi} \int_0^\infty F(\omega)\cos(\omega t)d\omega. \qquad (2.415)$$

Evidently with $F_c(\omega) = F(\omega)/2$ (2.414) and (2.415) correspond to (2.404) and (2.411), respectively.

In a similar manner, using the sine kernel, one can define the Fourier Sine Transform (FST):

$$F_s(\omega) = \int_0^\infty \sin(\omega t)f(t)\,dt. \qquad (2.416)$$

The corresponding inversion formula (which can be established either formally in terms of the normal equation as above or derived directly from the FT representation of an odd function) reads

$$f(t) = \frac{2}{\pi} \int_0^\infty F_s(\omega)\sin(\omega t)d\omega. \qquad (2.417)$$

Upon combining (2.416) and (2.417) we get the corresponding completeness relationship

$$\delta(t - t') = \int_0^\infty \sqrt{\frac{2}{\pi}} \sin(\omega t) \sqrt{\frac{2}{\pi}} \sin(\omega t')d\omega. \qquad (2.418)$$

Note that (2.417) and (2.418) require that $f(0) = 0$ so that only for such functions a pointwise convergent FST representation is possible.

Problems

1. Using (2.37) compute the limit as $M \to \infty$, thereby verifying (2.39).

2. Prove (2.48).

3. Derive the second-order Fejer sum (2.42).

4. For the periodic function shown in the following sketch:

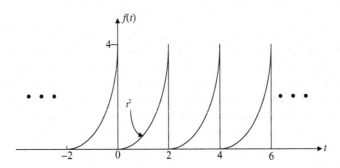

Figure P4: Periodic function with step discontinuities

(a) Compute the FS coefficients $\hat{f}_n$.

(b) Compute and plot the partial sum $f^N(t)$ for $N = 5$ and $N = 20$. Also compute the corresponding LMS errors.

(c) Repeat (b) for the first-order Fejer sum.

(d) Repeat (c) for the second-order Fejer sum.

5. Derive the interpolation formula (2.82)

6. Derive the interpolation formula (2.88)

7. Approximate the signal $f(t) = te^{-t}$ in the interval $(0, 4)$ by the first five terms of a Fourier sine series and an anharmonic Fourier series with expansion functions as in (2.101) assuming (a) $\beta = -1/3$ and (b) $\beta = -1$. Plot $f^5(t)$ for the three cases together with $f(t)$ on the same set of axes. Account for the different values attained by the three approximating sums at $t = 4$.

8. The integral

$$I = P \int_{-2}^{2} \frac{tdt}{(t-1)(t^2+1)\sin t}$$

is defined in the CPV sense. Evaluate it numerically.

9. Derive formulas (2.137)(2.137*)(2.141) and (2.142).

10. The differential equation

$$x''(t) + x'(t) + 3x(t) = 0$$

is to be solved for $t \geq -2$ using the FT. Assuming initial conditions $x(-2) = 3$ and $x'(-2) = 1$ write down the general solution in terms of the FT inversion formula.

11. Derive formula (2.73).

12. Prove that $K_\Omega^{(1)}(t)$ in (2.200) is a delta function kernel.

13. Prove the asymptotic form (2.201).

14. Compute the Fourier transform of the following signals:

$$a) \ \left(e^{-3t}\cos 4t\right)U(t) \quad b) e^{-4|t|}\sin 7t$$

$$c) \ \left(te^{-5t}\sin 4t\right)U(t) \quad d) \ \sum_{n=0}^{\infty} 4^{-n}\delta(t-nT)$$

$$e) \ \left(\frac{\sin at}{at}\right)\left(\frac{\sin 2a\,(t-1)}{a\,(t-1)}\right) \quad f) \ \sum_{n=-\infty}^{\infty} e^{-|t-4n|}$$

15. Compute the Fourier transform of the following signals:

(a) $f(t) = \frac{\sin at}{\pi t} U(t)$

(b) $f(t) = \int_{-\infty}^{\infty} g(t+x)g^*(x)dx$ with $g(t) = e^{-at}U(t-2)$

(c) $f(t) = w(t)$ with $w(t)$ defined in the following sketch.

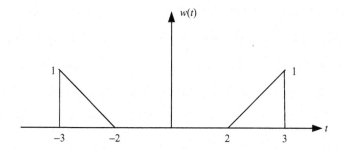

16. With the aid of Parseval's theorem evaluate $\int_{-\infty}^{\infty} \frac{\sin^4 x}{x^4}dx$.

17. With $F(\omega) = R(\omega)+iX(\omega)$ the FT of a causal signal find $X(\omega)$ when a) $R(\omega) = 1/(1+\omega^2)$ b) $R(\omega) = \frac{\sin^2 2\omega}{\omega^2}$.

18. Given the real signal $1/(1+t^2)$ construct the corresponding analytic signal and its FT.

19. Derive (2.217).

20. For the signal $z(t) = \frac{\cos 5t}{1+t^2}$ compute and plot the spectra of the inphase and quadrature components $x(t)$ and $y(t)$ for $\omega_0 = 5, 10, 20$. Interpret your results in view of the constraint (2.226).

21. The amplitude of a minimum phase FT is given by $|F(\omega)| = \frac{1}{1+\omega^{2n}}, n > 1$. Compute the phase.

Chapter 3

Linear Systems

3.1 Fundamental Properties

3.1.1 Single-valuedness, Reality, and Causality

At the beginning of Chap. 1 we defined a system as a mapping of an input signal into an output signal, as expressed by (3.3). The nature of the mathematical operation or operations that the system performs on the input vector $\mathbf{f}(t)$ to yield the output vector $\mathbf{y}(t)$ defines the (system) operator $\mathfrak{T}$. Of course, of the infinite variety of possible mathematical operations that could be carried out on the input signal only certain classes could represent models of physical systems. For example, given a specific input signal $\mathbf{f}(t)$ a reasonable expectation is for the output to be represented by a unique time function $\mathbf{y}(t)$ (rather than two or more different time functions, say $\mathbf{y}_1(t)$ and $\mathbf{y}_2(t)$). In other words, from a mathematical standpoint, we would like the operator $\mathfrak{T}$ to be single valued. Another necessary attribute of a physical system is that it be real in the sense that an input represented by a real $\mathbf{f}(t)$ should always result in a real $\mathbf{y}(t)$.

It should be emphasized that the concept of the systems operator $\mathfrak{T}$ as used herein requires the specification of the input for $-\infty < t < \infty$ enabling the determination of the output for $-\infty < t < \infty$. Next let us consider the system response (output) due to two possible inputs $\mathbf{f}_1(t)$ and $\mathbf{f}_2(t)$ related as follows:

$$\mathbf{f}_1(t) = \mathbf{f}_2(t) \ ; \ t < T,$$
$$\mathbf{f}_1(t) \neq \mathbf{f}_2(t) \ ; \ t \geq T.$$

What we have postulated here are two inputs that are identical over the infinite past up to $t = T$ but differ subsequently. The corresponding outputs may be represented symbolically as follows:

$$\mathbf{y}_1(t) = \mathfrak{T}\{\mathbf{f}_1(t)\},$$
$$\mathbf{y}_2(t) = \mathfrak{T}\{\mathbf{f}_2(t)\}.$$

W. Wasylkiwskyj, *Signals and Transforms in Linear Systems Analysis*,
DOI 10.1007/978-1-4614-3287-6_3, © Springer Science+Business Media, LLC 2013

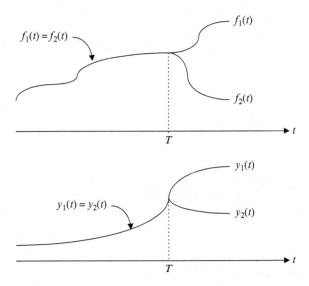

Figure 3.1: Illustration of causality

What is the relationship between $\mathbf{y}_1(t)$ and $\mathbf{y}_2(t)$? Clearly the assumption that $\mathbf{y}_1(t) \neq \mathbf{y}_2(t)$ for $t < T$ with identical inputs would be untenable for it would mean that the system could *predict* future changes in the input. Were such a prediction possible the state of affairs would be bizarre indeed for we could never pull the plug on such a system: it would anticipate our action! We say that such a system is *noncausal* and physically unrealizable. The definition of causality can be phrased as follows. A system is said to be causal if, whenever two inputs are identical for $t < T$ the corresponding two outputs are also identical for $t < T$. The relationship between the input and output signals of a causal system is illustrated in Fig. 3.1.

It is important to keep in mind that the notion of causality is forced upon us only when the input/output functional domains involve real time. System input/output models where the functional domains involve other variables (e.g., space) are not governed by an equivalent constraint.

We now confine our attention to a very special class of systems: systems that are said to be linear. We define a system as linear if, given

$$\mathbf{y}_1(t) = \mathfrak{T}\{\mathbf{f}_1(t)\},$$
$$\mathbf{y}_2(t) = \mathfrak{T}\{\mathbf{f}_2(t)\},$$

and any two constants α_1 and α_2, the following holds:

$$\mathfrak{T}\{\alpha_1\mathbf{f}_1(t) + \alpha_2\mathbf{f}_2(t)\} = \alpha_1\mathbf{y}_1(t) + \alpha_2\mathbf{y}_2(t). \qquad (3.1)$$

This principle of superposition generalizes to any number of input vector functions. Thus if (3.1) holds then given any number of relations of the form $\mathbf{y}_n(t) = \mathfrak{T}\{\mathbf{f}_n(t)\}$ and constants α_n we have

$$\mathfrak{T}\left\{\sum_n \alpha_n \mathbf{f}_n(t)\right\} = \sum_n \alpha_n \mathfrak{T}\{\mathbf{f}_n(t)\}$$

$$= \sum_n \alpha_n \mathbf{y}_n(t). \tag{3.2}$$

These sums may extend over a finite or an infinite number of members. In the latter case it is presupposed that the necessary convergence criteria hold. In fact if we interpret an integral as a limit of a sum (3.2) can even be extended to integrals. Thus suppose

$$\mathbf{y}(\xi, t) = \mathfrak{T}\{\mathbf{f}(\xi, t)\}, \tag{3.3}$$

wherein ξ takes on a continuum of values within some prescribed range. If this range encompasses the entire real number line, the extended superposition principle reads

$$\mathfrak{T}\left\{\int_{-\infty}^{\infty} \alpha(\xi)\mathbf{f}(\xi, t)\, d\xi\right\} = \int_{-\infty}^{\infty} \alpha(\xi)\mathfrak{T}\{\mathbf{f}(\xi, t)\}\, d\xi$$

$$= \int_{-\infty}^{\infty} \alpha(\xi)\mathbf{y}(\xi, t)\, d\xi. \tag{3.4}$$

We shall find the last form of the superposition principle particularly useful in the application of integral transform techniques to linear systems. A system operator $\mathfrak{T}$ that obeys the superposition principle is called a linear operator. We recall that both the Fourier transform and the Hilbert transform are linear operators.

3.1.2   Impulse Response

In the following we specialize to single input/single output systems so that

$$y(t) = \mathfrak{T}\{f(t)\}. \tag{3.5}$$

Suppose the input is the impulse $\delta(t - \tau)$. The corresponding output is called the system impulse response $h(t, \tau)$. The linear operator in (3.5) provides the formal identity

$$h(t, \tau) = \mathfrak{T}\{\delta(t - \tau)\}. \tag{3.6}$$

Note that the impulse response depends on two variables: the time of observation t and the time τ at which the impulse is applied. Since

$$f(t) = \int_{-\infty}^{\infty} f(\tau)\delta(t - \tau)\, d\tau$$

we can substitute it into (3.5), and using the superposition principle, bring the linear operator into the integrand to operate on the impulse function directly. Thus

$$y\left(t\right) = \Im\left\{\int_{-\infty}^{\infty} f\left(\tau\right)\delta\left(t-\tau\right)d\tau\right\} = \int_{-\infty}^{\infty} f\left(\tau\right)\Im\left\{\delta\left(t-\tau\right)\right\}d\tau.$$

Substituting (3.6) into the last integral gives

$$y\left(t\right) = \int_{-\infty}^{\infty} f\left(\tau\right)h\left(t,\tau\right)d\tau, \tag{3.7}$$

which is perhaps the most important single relationship in linear systems theory. It shows explicitly that the impulse response completely characterizes the input/output properties of a linear system in the sense that if we know the impulse response the response due to any input $f\left(t\right)$ is uniquely determined. It is important to note that in accordance with (3.7) this input must be known for all the infinite past and future. In practice this is of course neither possible nor necessary. We postpone until a later juncture the discussion of the important practical consequences of specifying $f\left(t\right)$ only over a finite segment of the time axis.

Note that the linear operator $\Im$ defined previously only in terms of its superposition properties now assumes the specific form of an integral operator with kernel $h\left(t,\tau\right)$. The question may well be asked at this point whether all single input/single output linear systems are governed by integral operators. An affirmative answer can be given provided $h\left(t,\tau\right)$ is allowed to encompass singularity functions. For example, suppose

$$h\left(t,\tau\right) = a_0\delta\left(t-\tau\right) + a_1\delta^{(1)}\left(t-\tau\right), \tag{3.8}$$

where a_0 and a_1 are constants. Then (3.7) gives

$$y\left(t\right) = a_0 f\left(t\right) + a_1\frac{df\left(t\right)}{dt}, \tag{3.9}$$

so that in this case an alternative representation of the integral operator is

$$\Im\left\{f\left(t\right)\right\} = \left\{a_0 + a_1\frac{d}{dt}\right\}f\left(t\right) = a_0 f\left(t\right) + a_1\frac{df\left(t\right)}{dt}, \tag{3.10}$$

i.e., a differential operator. An important feature of a differential operator not shared by an integral operator (with a non symbolic kernel) is that it is memoryless: the output at a specific time, say $t = t_1$, is dependent entirely on the properties of the input at $t = t_1$ only. Clearly we can generate differential operators of any order by appending terms with higher order singularity functions to (3.8).

We have associated the notion of causality with physical realizability of any system. Let us now examine the implications of this concept with respect to a linear system. Thus starting with two possible inputs $f_1(t)$ and $f_2(t)$ the corresponding outputs are $\mathfrak{T}\{f_1(t)\} = y_1(t)$ and $\mathfrak{T}\{f_2(t)\} = y_2(t)$ we suppose that $f_1(t) = f_2(t)$ for $t < T$. If the system is causal, then in accordance with the previous definition $y_1(t) = y_2(t)$ for $t < T$. However since the system is also linear it follows by superposition that the input $f_1(t) - f_2(t) = 0$ for $t < T$ results in the output $y_1(t) - y_2(t) = 0$ for $t < T$. In other words, for a linear system the definition of causality may be modified to read: a system is causal if for any nontrivial input which is zero for $t < T$ the output is also zero for $t < T$.

We now prove that a linear system represented by the linear operator (3.7) is causal if and only if

$$h(t, \tau) = 0 \text{ for } t < \tau. \qquad (3.11)$$

First suppose (3.11) holds. Then from (3.7) it follows that $y(t) = \int_{-\infty}^{t} f(\tau) h(t, \tau) d\tau$ from which we see directly that if $f(t) = 0$ for $t < T$ then also $y(t) = 0$ for $t < T$. Hence the system is causal. On the other hand, suppose the system is causal. Then (3.7) gives with $f(t) = 0$ for $t < T$ $y(t) = \int_{T}^{\infty} f(\tau) h(t, \tau) d\tau = 0$ for $t < T$. The last expression must be satisfied for arbitrary $f(\tau)$ for $\tau > T$. This is possible only if $h(t, \tau) = 0$ for $\tau > T$, but since $T > t$ this implies (3.11).

In summary the input/output relationship for a causal linear system is

$$y(t) = \int_{-\infty}^{t} f(\tau) h(t, \tau) d\tau. \qquad (3.12)$$

The relationship between the position of the impulse in time and the response of a causal linear system is shown in Fig. 3.2.

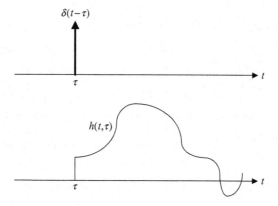

Figure 3.2: Impulse response of a causal linear system

Even though only causal systems are physically realizable, it is frequently convenient to use simplified system models that are not causal so that the general form of the input/output relationship (3.7) is still of value.

3.1.3   Step Response

Certain features of a linear system are better represented by its response to a unit step rather than to a unit impulse. We define the step response $a(t, \tau)$ as the response to $U(t - \tau)$, i.e.,

$$a(t, \tau) = \mathfrak{T}\{U(t - \tau)\}. \tag{3.13}$$

In view of (3.7) we get the relationship

$$
\begin{aligned}
a(t, \tau) &= \int_{-\infty}^{\infty} h(t, \tau')U(\tau' - \tau)\, d\tau' \\
&= \int_{\tau}^{\infty} h(t, \tau')d\tau'.
\end{aligned} \tag{3.14}
$$

Differentiation yields

$$\frac{da(t, \tau)}{d\tau} = -h(t, \tau). \tag{3.15}$$

Thus we can replace (3.7) by the equivalent form

$$y(t) = -\int_{-\infty}^{\infty} f(\tau)\frac{da(t, \tau)}{d\tau}d\tau. \tag{3.16}$$

When the system is causal $a(t, \tau) = 0$ for $t < \tau$ so that the preceding becomes

$$y(t) = -\int_{-\infty}^{t} f(\tau)\frac{da(t, \tau)}{d\tau}d\tau. \tag{3.17}$$

3.1.4   Stability

An important attribute of a system is that of stability, a term that generally refers to some measure of finiteness of the system response. Among the several possible definitions of stability we shall mention only one: the so-called bounded input/bounded output (BIBO) stability. Its formal definition is as follows. A system is said to be BIBO stable if for any bounded input $f(t)$, i.e., $|f(t)| < B = $ constant, there exists a constant I such that $|y(t)| < BI$. We shall prove the following [16]. A linear system is BIBO stable if and only if

$$\int_{-\infty}^{\infty} |h(t, \tau)|\, d\tau \le I < \infty \text{ for all } |t| < \infty. \tag{3.18}$$

To prove sufficiency we suppose that (3.18) holds. Then if $|f(t)| < B$ we have $|y(t)| = \left|\int_{-\infty}^{\infty} f(\tau)h(t, \tau)\, d\tau\right| \le B\int_{-\infty}^{\infty} |h(t, \tau)|\, d\tau \le BI$. To prove the necessity consider an input of the form

$$f(\xi, t) = \begin{cases} \frac{h(\xi, t)}{|h(\xi, t)|} & ; h(\xi, t) \ne 0, \\ 1 & ; h(\xi, t) = 0. \end{cases}$$

where ξ may be regarded as fixed parameter. Clearly $f(\xi, t)$ is a bounded function for all real ξ and the system output is

$$y(\xi, t) = \int_{-\infty}^{\infty} \frac{h(\xi, \tau)}{|h(\xi, \tau)|} h(t, \tau) \, d\tau.$$

Now at $t = \xi$ this becomes

$$y(\xi, \xi) = \int_{-\infty}^{\infty} \frac{h^2(\xi, \tau)}{|h(\xi, \tau)|} d\tau = \int_{-\infty}^{\infty} |h(\xi, \tau)| \, d\tau$$

so that for $y(\xi, \xi)$ to be bounded requires the boundedness of the last integral.

3.1.5   Time-invariance

A very important special class of linear systems are those termed time-invariant. We say a linear system is time-invariant if the input/output relationship

$$y(t) = \mathfrak{T}\{f(t)\} \tag{3.19a}$$

implies

$$y(t - T) = \mathfrak{T}\{f(t - T)\} \tag{3.19b}$$

for all real-time shifts T. In other words, the absolute time of initialization of the input has no effect on the relative relationship between the input and the output. For example, if we were to subject a linear time-invariant (LTI) system to inputs with different time delays, we would observe a sequence such as depicted in Fig. 3.3.

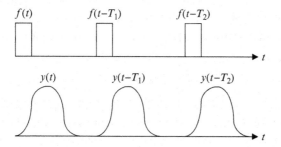

Figure 3.3: Input/output relationship for a time-invariant system

Physically the property of time-invariance means that the system input/output relationship is defined by a set of parameters that do not vary with time. For example, as we shall see in the sequel, if the input/output relationship is described by a differential equation, then time-invariance requires that the coefficients of this differential equation be constants (i.e., not functions of time).

To see what constraint is imposed by time-invariance on the system impulse response we return to (3.7) and compute the output due to a time shifted input

$$y(t) = \int_{-\infty}^{\infty} f(\tau - T) h(t, \tau) d\tau. \tag{3.20}$$

By time-invariance the same result must follow by simply displacing the time variable t in (3.7) by T. Thus

$$
\begin{aligned}
y(t - T) &= \int_{-\infty}^{\infty} f(\tau) h(t - T, \tau) d\tau \\
&= \int_{-\infty}^{\infty} f(\tau - T) h(t - T, \tau - T) d\tau.
\end{aligned}
\tag{3.21}
$$

Since (3.20) and (3.21) must be identical for arbitrary $f(\tau - T)$ we have $h(t, \tau) = h(t - T, \tau - T)$, independent of the choice of T. Clearly this implies that for a time-invariant system the impulse response is a function only of the difference between the time of the application of the impulse and the time of observation. Therefore we can write for the impulse response

$$h(t, \tau) = h(t - \tau). \tag{3.22}$$

The property of time-invariance allows us to specify the system impulse response in terms of only a single independent variable. This follows from the observation that $h(t - \tau)$ can be determined for any τ from the specification $h(t, 0) \equiv h(t)$. In virtue of (3.22) the input/output relationship (3.7) can now be rewritten as the convolution

$$y(t) = \int_{-\infty}^{\infty} f(\tau) h(t - \tau) d\tau \equiv \int_{-\infty}^{\infty} f(t - \tau) h(\tau) d\tau. \tag{3.23}$$

Also the condition for BIBO stability (3.18) can be restated in terms of only one variable, viz.,

$$\int_{-\infty}^{\infty} |h(t)| dt < \infty. \tag{3.24}$$

Note that the concepts of causality and time invariance are independent of each other. If a time-invariant system is also causal, the added constraint (3.22) in (3.11) requires that the impulse response vanishes for negative arguments, i.e.,

$$h(t) = 0 \text{ for } t < 0. \tag{3.25}$$

Observe that for a causal LTI system the convolution (3.23) may be modified to read

$$
\begin{aligned}
y(t) &= \int_{-\infty}^{t} f(\tau) h(t - \tau) d\tau \\
&= \int_{0}^{\infty} f(t - \tau) h(\tau) d\tau.
\end{aligned}
\tag{3.26}
$$

For an LTI system the step response (3.13) is likewise a function of $t - \tau$ and is completely defined in terms of the single variable t. Accordingly (3.14) now becomes

$$a(t) = \int_{-\infty}^{t} h(\tau)\, d\tau, \qquad (3.26^*)$$

which for a causal system reads

$$a(t) = \int_{0}^{t} h(\tau)\, d\tau. \qquad (3.27)$$

Thus unless $h(t)$ has a delta function at the origin, $a(0) = 0$. In view of (3.27) we have $da(t)/dt = h(t)$ and if the LTI system is causal we may replace (3.17) by

$$y(t) = \int_{-\infty}^{t} f(\tau)\, a'(t - \tau)\, d\tau, \qquad (3.28)$$

where the $'$ denotes the derivative with respect to the argument. When the input is zero for $t < 0$ $y(t)$ exists only for positive t in which case we may replace (3.28) by

$$y(t) = \int_{0}^{t} f(\tau)\, a'(t - \tau)\, d\tau. \qquad (3.29)$$

Formula (3.29) is referred to as the Duhamel integral and has played an important role in the older literature on linear systems (before the popularization of the delta function).

3.2 Characterizations in terms of Input/Output Relationships

3.2.1 LTI Systems

The exponential function plays a unique role in the analysis of LTI systems. The special feature that accounts for this prominence is the replication by an LTI system of an exponential input in the output. Even though this follows almost by inspection from the basic convolution representation (3.31) it is instructive to derive it by invoking explicitly the postulates of linearity and time-invariance.[1] Consider the input $e^{i\omega t}$ where ω is a real parameter. Then

$$\mathfrak{T}\left\{e^{i\omega t}\right\} = q(t, \omega)$$

and introducing a time delay τ we have by virtue of time-invariance

$$\mathfrak{T}\left\{e^{i\omega(t-\tau)}\right\} = q(t - \tau, \omega)$$

---

[1]This proof is believed to have been first presented by Norbert Wiener in "The Fourier Integral and Some of its Applications," Cambridge University Press, New York, 1933, where the LTI operator is referred to as "the operator of the closed cycle."

Furthermore linearity permits us to factor out $e^{-i\omega\tau}$ so that

$$\mathfrak{T}\left\{e^{i\omega t}\right\} = e^{i\omega\tau}q\left(t - \tau, \omega\right).$$

But τ is arbitrary so that in the last expression we may set it equal to t resulting in

$$\mathfrak{T}\left\{e^{i\omega t}\right\} = e^{i\omega t}q\left(0, \omega\right). \tag{3.30}$$

Equation (3.30) may be interpreted as an eigenvalue problem for the linear operator $\mathfrak{T}$ wherein $e^{i\omega t}$ is an eigenfunction and $q\left(0, \omega\right)$ an eigenvalue. In other words the exponential $e^{i\omega t}$ is an eigenfunction of every LTI system. Interpreting the real parameter ω as the frequency (3.30) states that an LTI system is incapable of generating frequencies at the output not present at the input. Clearly if the input is comprised of a sum of sinusoids say $\sum \alpha_n e^{i\omega_n t}$, then superposition gives

$$\mathfrak{T}\left\{\sum \alpha_n e^{i\omega_n t}\right\} = \sum e^{i\omega_n t}\alpha_n q\left(0, \omega_n\right). \tag{3.31}$$

Note that here and in (3.30) the sinusoidal inputs are presumed to be of infinite duration so that (3.30) and (3.31) are not valid for inputs initiated at some finite time in the past. Not all LTI systems can be characterized by a finite $q\left(0, \omega\right)$. As we shall see in the following for a finite exponential response to exist it is sufficient (but not necessary) that the system be BIBO stable. How is the constant $q\left(0, \omega\right)$ related to the system impulse response? The answer is provided at once by (3.23) with $f\left(t\right) = e^{i\omega t}$. After a simple change of variable we obtain

$$y\left(t\right) = e^{i\omega t}\int_{-\infty}^{\infty} h\left(\tau\right)e^{-i\omega\tau}d\tau = e^{i\omega t}H\left(\omega\right) \tag{3.32}$$

so that $q\left(0, \omega\right) = H\left(\omega\right)$ is identified as the Fourier Transform of the impulse response, or, using our compact notation

$$h\left(t\right) \overset{\mathcal{F}}{\Longleftrightarrow} H\left(\omega\right). \tag{3.33}$$

Hence a finite exponential response (3.30) will exist for any LTI system whose impulse response possesses a Fourier Transform. In virtue of (3.24) this is the case for all BIBO stable systems. On the other hand we know that absolute integrability is not a necessary requirement for the existence of the Fourier Transform. Thus even though the system with the impulse response

$$h\left(t\right) = \frac{\sin at}{\pi t}U\left(t\right) \tag{3.34}$$

is not BIBO stable its Fourier Transform exists. Indeed,

$$H\left(\omega\right) = \frac{1}{2}p_a\left(\omega\right) + i\frac{1}{2\pi}\ln\left|\frac{a - \omega}{a + \omega}\right|. \tag{3.35}$$

On the other hand the impulse response $h\left(t\right) = e^{\alpha t}U\left(t\right)$ with $\alpha > 0$ is neither BIBO stable nor does it possess a Fourier Transform.

We shall refer to $H(\omega)$ as the transfer function of the LTI system, a term justified by the frequency domain representation of the convolution (3.23):

$$Y(\omega) = F(\omega) H(\omega). \tag{3.36}$$

Let us now assume an input of the form

$$f(t) = e^{i\omega t} U(t - T),$$

i.e., an exponential that has been initialized at some finite $t = T$. Assuming the existence of the FT, the convolution integral for the response becomes

$$\begin{aligned}
y(t) &= \int_{-\infty}^{\infty} e^{i\omega\tau} U(\tau - T) h(t - \tau)\, d\tau \\
&= \int_{T}^{\infty} e^{i\omega\tau} h(t - \tau)\, d\tau \\
&= e^{i\omega t} \int_{-\infty}^{t-T} e^{-i\omega\tau} h(\tau)\, d\tau \\
&= e^{i\omega t} H(\omega) - e^{i\omega t} \int_{t-T}^{\infty} h(\tau) e^{-i\omega\tau}\, d\tau.
\end{aligned} \tag{3.37}$$

The last expression may be interpreted as follows. The first term on the right is the output to an exponential, initialized in the infinite past as in (3.32), while the second term may be considered a correction that accounts for the finite initialization time. Since this term vanishes as $t - T \longrightarrow \infty$ it is referred to as the transient response of the system. Thus after a sufficiently long time has elapsed the system output approaches $e^{i\omega t} H(\omega)$ which is defined as the steady state system response. Evidently if we know the transfer function then, by superposition, we can write down the steady response to any linear combination of sinusoids. For example, the steady state response to $A\cos(\omega_0 t - \theta)$ is

$$\Im\{A\cos(\omega_0 t - \theta)\} = A/2[H(\omega_0)\, e^{i(\omega_0 t - \theta)} + H(-\omega_0)\, e^{-i(\omega_0 t - \theta)}], \tag{3.38}$$

which is just another combination of sinusoids at the input frequency ω_0. Since we know $H(\omega)$, we can determine $h(t)$ from (3.33) so that the transient response could be determined using the second term on the right of (3.37), a calculation requiring, of course, rather more effort.

3.2.2 Time-varying Systems

Relationship between the Impulse Response and the Response to an Exponential Input

It is instructive to see how the preceding results are modified if the system is time-variant. Thus suppose we again attempt to characterize the system by means of the response to an exponential input. Such a response will, of course, no longer be representable by the product of a time-independent parameter with the exponential input but will in general be comprised of a time-varying amplitude and phase. Resorting again to the linear operator symbolism, the input/output relationship reads

$$\mathfrak{T}\left\{e^{i\omega t}\right\} = A\left(\omega, t\right) e^{i\theta(\omega, t)}, \tag{3.39}$$

wherein $A\left(\omega, t\right)$ and $\theta\left(\omega, t\right)$ are, respectively, the modulus and argument of a complex number and therefore real functions. As in case of an LTI system it is sufficient for the existence of such a representation to assume that the system is BIBO stable. What can we say about the impulse response of such a system? It is not hard to see that just like for an LTI system the response to an exponential defines the system impulse uniquely. Thus, since by definition the impulse response is a linear operation on a time-delayed delta function, we write $\mathfrak{T}\left\{\delta\left(t - \tau\right)\right\} = h\left(t, \tau\right)$ and represent this delta function as a superposition of exponentials (see 2.119 in 2.2). We linear operator symbol into the integrand. This results in

$$\mathfrak{T}\left\{\delta\left(t - \tau\right)\right\} = \mathfrak{T}\left\{\frac{1}{2\pi}\int_{-\infty}^{\infty} e^{i\omega(t-\tau)} d\omega\right\} = \frac{1}{2\pi}\int_{-\infty}^{\infty}\mathfrak{T}\left\{e^{i\omega(t-\tau)}\right\} d\omega$$

$$= \frac{1}{2\pi}\int_{-\infty}^{\infty}\mathfrak{T}\left\{e^{i\omega t}\right\} e^{-i\omega\tau} d\omega.$$

Finally substituting from (3.39) into the last integrand we obtain

$$h\left(t, \tau\right) = \frac{1}{2\pi}\int_{-\infty}^{\infty} A\left(\omega, t\right) e^{i\theta(\omega, t)} e^{-i\omega\tau} d\omega. \tag{3.40}$$

If the system is LTI, then $A\left(\omega, t\right) e^{i\theta(\omega, t)} = A\left(\omega\right) e^{i\theta(\omega)} e^{i\omega t}$ with $A\left(\omega\right) e^{i\theta(\omega)} = H\left(\omega\right)$ independent of t, so that, as expected, (3.40) gives $h\left(t, \tau\right) = h\left(t - \tau\right)$. Thus, by analogy with an LTI system, we may regard the quantity

$$A\left(\omega, t\right) e^{i[\theta(\omega, t) - \omega t]} \tag{3.41}$$

as the transfer function of a time-varying system, provided we take note of the fact that unlike in the LTI system this transfer function is *not the FT of the impulse response*.

As an example, suppose we want to determine the impulse response of the linear system defined by

$$\mathfrak{T}\left\{e^{i\omega t}\right\} = \frac{1}{(\alpha t)^2 + i\omega} e^{i\omega t}, \tag{3.42}$$

where α is a constant. Clearly the system is not time-invariant because the output is not of the form $H(\omega)e^{i\omega t}$. By superposition the impulse response is

$$h\left(t, \tau\right) = \frac{1}{2\pi}\int_{-\infty}^{\infty}\frac{e^{i\omega(t-\tau)}}{(\alpha t)^2 + i\omega} d\omega = e^{-(\alpha t)^2(t-\tau)} U\left(t - \tau\right). \tag{3.43}$$

From this we note that the system is causal, and, in accordance with (3.18), BIBO stable for $\alpha \neq 0$. On the other hand, if the exponential $e^{i\omega t}$ is initialized at $t = 0$ we get

$$\int_0^t e^{-(\alpha t)^2(t-\tau)} e^{i\omega \tau} d\tau = -\frac{e^{-(\alpha t)^2 t}}{(\alpha t)^2 + i\omega} + \frac{1}{(\alpha t)^2 + i\omega} e^{i\omega t}. \qquad (3.44)$$

The first term on the right is the transient response since it vanishes as $t \to \infty$ while the second term represents the steady state response. As expected, it is identical with the right side of (3.42). Note that this response is not purely sinusoidal but rather comprised of sinusoids with slowly varying amplitude and phase factors. Its deviation from the pure single spectral line of the input is best appreciated by viewing its FT which is easily found to be

$$Y(\omega) = -\frac{\pi}{K_1} e^{-|1-\omega/\omega_0|K_2} e^{i\pi/4} e^{-i|1-\omega/\omega_0|K_2}. \qquad (3.45)$$

We have relabeled the fixed input frequency in (3.42) by ω_0 and introduced the dimensionless parameters $K_1 = \alpha \omega_0^{1/2}$ and $K_2 = \alpha^{-1} \omega_0^{3/2}/2$. A plot of magnitude of $Y(\omega)$ as a function of ω/ω_0 is shown in Fig. 3.4.

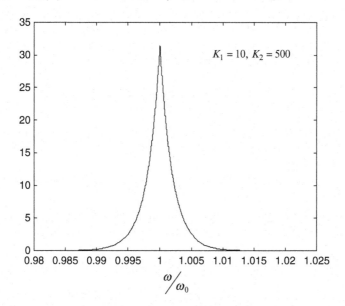

Figure 3.4: Plot of $|Y(\omega)|$ in Eq. (3.45)

Very often instead of the impulse response the transfer function (3.41) of a system constitutes the given data (either experimentally or through an analytical model). It is then more efficient to use this transfer function directly rather than going through the intermediate step of finding the system impulse response. We get the required formula by putting (3.40) into (3.7) in Sect. 3.1 and obtain

$$
y(t) = \frac{1}{2\pi} \int_{-\infty}^{\infty} A(\omega, t) e^{i\theta(\omega, t)} \left\{ \int_{-\infty}^{\infty} f(\tau) e^{-i\omega\tau} d\tau \right\} d\omega
$$

$$
= \frac{1}{2\pi} \int_{-\infty}^{\infty} A(\eta, t) e^{i\theta(\eta, t)} F(\eta) d\eta, \tag{3.46}
$$

where $F(\eta)$ is the FT of $f(t)$. To emphasize that the indicated integration is not being carried out directly over the frequency spectrum of the system output we have replaced the integration variable ω with η. In fact to find the spectrum of the output we must still compute the FT of the right side of (3.46). The result may be written as follows

$$
Y(\omega) = \frac{1}{2\pi} \int_{-\infty}^{\infty} H(\eta, \omega) F(\eta) d\eta, \tag{3.47}
$$

where we have introduced the transform pair

$$
A(\eta, t) e^{i\theta(\eta, t)} \overset{\mathcal{F}}{\Longleftrightarrow} H(\eta, \omega). \tag{3.48}
$$

Equation (3.47) highlights the fact that a time-varying system does not merely deemphasize(attenuate) and (or) enhance the frequency components already present in the input signal (as in an LTI system) but can generate frequency components not present in the input as we have already seen in the example in Fig. 3.4. Evidently for an LTI system

$$
H(\eta, \omega) = 2\pi H(\omega) \delta(\omega - \eta), \tag{3.49}
$$

where $H(\omega) = A(\omega) e^{i\theta(\omega)}$ so that we recover (3.32).

There are also cases where the transfer function (3.41) does not provide a useful characterization of a time-varying system. For example, for a system with the impulse response

$$
h(t, \tau) = \cos[\alpha(T - t)\tau] U(t - \tau) \tag{3.50}
$$

the transfer function does not exist. Nevertheless this impulse response is a useful approximation to a pulse compression circuit. Taking the rectangular pulse $p_{T/2}(t)$ as the input we find

$$
y(t) = \begin{cases} 0 \; ; \; t \leq -T/2, \\ \frac{\sin[\alpha(t-T)t]}{\alpha(T-t)} \; ; \; -T/2 \leq t \leq T/2, \\ \frac{2\sin\left[\frac{\alpha}{2}(T-t)T\right]}{\alpha(T-t)} \; ; \; t \geq T/2. \end{cases} \tag{3.51}
$$

A plot of (3.51) is shown in Fig. 3.5.

The rectangular pulse of duration T is compressed to a "pulse" of nominal duration $4\pi/\alpha T$. In view of the functional form of the output we see that this compression is accomplished by the system by effectively performing an FT on the input.

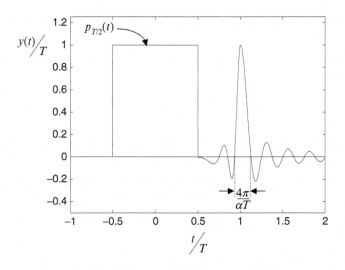

Figure 3.5: Pulse compression

Communications Channel Modeling

Mathematical representations of time-varying linear system play a key role in the modeling of so-called fading communications channels. This includes mobile radio communications, where signal fading is engendered by the motion of receivers and (or) transmitters relative to fixed scatterers and reflectors, and communications wherein the propagation paths between receivers and transmitters traverse intrinsically time-varying media as, e.g., in case of the ionosphere or the troposphere. In these applications it turns out to be more convenient to use a somewhat different form of the system response function than presented in the preceding. In the following we present a brief account of this alternative formulation.

To begin with, we introduce a new variable $\hat{\tau} = t - \tau$ and define a modified impulse response function $c(\hat{\tau}; t)$ by

$$c(\hat{\tau}; t) = h(t, t - \hat{\tau}). \tag{3.52}$$

Since t is the time of observation and τ the time at which the impulsive input is applied, $\hat{\tau}$ is just the time interval between the observation and the application of the impulse. Clearly if the system is LTI, then $c(\hat{\tau}; t)$ is only a function of $\hat{\tau}$ so that any explicit dependence of $c(\hat{\tau}; t)$ on the second variable (i.e., the observation time t) is a direct measure of the extent to which the system characteristics deviate from an LTI model. In terms of the modified impulse response the system output in (3.7) becomes

$$y(t) = \int_{-\infty}^{\infty} c(\hat{\tau}; t) f(t - \hat{\tau}) d\hat{\tau}. \tag{3.53}$$

Note also that from the definition of $\hat{\tau}$ it follows that for a causal system

$$c(\hat{\tau}; t) = 0 \; ; \; \hat{\tau} < 0 \qquad (3.54)$$

in which case the expression for the response is

$$y(t) = \int_0^\infty c(\hat{\tau}; t) f(t - \hat{\tau}) \, d\hat{\tau}. \qquad (3.55)$$

The dependence of the modified impulse response on the first variable is a measure of the frequency dispersive property of the channel whereas the dependence on the second variable is a measure of the Doppler spread. For example, the modified impulse response for a mobile link with terrain reflections might be

$$c(\hat{\tau}; t) = h_0(\hat{\tau}) + \beta(t) \, \delta[\hat{\tau} - \hat{\tau}_0(t)], \qquad (3.56)$$

where the first term on the right is the direct path contribution and the second term represents the time-varying multipath signal component resulting, e.g., from terrain reflections. Inserting this into (3.55) we see that the response resulting from the first term is that of an LTI system with some fixed frequency dispersive properties (depending on the nature of the direct path). The contribution from the second term is $\beta(t) f[t - \hat{\tau}_0(t)]$ which is a combination of an amplitude fluctuation (amplitude fading) and a variable time delay resulting in a Doppler spread of the signal.

By evaluating the FT with respect to the time delay variable and the time of observation the effects of dispersion and Doppler spread can be viewed in their respective frequency domains. To represent the spectrum of the dispersive effects we employ (3.40)

$$c(\hat{\tau}; t) = \frac{1}{2\pi} \int_{-\infty}^\infty A(\omega, t) \, e^{i\theta(\omega, t) - i\omega t} e^{i\omega\hat{\tau}} \, d\omega, \qquad (3.57)$$

which yields the transform pair

$$c(\hat{\tau}; t) \underset{(\hat{\tau}, \omega)}{\overset{\mathfrak{F}}{\longleftrightarrow}} A(\omega, t) \exp\{i[\theta(\omega, t) - \omega t]\}. \qquad (3.58)$$

The transform of (3.57) with respect to t is

$$c(\hat{\tau}; t) \underset{(t, \eta)}{\overset{\mathfrak{F}}{\longleftrightarrow}} \frac{1}{2\pi} \int_{-\infty}^\infty C(\omega, \eta) \exp(i\omega\hat{\tau}) \, d\omega, \qquad (3.59)$$

where we have used (3.48) with an interchange of the variables ω and η so that

$$C(\omega, \eta) = H(\omega, \omega + \eta) = \int_{-\infty}^\infty A(\omega, t) \, e^{i\theta(\omega, t)} e^{-i\omega t} e^{-i\eta t} \, dt. \qquad (3.60)$$

Taking account of (3.58) we can rewrite (3.60) as a 2-D FT:

$$C(\omega, \eta) = \int_{-\infty}^\infty \int_{-\infty}^\infty c(\hat{\tau}; t) \, e^{-i\omega\hat{\tau} - i\eta t} \, d\hat{\tau} \, dt. \qquad (3.61)$$

As an example consider a moving source of radiation with a fixed wave propagation speed v excited by the exponential input signal $e^{i\omega t}$. The linear transformation of this exponential into the RF signal at the antenna terminals of a fixed receiver located at range $r(t)$ can be represented as

$$\mathfrak{T}\left\{e^{i\omega t}\right\} = \frac{H_0\left(\omega\right) G\left(\omega, t\right)}{r\left(t\right)} e^{-i\frac{\omega r(t)}{v}} e^{i\omega t}, \tag{3.62}$$

where $H_0\left(\omega\right)$ is the transfer function of an equivalent stationary source (which for our purposes may be considered an LTI system) and the factor $G\left(\omega, t\right)$ is proportional to the square root of the product of the transmitter and receiver antenna gains at the instantaneous position of the trajectory defined by $r\left(t\right)$. As in (3.43) we can find the impulse response by superposition with the result

$$c\left(\hat{\tau}; t\right) = \frac{1}{2\pi} \int_{-\infty}^{\infty} \frac{H_0\left(\omega\right) G\left(\omega, t\right)}{r\left(t\right)} e^{i\omega\left(\hat{\tau} - \frac{r(t)}{v}\right)} d\omega. \tag{3.63}$$

Generally the variation of the gain with frequency is negligible within the signal information band so that we may set $G\left(\omega, t\right) \approx G\left(t\right)$. In that case (3.63) becomes

$$c\left(\hat{\tau}; t\right) = \left(\frac{G\left(t\right)}{r\left(t\right)}\right) \frac{1}{2\pi} \int_{-\infty}^{\infty} H_0\left(\omega\right) e^{i\omega\left(\hat{\tau} - \frac{r(t)}{v}\right)} d\omega \tag{3.64}$$

and the 2-D spectrum (3.61) is

$$C\left(\omega, \eta\right) = H_0\left(\omega\right) \int_{-\infty}^{\infty} \frac{G\left(t\right)}{r\left(t\right)} e^{-i\frac{\omega r(t)}{v}} e^{-i\eta t} dt. \tag{3.65}$$

The integral gives the spectrum of the Doppler spread. For a stationary source it reduces to $2\pi H_0\left(\omega\right) \delta\left(\eta\right) \left[G/r\right] e^{-i\frac{\omega r}{v}}$

3.3 Linear Systems Characterized by Ordinary Differential Equations

3.3.1 First-Order Differential Equations

In the preceding discussion we dealt only with the input/output characterization of a linear system. Such a description does not reveal anything about the physical processes that give rise to the system impulse response. The relationship between the impulse response and the dynamical variables governing the system must be established with the aid of the underlying differential/integral equations. For example, the following first-order linear differential equation

$$\frac{dy\left(t\right)}{dt} + a\left(t\right) y\left(t\right) = f\left(t\right), \tag{3.66}$$

where $a\left(t\right)$ is a given function of time, characterizes a linear system with input $f\left(t\right)$ and output $y\left(t\right)$. We can represent this system by the feedback network in

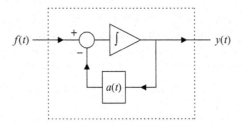

Figure 3.6: Feeedback representation of first-order system

Fig. 3.6 consisting of an integrator, a time-varying transducer and a differencing network.

Although much of the theory can be developed for quite general $a(t)$ for our purposes it will be sufficient to assume that is a smooth bounded function of time. Using our previous notation for a linear operator, $y(t) = \mathfrak{T}\{f(t)\}$. Moreover if the input is prescribed for $-\infty < t < \infty$ the output can be represented in terms of the impulse response as in (3.7). Let us determine the impulse response $h(t,\tau)$ of this system. Formally we are required to solve (3.66) when the input equals $\delta(t-\tau)$. Thus

$$\frac{dh(t,\tau)}{dt} + a(t)h(t,\tau) = \delta(t-\tau). \tag{3.67}$$

This first-order equation must be solved subject to an initial condition. For example, if we impose the requirements that $h(t,\tau)$ be causal, then the initial condition is given by (3.11) in Sect. 3.1. When $t \neq \tau$ the right side of (3.67) is zero so that $h(t,\tau)$ satisfies the homogeneous equation whose solution is

$$h(t,\tau) = C(\tau)\exp - \int_0^t a(t')\,dt', \tag{3.68}$$

where $C(\tau)$ is an integration constant that may depend on τ. To find this constant we first integrate both sides of (3.67) with respect to t between the limits $t = \tau - \epsilon \equiv \tau^-$ and $t = \tau + \epsilon \equiv \tau^+$, where ϵ is an arbitrarily small positive quantity. Thus

$$\int_{\tau^-}^{\tau^+} \frac{dh(t,\tau)}{dt}dt + \int_{\tau^-}^{\tau^+} a(t)h(t,\tau)\,dt = \int_{\tau^-}^{\tau^+} \delta(t-\tau)\,dt = 1 \tag{3.69}$$

and integrating the first member yields

$$h(\tau^+,\tau) - h(\tau^-,\tau) + \int_{\tau^-}^{\tau^+} a(t)h(t,\tau)\,dt = 1. \tag{3.70}$$

In virtue of the causality constraint $h(\tau^-,\tau) = 0$, and since $a(t)$ is bounded we have in the limit of vanishingly small ϵ

$$\lim_{\epsilon \to 0} \int_{\tau^-}^{\tau^+} a(t)h(t,\tau)\,dt = 0.$$

Consequently (3.70) yields

$$h(\tau,\tau) = 1. \tag{3.71}$$

Setting $t = \tau$ in (3.68) and using (3.71) we find the constant $C(\tau) = \exp \int_0^\tau a(t')\,dt'$ and taking account of causality we obtain

$$h(t,\tau) = \begin{cases} \exp - \int_\tau^t a(t')\,dt' & ; \ t \geq \tau, \\ 0 & ; \ t < \tau. \end{cases} \tag{3.72}$$

Alternatively and more compactly the preceding may be written

$$h(t,\tau) = \left(\exp - \int_\tau^t a(t')\,dt' \right) U(t - \tau). \tag{3.73}$$

Once the impulse response has been determined the response to an arbitrary input $f(t)$ is given in terms of the superposition integral (3.12 in Sect. 3.1). From this superposition integral it would appear that to determine the response at any time t requires the knowledge of the input over the entire infinite past up to the present instant t. This is more information than would be normally available for practically the input $f(t)$ would be known only for t at and past some time t_0. How does one determine the output under these circumstances? It turns out that because we are dealing here with a causal system the knowledge of the input prior to t_0 is actually superfluous provided we know the value of the output (initial condition) at $t = t_0$. To show this explicitly let us select a point $t_0, y(t_0)$ and rewrite the system response (3.12 in Sect. 3.1) as a sum of two contributions one dependent on the input prior to t_0 and the other dependent on the input for $t > t_0$. Thus we have for $t \geq t_0$

$$y(t) = \int_{-\infty}^{t_0} f(\tau) h(t,\tau)\,d\tau + \int_{t_0}^t f(\tau) h(t,\tau)\,d\tau. \tag{3.74}$$

Setting $t = t_0$ this gives

$$y(t_0) = \int_{-\infty}^{t_0} f(\tau) h(t_0,\tau)\,d\tau \tag{3.75}$$

so that the initial condition, i.e., $y(t_0)$, is completely determined by the summation of the input over the infinite past up to $t = t_0$. Defining $\hat{y}(t) = \exp - \int_0^t a(t')\,dt'$ we obtain the identity $\exp - \int_{t_0}^t a(t')\,dt' = \hat{y}(t)[\hat{y}(t_0)]^{-1} = h(t,t_0)$ so that $h(t,\tau) = h(t,t_0) h(t_0,\tau)$. Inserting this in the first integral in (3.74) and using (3.75) we get

$$y(t) = h(t,t_0) y(t_0) + \int_{t_0}^t f(\tau) h(t,\tau)\,d\tau \ ; t \geq t_0. \tag{3.76}$$

Note that

$$y_{0i}(t) \equiv h(t,t_0) y(t_0) \tag{3.77}$$

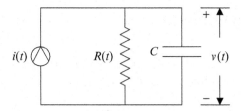

Figure 3.7: Time-varying resistor circuit

satisfies the homogeneous differential equation and when evaluated at $t = t_0$
yields $y(t_0)$. In virtue of (3.75) the latter is determined by the portion of the
input signal defined for $-\infty < t < t_0$. Thus for $t \geq t_0$ the net effect on the output
of this portion of the input is captured by the solution of the homogeneous
equation subject to the initial condition $y_{0i}(t_0) = y(t_0)$. Because $y_{0i}(t)$ is
obtained with $f(t) = 0$ for $t \geq t_0$ it is frequently referred to as the zero input
response.[2] The output engendered by segment of $f(t)$ for $t \geq t_0$ is

$$y_{0s}(t) \equiv \int_{t_0}^{t} f(\tau) h(t, \tau) d\tau \tag{3.78}$$

and corresponds to the output with zero initial condition at $t = t_0$. It is usually
referred to as the zero state response. For an LTI system $a(t) = a = $ constant so
that (3.73) takes on the form $h(t - \tau) = e^{-a(t-\tau)} U(t - \tau)$ which we may replace
it with the simpler statement

$$h(t) = e^{-at} U(t). \tag{3.79}$$

The representation of the solution in terms of a zero state and a zero input
response in (3.76) now reads

$$y(t) = e^{-a(t-t_0)} y(t_0) + \int_{t_0}^{t} f(\tau) e^{-a(t-\tau)} d\tau; \quad t \geq t_0. \tag{3.80}$$

Example. As an illustration of a time-varying first-order linear system consider the circuit in Fig. 3.7 where input is the current $i(t)$ supplied by an ideal
current source and the output is the voltage $v(t)$ taken across a fixed capacitor
C in parallel with the time-varying resistor $R(t)$.

Using Kirchhoff's circuit laws we get

$$\frac{dv}{dt} + \frac{1}{R(t)C} v = \frac{1}{C} i(t). \tag{3.81}$$

---

[2]Note that the zero input response (3.77) can also be obtained by exciting the system
with $f(t) = y(t_0)\delta(t - t_0)$. This is just a particular illustration of the general equivalence
principle applicable to linear differential equations permitting the replacement of a homogeneous system with specified nonzero initial conditions by an inhomogeneous system with zero
initial conditions using as inputs singularity functions whose coefficients incorporate the initial
conditions of the original problem.

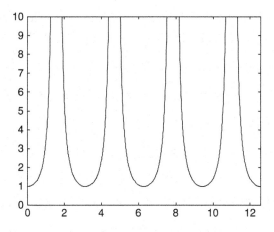

Figure 3.8: Plot of $\alpha CR(t)$ vs $\omega_0 t$

We suppose that the resistor varies sinusoidally as follows:

$$1/R(t) \equiv G(t) = \alpha C \cos^2 \omega_0 t = \frac{\alpha C}{2}[1 + \cos 2\omega_0 t], \qquad (3.82)$$

where α is a positive constant. Figure 3.8 shows a plot of $\alpha CR(t)$ as a function of $\omega_0 t$.

We see that the resistor acts approximately like an on–off switch, presenting an open circuit at $t = (2n + 1)\pi/2\omega_0$ and a resistance of $1/\alpha C$ at $t = n\pi/\omega_0, n = 0, \pm 1, \pm 2, \dots$. Physically, a time-varying resistance of this sort can be approximated by applying a sinusoidal excitation (local oscillator) to a circuit comprised of a suitable nonlinear device (e.g., a diode). If the magnitude of the input current is sufficiently small, the relationship between $v(t)$ and $i(t)$ may be modeled as a linear system which is generally referred to as small signal analysis of RF mixers or frequency converters. For example, in the context of a heterodyne receiver we may consider the current $i(t)$ as the information bearing RF input signal in which case the voltage $v(t)$ represents the desired downconverted (baseband) signal, and (since the circuit in Fig. 3.7 includes no band-pass filter) spurious responses in form of beats and their harmonics.

To maintain consistency with the parameters introduced in (3.66) we set $f(t) = i(t)/C$. Upon substituting $a(t) = 1/R(t)C = \alpha \cos^2 \omega_0 t$ in (3.73) we obtain the impulse response

$$h(t, \tau) = e^{-\frac{\alpha}{2}(t-\tau)} e^{-\frac{\alpha}{4\omega_0}[\sin 2\omega_0 t - \sin 2\omega_0 \tau]} U(t - \tau). \qquad (3.83)$$

If we specify the initial conditions at $t = 0$, the complete response for $t \geq 0$ reads

$$v(t) = v(0)e^{-\frac{\alpha}{2}t} e^{-\frac{\alpha}{4\omega_0} \sin 2\omega_0 t}$$

$$+ \frac{1}{C} \int_0^t e^{-\frac{\alpha}{2}(t-\tau)} e^{-\frac{\alpha}{4\omega_0}[\sin 2\omega_0 t - \sin 2\omega_0 \tau]} i(\tau) \, d\tau, \qquad (3.84)$$

wherein, in accordance with our previous designation, the first term on the right is the zero input response and the second term the zero state response. The numerical evaluation of the integral in (3.84) is facilitated by the introduction of the Fourier series representation

$$e^{\pm\frac{\alpha}{4\omega_0}\sin 2\omega_0 t} = \sum_{n=-\infty}^{\infty} q_n^{\pm} e^{i2n\omega_0 t} \tag{3.85}$$

with[3]

$$q_n^{\pm} = \frac{\omega_0}{\pi} \int_{-\frac{\pi}{\omega_0}}^{\frac{\pi}{\omega_0}} e^{\pm\frac{\alpha}{4\omega_0}\sin 2\omega_0 t} e^{-i2n\omega_0 t} dt. \tag{3.86}$$

Let us now consider a sinusoidal current of the form

$$i(t) = A\cos(\omega_1 t + \varphi), \tag{3.87}$$

which may be taken as the carrier of an RF signal (φ is a fixed phase reference). Substituting this in (3.84) and carrying out the simple integration term-by-term and letting $\alpha t \sim \infty$ we obtain the steady state response

$$v(t) \sim \frac{A}{2C} \sum_{\ell=-\infty}^{\infty} \left\{ \beta_\ell^+ e^{i(2\ell\omega_0+\omega_1)t} + \beta_\ell^- e^{i(2\ell\omega_0-\omega_1)t} \right\}, \tag{3.88}$$

where

$$\beta_\ell^+ = e^{i\varphi} \sum_{n=-\infty}^{\infty} \frac{q_{\ell-n}^- q_n^+}{2n\omega_0 + \omega_1 + \alpha/2}, \tag{3.89a}$$

$$\beta_\ell^- = e^{-i\varphi} \sum_{n=-\infty}^{\infty} \frac{q_{\ell-n}^- q_n^+}{2n\omega_0 - \omega_1 + \alpha/2}. \tag{3.89b}$$

These coefficients give the amplitude and phase of the sinusoidal outputs at various frequencies. For example, for the fundamental (input) frequency we have

$$v_0(t) = \frac{A}{2C}(\beta_0^+ e^{i\omega_1 t} + \beta_0^- e^{-i\omega_1 t}). \tag{3.90}$$

The amplitude and phase of the upper sideband (sum of the carrier and the LO frequencies) is

$$v_1^+(t) = \frac{A}{2C}(\beta_1^+ e^{i(2\omega_0+\omega_1)t} + \beta_{-1}^- e^{-i(2\omega_0+\omega_1)t}) \tag{3.91}$$

and similarly for the lower sideband (difference of the carrier and the LO frequencies)

$$v_1^-(t) = \frac{A}{2C}(\beta_{-1}^+ e^{i(-2\omega_0+\omega_1)t} + \beta_1^- e^{i(2\omega_0-\omega_1)t}). \tag{3.92}$$

---

[3]These Fourier coefficients can be expressed in terms of the modified Bessel functions I_n as follows: $q_n^{\pm} = (\pm i)^n I_n(\frac{\alpha}{4\omega_0})$.

An important design parameter of a frequency converter is its conversion efficiency. In the present model it is numerically equal to the squared magnitude of the coefficients $\beta_\ell^\pm$. For example, the conversion efficiency for the lower sideband is $\left|\beta_{-1}^+\right|^2 = \left|\beta_1^-\right|^2$.

In a practical circuit the converter output frequency (e.g., the lower sideband) would be selected by means of a band-pass filter so that our analysis is only approximate. The more accurate approach requires the inclusion of the effects of the reactances introduced by the filter which in turn leads to a differential equation of order higher than the first.

3.3.2 Second-Order Differential Equations

Time-varying Systems

Next we consider a linear system described by the second-order differential equation

$$\frac{d^2y\,(t)}{dt^2} + a(t)\frac{dy\,(t)}{dt} + b(t)y\,(t) = f(t), \tag{3.93}$$

where we suppose that the coefficients $a(t)$ and $b(t)$ are continuous functions of t. As in the first-order system $f(t)$ is the input and $y\,(t)$ the output. We can again represent the system by a feedback network. To account for the additional derivative requires two loops, as shown in Fig. 3.9.

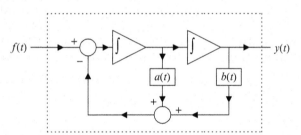

Figure 3.9: Feedback representation of second-order system

Let us first consider the homogeneous equation.

$$\frac{d^2y\,(t)}{dt^2} + a(t)\frac{dy\,(t)}{dt} + b(t)y\,(t) = 0. \tag{3.94}$$

It will be demonstrated in (3.98), that (3.95) has two linearly independent solutions which we presently denote by $y_1(t)$ and $y_2\,(t)$. A linear superposition of these can be used to construct the zero input response for $t \geq t_0$ which we define as in the case of the first-order system. However now two constants are required instead of one for its specification. Thus

$$y_{0i}(t) = \alpha\,(t_0)\,y_1(t) + \beta\,(t_0)\,y_2(t); \quad t \geq t_0, \tag{3.95}$$

where the two constants $\alpha(t_0)$ and $\beta(t_0)$ are to be determined from the initial conditions $y(t_0)$ and $y'(t_0)$. We find

$$\alpha(t_0) = \frac{y_2'(t_0)y(t_0) - y_2(t_0)y'(t_0)}{W[y_1(t_0), y_2(t_0)]} \tag{3.96a}$$

$$\beta(t_0) = \frac{-y_1'(t_0)y(t_0) + y_1(t_0)y'(t_0)}{W[y_1(t_0), y_2(t_0)]}, \tag{3.96b}$$

where $W[y_1(t_0), y_2(t_0)] \equiv W(t_0)$ is the Wronskian (see 1.35 in Sect. 1.2) evaluated at $t = t_0$. It is not hard to show that the Wronskian $W(t)$ satisfies the first-order differential equation

$$\frac{dW(t)}{dt} + a(t)W(t) = 0. \tag{3.97}$$

Hence it can be determined to within a multiplicative constant from the coefficient $a(t)$, viz.,

$$W(t) = W(t_0)\, e^{-\int_{t_0}^{t} a(\tau)d\tau}. \tag{3.98}$$

As long as the coefficient $a(t)$ is bounded so that the integral exists for all finite values of t (3.98) guarantees that if the Wronskian is not zero at one particular instant of time it will not be zero for all finite time. This means that if the two solutions of the homogeneous equation can be shown to be linearly independent at some fixed time $t = t_0$ they must be linearly independent for arbitrary finite t. Also, from the definition of $W(t)$ it follows that given the Wronskian and one solution of (3.94), say $y_1(t)$, of the homogeneous equation (3.94) we can find a second solution, linearly independent from the first, by solving the first-order differential equation $y_2'(t) - [y_1'(t)/y_1(t)]y_2(t) = W(t)/y_1(t)$ for $y_2(t)$. We get

$$y_2(t) = \frac{y_1(t)}{y_1(t_0)}y_2(t_0) + y_1(t)\int_{t_0}^{t} \frac{W(\tau)}{y_1^2(\tau)}d\tau. \tag{3.99}$$

As in the case of the first-order system we can construct the complete solution to (3.93) in terms of the impulse response. Presently we have

$$\frac{d^2 h(t,\tau)}{dt^2} + a(t)\frac{dh(t,\tau)}{dt} + b(t)h(t,\tau) = \delta(t-\tau), \tag{3.100}$$

which we solve subject to the causality constraint $h(t,\tau) = 0$; $t < \tau$. Now Eq. (3.100) is presumed to be valid for $|t| < \infty$ and, in particular, at $t = \tau$. Therefore the delta function source appearing on the right side must be balanced by a delta function on the left side. Because the derivative of a delta function is a singularity function of a higher order (doublet), to achieve a balance, the delta function on the left side can be contained only in the second derivative. It follows then that the first derivative of $h(t,\tau)$ has a step discontinuity at $t = \tau$ so that $h(t,\tau)$ is a continuous function of t at $t = \tau$. Coupled with the causality constraint this requires that

$$h(t,\tau) = 0 \; ; t \le \tau. \tag{3.101}$$

Because $h(t, \tau)$ vanishes identically for $t < \tau$ we also have

$$\frac{dh(t, \tau)}{dt} = 0 \quad ; t < \tau. \tag{3.102}$$

For $t \neq \tau$ the impulse response satisfies the homogeneous equation. Therefore we can represent it as a linear combination of $y_1(t)$ and $y_2(t)$. Since it is also continuous at $t = \tau$ the representation valid for all time reads

$$h(t, \tau) = \begin{cases} A \, y_1(t) + B \, y_2(t) & ; t \geq \tau, \\ 0 & ; t \leq \tau. \end{cases} \tag{3.103}$$

The coefficients A and B are functions of τ. One equation follows immediately from (3.101), for at $t = \tau$ we must have

$$A \, y_1(\tau) + B \, y_2(\tau) = 0. \tag{3.104}$$

The second equation follows from an integration of both sides of (3.100) between the limits $t = \tau + \epsilon \equiv \tau^+$ and $t = \tau - \epsilon \equiv \tau^-$ as in (3.69). We get

$$\int_{\tau^-}^{\tau^+} \frac{d^2 h(t, \tau)}{dt^2} dt + \int_{\tau^-}^{\tau^+} a(t) \frac{dh(t, \tau)}{dt} dt + \int_{\tau^-}^{\tau^+} b(t) h(t, \tau) \, dt = \int_{\tau^-}^{\tau^+} \delta(t - \tau) \, dt = 1. \tag{3.105}$$

Because $h(t, \tau)$ and its derivative are integrable over any finite interval we have

$$\lim_{\epsilon \to 0} \int_{\tau^-}^{\tau^+} a(t) \frac{dh(t, \tau)}{dt} dt \quad \to \quad 0$$

$$\lim_{\epsilon \to 0} \int_{\tau^-}^{\tau^+} b(t) h(t, \tau) \, dt \quad \to \quad 0$$

and taking account of (3.102) the integration of the first member on the left of (3.105) is

$$\lim_{\epsilon \to 0} \int_{\tau^-}^{\tau^+} \frac{d^2 h(t, \tau)}{dt^2} dt = \frac{dh(t, \tau)}{dt}\Big|_{t=\tau} = 1. \tag{3.106}$$

As a result the second relationship between A and B is

$$A \, y_1'(\tau) + B \, y_2'(\tau) = 1. \tag{3.107}$$

Solving (3.104) and (3.107) for A and B and substituting in (3.103) we obtain

$$h(t, \tau) = \frac{y_1(\tau) y_2(t) - y_2(\tau) y_1(t)}{W[y_1(\tau), y_2(\tau)]} U(t - \tau). \tag{3.108}$$

In parallel with (3.76) the complete solution for $t \geq t_0$ comprising the zero input as well as the zero state response is

$$y(t) = \alpha(t_0) y_1(t) + \beta(t_0) y_2(t) + \int_{t_0}^{t} h(t, \tau) f(\tau) \, d\tau, \tag{3.109}$$

where $\alpha(t_0)$ and $\beta(t_0)$ are given by (3.96).

Example. With the choice of the coefficients $a(t) = (t^2 + 4t + 2)(t^2 + 3t + 2)^{-1}$ and $b(t) = t(t^2 + 3t + 2)^{-1}$ we can validate by direct substitution that $y_1(t) = e^{-t}$ and $y_2(t) = (t + 2)^{-1}$ are solutions of (3.94). The corresponding Wronskian is $W(t) = e^{-t}(t + 2)^{-2}(t + 1)$. We note that this Wronskian equals zero at $t = -1$. This is not surprising since the coefficient $a(t)$ becomes infinite at this point so that the integral in (3.98) will diverge whenever $t = -1$ falls within the integration interval. Using (3.108) we get the impulse response

$$h(t, \tau) = \frac{1}{\tau + 1} \left[\frac{\tau + 2}{t + 2} - e^{-(t - \tau)} \right] U(t - \tau). \tag{3.110}$$

Although this impulse response is finite provided t and τ exceed -1, it is not BIBO stable since (3.110) fails to satisfy (3.18).

Equivalent First-Order Vector System. The second-order system (3.93) can also be reformulated as a two-dimensional vector first-order system by introducing the change of variables $x_1 = y$ and $x_2 = y'$. Then (3.93) becomes $x_2' + ax_2 + bx_1 = f$ which together with $x_1' = x_2$ may be written in the following form

$$\begin{bmatrix} x_1' \\ x_2' \end{bmatrix} + \begin{bmatrix} 0 & -1 \\ b & a \end{bmatrix} \begin{bmatrix} x_1 \\ x_2 \end{bmatrix} = \begin{bmatrix} 0 \\ f \end{bmatrix}. \tag{3.111}$$

Setting

$$\mathbf{x} = \begin{bmatrix} x_1 & x_2 \end{bmatrix}^T, \mathfrak{A} = \begin{bmatrix} 0 & -1 \\ b & a \end{bmatrix}, \mathbf{f} = \begin{bmatrix} 0 & f \end{bmatrix}^T \tag{3.112}$$

we achieve the compact matrix form

$$\frac{d\mathbf{x}}{dt} + \mathfrak{A}\mathbf{x} = \mathbf{f}, \tag{3.113}$$

which constitutes the so-called state space representation of the differential equation (3.93). Let us first consider the homogeneous equation

$$\frac{d\mathbf{x}}{dt} + \mathfrak{A}\mathbf{x} = \mathbf{0}, \tag{3.114}$$

which is, of course, completely equivalent to (3.94). Using the two linearly independent solutions y_1 and y_2 we construct the matrix

$$\mathbf{X}(t) = \begin{bmatrix} y_1(t) & y_2(t) \\ y_1'(t) & y_2'(t) \end{bmatrix}, \tag{3.115}$$

which necessarily satisfies (3.113). Thus

$$\frac{d\mathbf{X}(t)}{dt} + \mathfrak{A}\mathbf{X}(t) = \mathbf{0}. \tag{3.116}$$

Since $\mathbf{X}(t)$ is nonsingular $\mathbf{X}(t)\,\mathbf{X}^{-1}(t) = \mathbf{I}$ is satisfied for all t. Hence it can be differentiated to obtain

$$\frac{d\mathbf{X}}{dt}\mathbf{X}^{-1} + \mathbf{X}\frac{d\mathbf{X}^{-1}}{dt} = \mathbf{0}. \tag{3.117}$$

After substituting for the derivative from (3.116) and multiplying from the left by $\mathbf{X}^{-1}$ we get

$$\frac{d\mathbf{X}^{-1}}{dt} - \mathbf{X}^{-1}\mathfrak{A} = \mathbf{0}. \tag{3.118}$$

This so-called adjoint equation can be used to solve the inhomogeneous system (3.113). This is done by first multiplying both sides of (3.113) from the left by $\mathbf{X}^{-1}(t)$ and (3.118) from the right by $\mathbf{x}$ and then adding. As a result we obtain

$$\mathbf{X}^{-1}\frac{d\mathbf{x}}{dt} + \frac{d\mathbf{X}^{-1}}{dt}\mathbf{x} + \mathbf{X}^{-1}\mathfrak{A}\mathbf{x} - \mathbf{X}^{-1}\mathfrak{A}\mathbf{x} = \mathbf{X}^{-1}\mathbf{f},$$

which is obviously equivalent to

$$\frac{d[\mathbf{X}^{-1}(t)\,\mathbf{x}(t)]}{dt} = \mathbf{X}^{-1}(t)\,\mathbf{f}(t).$$

Integrating between a fixed limit t_0 and a variable limit t gives for $t \geq t_0$

$$\mathbf{X}^{-1}(t)\,\mathbf{x}(t) - \mathbf{X}^{-1}(t_0)\,\mathbf{x}(t_0) = \int_{t_0}^{t} \mathbf{X}^{-1}(\tau)\,\mathbf{f}(\tau)\,d\tau.$$

The general solution for $\mathbf{x}(t)$ now follows after multiplication from the left by $\mathbf{W}(t)$. We write it in the form

$$\mathbf{x}(t) = \mathbf{H}(t, t_0)\,\mathbf{x}(t_0) + \int_{t_0}^{t} \mathbf{H}(t, \tau)\,\mathbf{f}(\tau)\,d\tau, \tag{3.119}$$

where

$$\mathbf{H}(t, \tau) = \mathbf{X}(t)\,\mathbf{X}^{-1}(\tau). \tag{3.120}$$

Equation (3.119) is equivalent to (3.109). Structurally however there is a closer resemblance to (3.76). For example, it is not hard to see that $\mathbf{H}(t, \tau)$ is the 2X2 matrix counterpart to $h(t, \tau)$ and is the impulse response satisfying

$$\frac{d\mathbf{H}(t, \tau)}{dt} + \mathfrak{A}\mathbf{H}(t, \tau) = \mathbf{I}\delta(t - \tau). \tag{3.121}$$

In fact one can start with (3.121) and (3.116) and imposing causality obtain (3.120). Also, just as in the scalar first-order system, the zero input response can be related directly to the input prior for $t < t_0$, i.e., for $t \geq t_0$

$$\mathbf{H}(t, t_0)\,\mathbf{x}(t_0) = \int_{-\infty}^{t_0} \mathbf{H}(t, \tau)\,\mathbf{f}(\tau)\,d\tau. \tag{3.122}$$

LTI System

Time Domain Response. For an LTI system the coefficients a and b in (3.93) are constants and the homogenous equation (3.94) is satisfied by the exponential $e^{\lambda t}$. Inserting this in (3.94) we get the characteristic equation

$$\lambda_{1,2}^2 + a\lambda_{1,2} + b = 0. \tag{3.123}$$

When the two roots λ_1 and λ_2 are distinct, the two linearly independent solutions of (3.93) are $y_1 = e^{\lambda_1 t}$ and $y_2 = e^{\lambda_2 t}$. The corresponding Wronskian is

$$W(t) = (\lambda_2 - \lambda_1) e^{(\lambda_1 + \lambda_2)t} \tag{3.124}$$

and from (3.108) we get for the impulse response

$$h(t) = \frac{e^{\lambda_2 t} - e^{\lambda_1 t}}{\lambda_2 - \lambda_1} U(t), \tag{3.125}$$

where, as usual for an LTI system, we have set the variable τ to zero. We note that when λ_1 and λ_2 have negative real parts the impulse response will decay to zero at infinity. It will then also be absolutely integrable so that, in accordance with (3.24) in Sect. 3.1, the system will be BIBO stable. Physically this implies net energy dissipation (rather than generation) within the system and requires that $a > 0$. In a mechanical system comprising a mass and a compliant spring this coefficient generally represents viscous damping and in an electrical network ohmic resistance (or conductance). For a dissipative system three classes of solutions can be distinguished. When the damping is very strong both roots are real ($a^2 > 4b$) and the system is said to be overdamped. As the damping is reduced we reach the condition $a^2 = 4b$ which corresponds to the limiting case $\lambda_2 \to \lambda_1$ in (3.125) and constitutes the so-called critically damped case. If we reduce the damping further, $a^2 < 4b$ and both roots become complex. This is the so-called underdamped case for which the impulse response assumes the form

$$h(t) = e^{-(a/2)t} \frac{\sin \sqrt{b - a^2/4} t}{\sqrt{b - a^2/4}} U(t). \tag{3.126}$$

The waveform is a damped sinusoid oscillating at the frequency $\omega_{res} = \sqrt{b - a^2/4}$. As $a \to 0, \omega_{res} \to \sqrt{b} \equiv \omega_0$, which would be, e.g., the frequency of oscillation of an undamped mass/spring or inductor/capacitor combination. Clearly in this limiting case the system is no longer BIBO stable.

Using the formulas in (3.96a) for the constants we can get an explicit form for the zero input response. Adding it to the zero state response we get for $t \geq t_0$

$$y(t) = \frac{\lambda_2 y(t_0) - y'(t_0)}{\lambda_2 - \lambda_1} e^{\lambda_1(t-t_0)} + \frac{-\lambda_1 y(t_0) + y'(t_0)}{\lambda_2 - \lambda_1} e^{\lambda_2(t-t_0)}$$
$$+ \int_{t_0}^{t} \frac{e^{\lambda_2(t-\tau)} - e^{\lambda_1(t-\tau)}}{\lambda_2 - \lambda_1} f(\tau) \, d\tau. \tag{3.127}$$

We still have to examine the critically damped case $\lambda_1 = \lambda_2 = \lambda$. We can still choose $y_1 = e^{\lambda t}$ for one of the solutions and substitute in (3.99) to get the second linearly independent solution. Since $\lambda_1 + \lambda_2 = 2\lambda$, the Wronskian is $W(t) = W(t_0) e^{-a(t-t_0)} = W(t_0) e^{2\lambda(t-t_0)}$. We then obtain

$$
y_2(t) = e^{\lambda(t-t_0)} y_2(t_0) + e^{\lambda t} W(t_0) \int_{t_0}^{t} \frac{e^{2\lambda(\tau-t_0)}}{e^{2\lambda\tau}} d\tau
$$

$$
= e^{\lambda(t-t_0)} y_2(t_0) + (t - t_0) W(t_0) e^{\lambda t} e^{-2\lambda t_0}.
$$

After setting $t_0 = 0$ and $y_2(0) = 0$ we get, upon dropping the constant $W(0)$,

$$
y_2(t) = t e^{\lambda t}. \tag{3.128}
$$

With $e^{\lambda t}$ and $t e^{\lambda t}$ as the two linearly independent solutions we get for the Wronskian

$$
W(t) = e^{2\lambda t}. \tag{3.129}
$$

Using (3.108) the impulse response is

$$
h(t) = t e^{\lambda t} U(t) \tag{3.130}
$$

while the complete response with initial conditions specified at $t = t_0$ becomes

$$
y(t) = \{y(t_0) + (t - t_0)[y'(t_0) - \lambda y(t_0)]\} e^{\lambda(t-t_0)}
$$

$$
+ \int_{t_0}^{t} (t - \tau) e^{\lambda(t-\tau)} f(\tau) d\tau. \tag{3.131}
$$

Transfer Function. Recall that if the system is BIBO stable then with $f(t) = e^{i\omega t}$ the output $y(t) = H(\omega) e^{i\omega t}$ is identical with the steady state response when the same exponential is initiated at some finite time. If we know the impulse response, then its FT equals $H(\omega)$. However if we have the governing differential equation we can get the transfer function $H(\omega)$ more directly by simply substituting $f(t)$ and $y(t)$ into the DE, cancelling the common exponential, and solving for $H(\omega)$. Carrying out this operation on (3.93) (with a and b are constant) we get

$$
H(\omega) = \frac{1}{-\omega^2 + i\omega a + b}. \tag{3.132}
$$

Assuming $b > 0$ and setting $\alpha = -a/2, \beta = \sqrt{b - (a/2)^2}, \omega_0 = \sqrt{b} = \sqrt{\beta^2 + \alpha^2}$, $Q = \beta/\alpha$ we write the magnitude of (3.132) in the following normalized form:

$$
\omega_0^2 |H(\omega)| = \frac{1}{\sqrt{\left[\left(\frac{\omega}{\omega_0}\right)^2 - 1\right]^2 + \frac{4}{1+Q^2}\left(\frac{\omega}{\omega_0}\right)^2}}. \tag{3.133}
$$

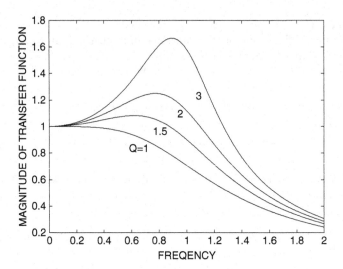

Figure 3.10: Plot of $\omega_0^2 \left| H\left(\omega\right)\right|$ vs ω/ω_0

In this parameterization we assume $\alpha < 0$ and $\omega_0^2 \geq \alpha^2$ which covers the under-damped range up to critical damping at $\omega_0^2 = \alpha^2$. The parameter Q is a relative measure of the dissipation in the system ($Q = 1$ for critical damping) and the width of the resonant peak which occurs at $\omega = \omega_{res} = \omega_0\sqrt{(Q^2 - 1)/(Q^2 + 1)}$. A plot of $\omega_0^2 \left| H\left(\omega\right)\right|$ as a function of ω/ω_0 is shown in Fig. 3.10 for several values of Q.

The relationship among $\omega_{res}, \omega_0, \alpha$, the position of the resonance peak along the frequency axis and the position of the pole[4] $\alpha \pm i\beta$ in the complex plane can be displayed geometrically as shown in Fig. 3.11.

Direct Solution Using the FT. An alternative technique for obtaining the general solution (3.127) or (3.131) of an LTI system is to apply the FT directly to the differential equation, an example of which we have already encountered in 2.2.6. Thus if the input signal $f\left(t\right)$ in (3.93) is specified for $t \geq t_0$ only and we are given the values $y\left(t_0\right)$ and $y'\left(t_0\right)$, the imposition of causality on the solution permits us to consider an equivalent problem wherein we set both $f\left(t\right)$ and $y\left(t\right)$ to zero for $t < t_0$. We then evaluate

$$F\left(\omega\right) = \int_{t_0}^{\infty} f\left(t\right) e^{-i\omega t} dt \tag{3.134}$$

while the implied discontinuity of $y\left(t\right)$ at $t = t_0$ forces us to interpret the derivatives in the differential equations as derivatives of "smooth" functions, as in (2.158 in 2.2) and (2.159 in 2.2).

---

[4]Here we adhere to the convention and define the pole in terms of the complex variable $s = i\omega$ so that the roots of the denominator of the transfer function are solutions of $s^2 - 2\alpha s + \omega_0^2 = 0$.

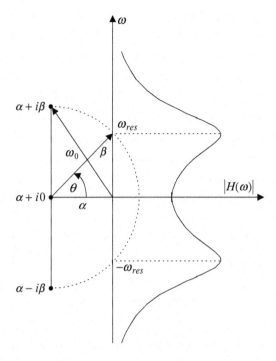

Figure 3.11: Relationship between $|H(\omega)|$ and Pole position

$$y'(t) \overset{\mathcal{F}}{\Longleftrightarrow} i\omega Y(\omega) - e^{-i\omega t_0} y(t_0) \tag{3.135a}$$

$$y''(t) \overset{\mathcal{F}}{\Longleftrightarrow} i\omega \left[i\omega Y(\omega) - e^{-i\omega t_0} y(t_0) \right] - e^{-i\omega t_0} y'(t_0)$$

$$= -\omega^2 Y(\omega) - i\omega e^{-i\omega t_0} y(t_0) - e^{-i\omega t_0} y'(t_0). \tag{3.135b}$$

Substituting (3.135a) and (3.135b) in (3.93), rearranging terms, and solving for $Y(\omega)$ we get

$$Y(\omega) = \frac{(a + i\omega) y(t_0) + y'(t_0)}{-\omega^2 + i\omega a + b} e^{-i\omega t_0} + \frac{F(\omega)}{-\omega^2 + i\omega a + b}. \tag{3.136}$$

Now $-\omega^2 + i\omega a + b = (i\omega - \lambda_1)(i\omega - \lambda_2)$ and assuming for definiteness that these roots are distinct, the inverse FT of the first term on the right of (3.136) is easily shown (e.g., using a residue evaluation as in 3.41) to be identical with the zero input response in (3.127). Also in virtue of (3.132) and the convolution theorem for the FT the second term on the right of (3.136) is just the zero state response in (3.127).

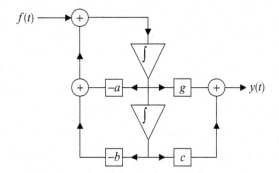

Figure 3.12: Feedforward representation of the right side of (3.137)

Input Transformed by a Differential Operator. Consider now an LTI system represented by (3.93) wherein the right side is replaced by a differential operator on the input. For example

$$\frac{d^2y}{dt^2} + a\frac{dy}{dt} + by = cf + g\frac{df}{dt}. \qquad (3.137)$$

Clearly this modification of the input leaves the zero input response unaffected. We can find the impulse response by setting $f = \delta(t)$ and convolving (3.125) or (3.130), which we presently denote by $h_0(t)$, with the input. We obtain

$$h(t) = \int_{-\infty}^{t} h_0(t - \tau)[c\delta(\tau) + g\delta^{(1)}(\tau)]d\tau$$
$$= ch_0(t) + gh_0'(t). \qquad (3.138)$$

In virtue of (3.102) and (3.106) this impulse response is discontinuous at $t = 0$, viz., $h(0^+) = g$. Alternatively, we can determine the impulse response by first finding the response of the system to $e^{i\omega t}$ we obtain the transfer function

$$H(\omega) = \frac{c + i\omega g}{-\omega^2 + i\omega a + b} \qquad (3.139)$$

and then compute $h(t)$ with the aid of the FT inversion formula. Note that in addition to the two poles at $\omega = \pm i\lambda_{1,2}$ this transfer function has a zero at $\omega = ic/g$. This zero is a direct consequence of the differentiation operation on the input. This differentiation operation can also be modeled by adding to the feedback loop in Fig. 4.9 that represents the left side of (3.137) a feedforward loop shown in Fig. 3.12.

Here the coefficients c and g represent gain settings. Note that it is this loop that is responsible for the creation of the zero in the numerator of the transfer function. One realization of this system as an electrical network is shown in Fig. 3.13 where we identify the parameters in (3.137) as $a = (R_1 + R_2)/L, b = 1/LC, c = 1/LC$, and $g = R_1/L$. When one of the factors in the denominator matches the zero in the numerator we obtain a second realization discussed in the following example.

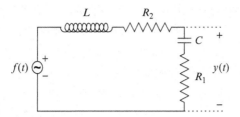

Figure 3.13: Electrical network realization of (3.137)

Example.   For the system

$$\frac{d^2y}{dt^2} + 3\frac{dy}{dt} + 2y = 2f + 2\frac{df}{dt} \tag{3.140}$$

let us find the output when $f = e^{-\alpha|t|}$, with $\alpha > 0$. Since the input is specified for $-\infty < t < \infty$ initial conditions at the output are redundant. (In this case the zero input response is in fact zero!) One way of determining the output is to find the impulse response and convolve it with the input. From the characteristic equation $\lambda^2 + 3\lambda + 2 = 0$ we find $\lambda_1 = -1, \lambda_2 = -2$. Instead of substituting in the general formulas we have derived let us determine the impulse response directly for the parameters of (3.140). First, with the right side of (3.140) set to $\delta(t)$ the response is $h_0(t) = \left(Ae^{-t} + Be^{-2t}\right)U(t)$. Setting $h_0(0) = 0$ and $h_0'(0^+) = 1$ we get $A + B = 0$ and $-A - 2B = 1$ so that $A = 1$ and $B = -1$. The system impulse response is then (see (3.138))

$$h(t) = 2\frac{dh_0(t)}{dt} + 2h_0(t) = 2e^{-2t}U(t). \tag{3.141}$$

Alternatively we can compute the impulse response from the transfer function. Because for the parameters at hand the zero at $\omega = i$ in the numerator coincides with the pole in the denominator they mutually cancel and we get

$$H(\omega) = \frac{2}{i\omega + 2}. \tag{3.142}$$

Thus in this simple case we see that the inverse FT of (3.142) gives (3.141). The system response to the specified input is

$$y(t) = \int_{-\infty}^{t} 2e^{-2(t-\tau)}e^{-\alpha|\tau|}d\tau = \begin{cases} \frac{2e^{\alpha t}}{2+\alpha} \;;\; t \le 0, \\ \frac{2e^{-2t}}{2+\alpha} + 2\frac{e^{-\alpha t}-e^{-2t}}{2-\alpha} \;;\; t \ge 0. \end{cases} \tag{3.143}$$

Another way of getting this result is to multiply the transfer function by the FT of the input and invert the FT. Thus

$$y(t) = \frac{1}{2\pi} \int_{-\infty}^{\infty} \frac{4\alpha e^{i\omega t}}{(i\omega + 2)(\omega^2 + \alpha^2)} d\omega. \tag{3.144}$$

We leave it as an exercise to show that (3.144) yields (3.143).

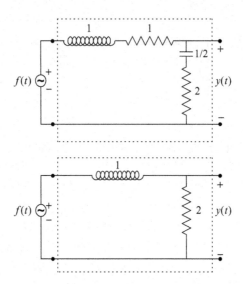

Figure 3.14: Two electrical networks with identical transfer functions

Note that (3.142) is the transfer function of a first-order system whereas the differential equation (3.140) describes a second-order system. In fact this transfer function holds for both electrical circuits in Fig. 3.14.

Apparently if only the input and output are known one cannot distinguish between the two networks. As mentioned above this is a direct result of the cancellation of factors in the numerator and the denominator. For differential equations of higher order the number of possible cancellations is increased.

This indeterminacy applies strictly to measurements of the transfer function alone, which in the present example is a voltage ratio. If additional or alternative measurements are permitted (e.g., the input impedance), then in most cases alternative realizations of same transfer function can be distinguished.

Transfer functions can be defined in various ways. For example, as an input voltage to current ratio, i.e., the input impedance. A famous problem in this class leading to indeterminacy is the so-called Slepian's Black Box Problem [22] for which the circuits are shown in Fig. 3.15.

The problem is to determine from the input impedance whether the "black box" contains the resistor R in (a) or the series arrangement of the parallel RL and RC circuits shown in (b). A simple calculation shows that when $R = \sqrt{L/C}$ the input impedance looking into the terminal equals R independent of which circuit is in the "box." Unlike indeterminacies caused by the cancellation of common factors, where they generally can be resolved by the addition of external circuit elements at the input of the output ports, here no such solutions are possible.[5]

---

[5]The problem has a long history. One of the early resolutions was on the thermal noise generated by the resistor. See, e.g., [11].

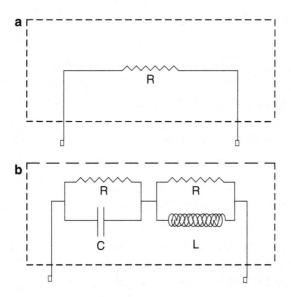

Figure 3.15: The Slepian black box problem

3.3.3 N-th Order Differential Equations

Time-Varying Systems

Standard Form. We now consider a linear system described by an N-th order differential equation. Represented as a general form of (3.93) that also includes the differentiation of the input as in (3.137) it reads

$$\sum_{k=0}^{N} a_{N-k}(t) \frac{d^k}{dt^k} y(t) = \sum_{k=0}^{M} b_{N-k}(t) \frac{d^k}{dt^k} f(t), \qquad (3.145)$$

where $M < N$. Here again $f(t)$ is the input and $y(t)$ the output. As in the two-dimensional case we will treat the differential operator on the right side of (3.145) as an equivalent input. If we wish to exclude singularity functions, the restrictions on $f(t)$ are more stringent than in two-dimensions since now it must now possess derivatives up to order N. When $N = 0$ only this equivalent input remains. In this case it is more reasonable to regard it as the differential operator

$$y(t) = \frac{1}{a_N(t)} \sum_{k=0}^{M} b_{M-k}(t) \frac{d^k f(t)}{dt^k}. \qquad (3.146)$$

Its impulse response is a sum of singularity functions

$$h(t, \tau) = \frac{1}{a_N(t)} \sum_{k=0}^{M} b_{M-k}(t) \delta^{(k)}(t - \tau). \qquad (3.147)$$

Retaining the first term $(k = 0)$ we have

$$\sum_{k=0}^{N} a_{N-k}(t) \frac{d^k y(t)}{dt^k} = b_M(t) f(t) \tag{3.148}$$

for which the impulse response is obtained by solving

$$\sum_{k=0}^{N} a_{N-k}(t) \frac{d^k h_0(t,\tau)}{dt^k} = b_M(t) \delta(t-\tau) \tag{3.149}$$

for $h_0(t,\tau)$ subject to the initial condition (causality)

$$h_0(t,\tau) = 0 \; ; \; t < \tau. \tag{3.150}$$

It will be notationally more convenient to deal with (3.149) if the coefficient of the highest derivative is normalized to unity or, equivalently, both sides of (3.149) are divided by $a_0(t)$. The restated problem is then

$$\frac{d^N h_0(t,\tau)}{dt^N} + \sum_{k=0}^{N-1} \alpha_{N-k}(t) \frac{d^k h_0(t,\tau)}{dt^k} = \beta_M(t) \delta(t-\tau), \tag{3.151}$$

where we have defined

$$\alpha_{N-k}(t) = \frac{a_{N-k}(t)}{a_0(t)}, \beta_M(t) = \frac{b_M(t)}{a_0(t)}. \tag{3.152}$$

We shall assume that $N > 1$ and that the $\alpha_{N-k}(t)$ and $\beta_M(t)$ are continuous functions for all t. As a consequence, just as we have seen in the case of $N = 2$, (3.151) requires that at $t = \tau$ the N-th derivative contain a delta function, the $N - 1$-st derivative a step discontinuity and all lower order derivatives including $h_0(t,\tau)$ be continuous. We now construct $h_0(t,\tau)$ from the N linearly independent solutions $y_n(t) n = 1, 2, \ldots N$ of the homogeneous equation

$$\frac{d^N y_n(t)}{dt^N} + \sum_{k=0}^{N-1} \alpha_{N-k}(t) \frac{d^k y_n(t)}{dt^k} = 0 \; ; \; n = 1, 2, \ldots N. \tag{3.153}$$

Clearly for $t \neq \tau h_0(t,\tau)$ must be a linear superposition of the solutions of (3.153). Since by continuity of $h_0(t,\tau)$ the same superposition must also hold at $t = \tau$ we may write

$$h_0(t,\tau) = \begin{cases} \sum_{n=1}^{N} A_n(\tau) y_n(t) \; ; \; t \geq \tau, \\ 0 \; ; \; t < \tau, \end{cases} \tag{3.154}$$

where we anticipate that the expansion coefficients $A_n(\tau)$ must be functions of τ. Employing the continuity of $h_0(t,\tau)$ and its derivatives of order lower than

$N - 1$ at $t = \tau$ in (3.154) provides us with $N - 1$ equations for the coefficients $A_n(\tau)$. Thus

$$\sum_{n=1}^{N} A_n(\tau) y_n^{(m)}(\tau) = 0 \; ; \; m = 0, 1, \ldots N - 2, \tag{3.155}$$

where $y_n^{(m)}(\tau) \equiv d^m y(\tau)/d\tau^m$ and $y_n^{(0)}(\tau) \equiv y_n(\tau)$. To solve for $A_n(\tau)$ we need one more equation which we obtain by integrating (3.151) between the limits $t = \tau + \varepsilon = \tau^+$ and $t = \tau - \varepsilon = \tau^-$:

$$\frac{d^{N-1}h_0(t,\tau)}{dt^{N-1}}\Big|_{t=\tau^+} - \frac{d^{N-1}h_0(t,\tau)}{dt^{N-1}}\Big|_{t=\tau^-}$$
$$+ \sum_{k=0}^{N-1} \int_{\tau^-}^{\tau^+} \alpha_{N-k}(t) \frac{d^k h_0(t,\tau)}{dt^k} dt$$
$$= \beta_M(\tau). \tag{3.156}$$

The causality condition (3.150) requires that $d^{N-1}h_0(t,\tau)/dt^{N-1}|_{t=\tau^-} = 0$ while by virtue of the continuity of the members in the integrands the contribution of the entire sum vanishes as $\varepsilon \to 0$. Hence (3.156) is equivalent to

$$\frac{d^{N-1}h_0(t,\tau)}{dt^{N-1}}\Big|_{t=\tau^+} = \beta_M(\tau) \tag{3.157}$$

and using (3.154) we obtain

$$\sum_{n=1}^{N} A_n(\tau) y_n^{(N-1)}(\tau) = \beta_M(\tau) . \tag{3.158}$$

Combining (3.155) and (3.158) into a single matrix equation gives

$$\begin{bmatrix} y_1(\tau) & y_2(\tau) & \cdots & y_N(\tau) \\ y_1^{(1)}(\tau) & y_2^{(1)}(\tau) & \cdots & y_N^{(1)}(\tau) \\ y_1^{(2)}(\tau) & y_2^{(2)}(\tau) & \cdots & y_N^{(2)}(\tau) \\ & & \cdots & \\ y_1^{(N-1)}(\tau) & y_2^{(N-1)}(\tau) & \cdots & y_N^{(N-1)}(\tau) \end{bmatrix} \begin{bmatrix} A_1(\tau) \\ A_2(\tau) \\ A_3(\tau) \\ \vdots \\ A_N(\tau) \end{bmatrix} = \begin{bmatrix} 0 \\ 0 \\ 0 \\ \vdots \\ \beta_M(\tau) \end{bmatrix}. \tag{3.159}$$

We recognize the determinant of this matrix as the Wronskian of the linearly independent solutions of the DE so that (3.159) always has a unique solution which, together with (3.154), yields the impulse response of (3.149).

We now return to the general case (3.145). Again it will be convenient to normalize the coefficients $b_{M-k}(t)$ as in (3.152)

$$\beta_{M-k}(t) = \frac{b_{M-k}(t)}{a_0(t)}; \; k = 0, 1, 2, \ldots M \tag{3.160}$$

so that (3.145) becomes

$$\sum_{k=0}^{N} \alpha_{N-k}(t) \frac{d^k y(t)}{dt^k} = \sum_{k=0}^{M} \beta_{M-k}(t) \frac{d^k f(t)}{dt^k}$$

$$= \beta_M(t) \left\{ \sum_{k=0}^{M} \frac{\beta_{M-k}(t)}{\beta_M(t)} \frac{d^k f(t)}{dt^k} \right\}. \tag{3.161}$$

Since the impulse response $h_0(t,\tau)$ is the causal solution of (3.161) when the sum in braces equals $\delta(t-\tau)$ we may use superposition and solve for $y(t)$ as follows:

$$y(t) = \int_{-\infty}^{t} d\tau' h_0(t,\tau') \sum_{k=0}^{M} \frac{\beta_{M-k}\left(\tau'\right)}{\beta_M\left(\tau'\right)} \frac{d^k f\left(\tau'\right)}{d\tau'^k}. \tag{3.162}$$

The impulse response of the system, as viewed from the actual port (i.e., $f(t)$) rather than the equivalent port, can be obtained by replacing $x\left(\tau'\right)$ in the preceding expression by $\delta\left(\tau'-\tau\right)$ and using the properties of the delta function and its derivatives. Thus we obtain

$$h(t,\tau) = h_0(t,\tau) + \sum_{k=1}^{M} (-1)^k \frac{d^k}{d\tau^k} \left\{ \frac{h_0(t,\tau) \beta_{M-k}(\tau)}{\beta_M(\tau)} \right\} \tag{3.163a}$$

and we can rewrite (3.162) in the standard form

$$y(t) = \int_{-\infty}^{t} h(t,\tau) f(\tau) d\tau. \tag{3.163b}$$

Note that (3.163a) assumes that $\beta_{M-k}(\tau)/\beta_M(\tau)$ possesses derivatives up to at least order M. In that case $h(t,\tau)$ will be continuous at $t = \tau$ provided $M < N - 1$ and will contain singularity functions whenever $M \geq N$.

Note also that in the preceding development the initial conditions were excluded (or, equivalently, were specified at $-\infty$.). Rather than exhibiting them explicitly as in (3.109) it is notationally simpler to include them in the state variable formulation discussed next.

Equivalent N-Dimensional First-Order Vector Form. The change of variables used to transform the second-order differential equation into the two-dimensional first-order system (3.111) generalizes in the N-dimensional case

$$\mathbf{x} = \begin{bmatrix} y & y^{(1)} & y^{(2)} & \cdots & y^{(N-2)} & y^{(N-1)} \end{bmatrix}^T$$

$$= \begin{bmatrix} x_1 & x_2 & x_3 & \cdots & x_{N-1} & x_N \end{bmatrix}^T, \tag{3.164}$$

where $y^{(k)} = d^k y/dt_k$. Next we set the right side of (3.145) to f_M and rewrite it in the compact form

$$\sum_{k=0}^{N} a_{N-k} \, y^{(k)} = f_M. \tag{3.165}$$

Using the correspondence between the derivatives $y^{(k)}$ and the vector components x_k in (3.164) we establish the chain

$$x_2 = x_1^{(1)} \; , \; x_3 = x_2^{(1)}, \; x_4 = x_3^{(1)} \ldots x_N = x_{N-1}^{(1)}$$

from which we construct the N-dimensional state space representation

$$
\begin{bmatrix} x_1^{(1)} \\ x_2^{(1)} \\ x_2^{(1)} \\ . \\ x_{N-1}^{(1)} \\ x_N^{(1)} \end{bmatrix}
=
\begin{bmatrix}
0 & 1 & 0 & . & 0 & 0 \\
0 & 0 & 1 & . & 0 & 0 \\
0 & 0 & 0 & . & . & . \\
. & . & . & 0 & 1 & 0 \\
0 & 0 & . & 0 & 0 & 1 \\
-a_N & -a_{N-1} & . & a_3 & a_2 & a_1
\end{bmatrix}
\begin{bmatrix} x_1 \\ x_2 \\ x_3 \\ . \\ x_{N-1} \\ x_N \end{bmatrix}
=
\begin{bmatrix} 0 \\ 0 \\ 0 \\ . \\ 0 \\ f_M \end{bmatrix}
$$

(3.166)

where we have set $a_0 = 1$. In the block matrix notation (3.166) takes on the form

$$\mathbf{x}^{(1)} = \mathfrak{A}\mathbf{x} + \mathbf{f}_M. \tag{3.167}$$

The procedure for solving (3.167) for the state vector $\mathbf{x} = \mathbf{x}(t)$ is exactly the same as used in the two-dimensional case (3.113) through (3.119) since it is independent of the dimension of $\mathbf{X}(t)$. Thus if we use

$$\mathbf{X}(t) = \begin{bmatrix} \mathbf{x}_1 & \mathbf{x}_2 & \mathbf{x}_3 & . & \mathbf{x}_N \end{bmatrix} \tag{3.168}$$

the solution is of (3.167) is again given by (3.119). Note that (3.168) is identical with the matrix (3.159). Its columns are the N linearly independent solutions of

$$\mathbf{x}_k^{(1)} = \mathfrak{A}\mathbf{x}_k \; , k = 1, 2, \ldots N. \tag{3.169}$$

Feedback Representation. The feedback diagram in Fig. 3.9 when generalized to the N-dimensional case is shown in Fig. 3.16. That this arrangement is consistent with (3.145) when $M = 0$ can be established by noting that starting with highest derivative $y(t)^N$ successive integrations bring us down to the output $y(t)$. (Recall that $y^{(0)}(t) \equiv y(t)$) On the other hand, the weighted sum of the derivatives

$$-\sum_{k=1}^{N} a_k y^{(N-k)}(t)$$

when added to $f(t)$, must match $y^{(N)}(t)$. As a result we obtain (3.145). When $M \neq 0$ the representation in Fig. 3.16 should be extended to include the dependence on the derivatives of $f(t)$, as required by (3.162). In the following we discuss this extension for the time-invariant case.

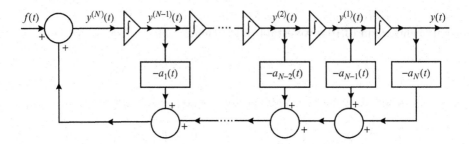

Figure 3.16: Feedback representation of a differential equation with time-varying coefficients

Time-invariance

In this case the coefficients a_{N-k} and b_{N-k} in (3.145) are constant and we can solve it using Fourier Transforms. With reference to (3.135) for initial conditions at $t = 0^+$ the Fourier Transform of the k-th derivative of $y(t)$ is

$$\mathcal{F}\left\{y(t)^{(k)}\right\} = (i\omega)^k Y(\omega) - \sum_{\ell=1}^{k} (i\omega)^{k-\ell} y\left(0^+\right)^{(\ell-1)}. \qquad (3.170)$$

After substituting into (3.145) and rearranging terms we get

$$\sum_{k=0}^{N} a_{N-k} (i\omega)^k Y(\omega) = \sum_{k=0}^{N} a_{N-k} \sum_{\ell=1}^{k} (i\omega)^{k-\ell} y\left(0^+\right)^{(\ell-1)}$$
$$+ F(\omega) \sum_{k=0}^{M} b_{M-k}(i\omega)^k, \qquad (3.171)$$

where the double summation on the right represents the total contribution from the initial conditions at $t = 0$, the second term the contribution from the equivalent source. Here, as in the two-dimensional case, we can identify in (3.171) the Fourier Transform of the zero state response

$$Y_{0s}(\omega) = \frac{\sum_{k=0}^{M} b_{M-k}(i\omega)^k}{\sum_{k=0}^{N} a_{N-k}(i\omega)^k} F(\omega) \qquad (3.172)$$

as well as the zero input response

$$Y_{0i}(\omega) = \frac{\sum_{k=0}^{N} a_{N-k} \sum_{\ell=1}^{k} (i\omega)^{k-\ell} y(0^+)^{(\ell-1)}}{\sum_{k=0}^{N} a_{N-k}(i\omega)^k}. \qquad (3.173)$$

As we know the former coincides with the time harmonic response so that we may designate the ratio

$$H_{N,M}(\omega) = \frac{\sum_{k=0}^{M} b_{M-k}(i\omega)^k}{\sum_{k=0}^{N} a_{N-k}(i\omega)^k} \qquad (3.174)$$

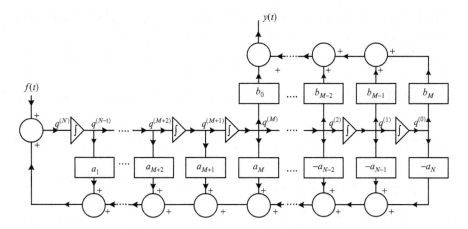

Figure 3.17: Feedback/feedforward representation of a transfer function

as the system transfer function. Its can be generated by adding a feed forward loop to the feedback loop in Fig. 3.16 resulting in Fig. 3.17. To trace the signal from input to output we start, as in Fig. 3.16, on the right side of the lower part of the diagram and sum the weighted derivatives of $q(t) \equiv q^{(0)}$ up to order $q^{(N-1)}$. Subtracting this from $f(t)$ yields $q^{(N)}$. As a result we get

$$f(t) = \sum_{k=0}^{N} a_{N-k} \, q^{(k)}(t),\qquad(3.175)$$

which is a differential equation for $q(t)$. This corresponds to the feedback portion of the loop in Fig. 3.17, where $y(t) \equiv q(t)$. Here $q(t)$ is fed "forward" and $y(t)$ is derived from the weighted sum

$$y(t) = \sum_{k=0}^{M} b_{M-k} q^{(k)}(t) \,.\qquad(3.176)$$

Evidently the solution for $y(t)$ in terms of $f(t)$ requires the elimination of $q(t)$ between (3.175) and (3.176). As long as the coefficients a_{N-k} and b_{M-k} are constant this is easily done with Fourier Transforms. Thus setting $\mathcal{F}\{q(t)\} = Q(\omega)$ we get from (3.176)

$$Y(\omega) = Q(\omega) \sum_{k=0}^{M} b_{M-k} (i\omega)^k \qquad(3.177)$$

and from (3.175)

$$F(\omega) = Q(\omega) \sum_{k=0}^{N} a_{N-k} (i\omega)^k \,. \qquad(3.178)$$

The ratio $Y(\omega)/F(\omega)$ is the transfer function (3.175).

When the coefficients in (3.175) and (3.176) depend on time a solution for $y(t)$ in terms of $f(t)$ may still exist but it will in general not satisfy (3.145) unless the feed forward coefficients b_{M-k} are constant.

Problems

1. The input $e^{i\omega t}$ to a linear system results in the output $\frac{e^{i4\omega t}}{1+2i\omega}$ for all real ω and t.

   (a) Is the system time-invariant? Justify your answer.
   (b) Find the impulse response.
   (c) Is the system causal?
   (d) Is the system stable?
   (e) Find the output when the input is

   $$f(t) = \begin{cases} 1 \; ; -1/2 \le t \le 1/2, \\ 0 \; ; \text{otherwise.} \end{cases}$$

2. The input $e^{i\omega t}$ to a linear system results in the output $\frac{e^{i\omega(1-2v/c)t}}{a^2+(vt)^2}$ for all real ω and t, where v, c and a are constants and such that

   $$0 < 2v/c < 1.$$

   (a) Find the impulse response.
   (b) Is the system causal? Justify your answer.
   (c) Find the output when the input is the pulse

   $$p_T(t) = \begin{cases} 1 \; ; -T \le t \le T, \\ 0 \; ; \text{otherwise.} \end{cases}$$

3. A linear system is defined by the differential equation

   $$\frac{dy(t)}{dt} + \left(\alpha + 3\beta t^2\right) y(t) = x(t),$$

   where α and β are real nonnegative constants and where $x(t)$ is the input and $y(t)$ the output. Find the system impulse response. With $x(t) = \cos 2t \exp -\beta t^3$ and initial condition $y(-3) = 2$ find the output for $t \ge -3$. In your solution identify the zero state, the zero input, the transient, and the steady state responses.

4. A linear system is defined by the differential equation

$$\frac{d^2y}{dt^2} + 2\frac{dy}{dt} + y = x + \frac{dx}{dt}$$

with $x(t)$ the input and $y(t)$ the output. Find

(a) The output when the input is $2\cos(6t)$.

(b) The impulse response.

(c) The output for $t \geq 0$ with initial conditions $y(0) = 1, y'(0) = -1$ when the input is $2\cos(6t)U(t)$. Identify the zero state, the zero input, and the steady state responses.

5. A linear system is defined by the differential equation

$$\frac{d^2y(t)}{dt^2} + 5\frac{dy(t)}{dt} + 6y(t) = 3x(t) + \frac{dx(t)}{dt},$$

where $x(t)$ is the input and $y(t)$ is the output.

(a) Find the system impulse response.

(b) Find the output when the input is $4\cos(3t)$.

(c) Find the output when the input is $2\sin(2t)U(t+2)$.

(d) Find the output when the input is $e^{-8|t|}$.

6. The input $e^{i\omega t}U(t)$ to a linear system results in the output

$$-\frac{e^{-t^3}}{t^2+i\omega} + \frac{1}{t^2+i\omega}e^{i\omega t}$$

for all real ω and t.

(a) Find the system impulse response.

(b) Find the output when the input is

$$f(t) = \begin{cases} 1 \, ; 0 \leq t \leq 1, \\ 0 \, ; \text{otherwise.} \end{cases}$$

7. Obtain the Fourier Transform of $Y(\omega)$ in (3.171) with initial conditions $t = t_0^+$.

Chapter 4

Laplace Transforms

4.1 Single-Sided Laplace Transform

4.1.1 Analytic Properties

The analysis of LTI system performance sometimes calls for signal models for which the FT does not exist. For example, the system defined by the DE

$$\frac{d^2 y}{dt^2} - 4\frac{dy}{dt} + 4y = f(t) \tag{4.1}$$

has a causal impulse response $te^{2t}U(t)$. This function has no FT and since it is not absolutely integrable it is not BIBO stable. Of course, an unstable system of this sort would generally not be the desired end product of a design effort but merely a pathological state of affairs requiring correction. On the other hand, to identify and apply design procedures that ensure stable system performance one must be able to quantify the unstable behavior of a system. This requires that the class of admissible signals includes signals that can be generated by unstable systems. These include functions that can become unbounded at infinity and of the generic form $t^n e^{qt}(q > 0)$ for which the FT does not exist. For such signals a convergent integral transform can be constructed by initially multiplying the signal by the exponential convergence factor $e^{-\sigma t}$ with $\sigma > 0$ and then computing the FT. Thus

$$\int_{-\infty}^{\infty} e^{-\sigma t} f(t) e^{-i\omega t} dt. \tag{4.2}$$

Clearly we can reduce the growth of the integrand as $t \to \infty$ by choosing a sufficiently large positive σ. Unfortunately this same factor will contribute to an exponential growth of the integrand for $t < 0$ turning the convergence factor effectively into a "divergence" factor. However if we confine ourselves to causal

W. Wasylkiwskyj, *Signals and Transforms in Linear Systems Analysis*,
DOI 10.1007/978-1-4614-3287-6_4, © Springer Science+Business Media, LLC 2013

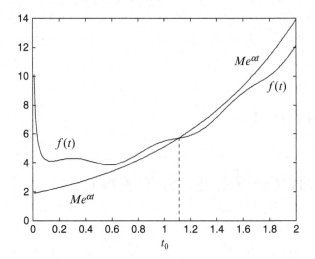

Figure 4.1: Growth of a function of exponential order

signals there is an easy remedy. We simply truncate the integral to nonnegative values of t and define an integral transform by

$$F(s) = \int_0^\infty e^{-st} f(t)\, dt, \tag{4.3}$$

where we have set $s = \sigma + i\omega$. Equation (4.3) defines the so-called one-sided (or unilateral) Laplace transform (LT). As in case of the FT we will at times find it convenient to use an abbreviation for the defining integral. Thus

$$\mathfrak{L}\{f(t)\} = F(s). \tag{4.4}$$

Having defined the transform our next task is to identify the class of functions for which (4.3) converges and delineate the resulting properties of $F(s)$. Even though we allow $f(t)$ to grow at infinity we cannot expect convergence for all possible functions that tend to infinity with t. In fact the use of the exponential convergence factor implies that our signals may not grow any faster than an exponential function. Formally we say that $f(t)$ is required to be of exponential order. Expressed in mathematical terminology this means that there exist real constants t_0, α, and M such that

$$|f(t)| < Me^{\alpha t}\ ;\ t > t_0. \tag{4.5}$$

Note that this definition does not imply that $f(t)$ may not exceed $Me^{\alpha t}$ for some values of t. It merely means that it may not exceed this exponential for $t > t_0$. This is illustrated graphically in Fig. 4.1.

The following notation is commonly employed to designate a function of exponential order:

$$\underset{t\sim\infty}{f(t)} \sim O\left(e^{\alpha t}\right). \tag{4.6}$$

Consistent with our previous assumption on the class of admissible signals we shall suppose that $f(t)$ is sectionally smooth. However, we shall relax our previous constraint on boundedness and replace it by the requirement that

$$g(t) \equiv \int_0^t f(\tau)\,d\tau \tag{4.7}$$

be finite for all finite $t \geq 0$. For example, within this framework we include functions of the form t^{-q} with $0 < q < 1$ which tend to infinity at $t = 0$. As we shall see signals with singularities of this type arise in the study of the transient behavior of distributed systems (e.g., transmission lines).

We shall prove the following. The LT of a sectionally smooth function satisfying (4.5) and (4.7) has the following properties:

$$F(s) \text{ is an analytic function of } s \text{ for } \operatorname{Re} s > \alpha \tag{4.8a}$$

and such that

$$\lim_{|s| \to \infty} F(s) \to 0 \text{ for } \operatorname{Re} s > \alpha. \tag{4.8b}$$

We first show that if $f(t)$ is of exponential order $g(t)$ is exponentially bounded for all nonnegative t. We start with (4.7) which implies that for $0 \leq t \leq t_0$

$$|g(t)| = \left| \int_0^t f(\tau)\,d\tau \right| \leq Q(t_0). \tag{4.9}$$

For $t > t_0$ we write the bound as a sum of two members as follows:

$$|g(t)| = \left| \int_0^{t_0} f(\tau)\,d\tau + \int_{t_0}^t f(\tau)\,d\tau \right| \leq Q(t_0) + \left| \int_{t_0}^t f(\tau)\,d\tau \right|. \tag{4.10}$$

In view of (4.5) we have $\left| \int_{t_0}^t f(\tau)\,d\tau \right| < M(e^{\alpha t} - e^{\alpha t_0})/\alpha$ so that the bound in (4.10) becomes

$$\begin{aligned}
|g(t)| &\leq e^{\alpha t}\left[Q(t_0)e^{-\alpha t} + \frac{M}{\alpha}(1 - e^{-\alpha(t - t_0)}) \right] \\
&\leq e^{\alpha t}\left[Q(t_0)e^{-\alpha t_0} + \frac{M}{\alpha} \right] = e^{\alpha t}P(t_0)
\end{aligned} \tag{4.11}$$

with $P(t_0) = Q(t_0)e^{-\alpha t_0} + M/\alpha$. If we now define $N = \max\{Q(t_0), P(t_0)\}$ then since for $0 \leq t \leq t_0$ the left side of (4.9) is also bounded by $Q(t_0)e^{\alpha t}$ we obtain in combination with (4.11)

$$|g(t)| \leq e^{\alpha t}N \quad ; \quad 0 \leq t < \infty. \tag{4.12}$$

Note that unlike $f(t)$ (see Fig. 4.1) $g(t)$ is bounded by the exponential for *all* nonnegative t. Next we integrate (4.3) by parts to obtain

$$F(s) = e^{-st}g(t)\big|_0^\infty + s\int_0^\infty e^{-st}\,g(t)\,dt. \tag{4.13}$$

With $s = \sigma + i\omega$ and choosing $\sigma > \alpha$ the first term on the right of (4.13) vanishes at the upper limit. It also vanishes at the lower limit because $g(0) = 0$. Therefore setting

$$G(s) = \int_0^\infty e^{-st} g(t) \, dt \qquad (4.14)$$

we obtain the relationship

$$F(s) = sG(s) \; ; \; \text{Re} \, s > \alpha. \qquad (4.15)$$

We now utilize (4.12) to bound (4.14) as follows:

$$|G(s)| \le \int_0^\infty \left| e^{-st} g(t) \right| dt \le N \int_0^\infty e^{-(\sigma-\alpha)t} dt = \frac{N}{\sigma - \alpha} \; ; \; \sigma > \alpha. \qquad (4.16)$$

The preceding states that $G(s)$ exists and is bounded for $\text{Re} \, s > \alpha$. We can similarly bound the derivative of $G(s)$. Thus

$$|G'(s)| \le \int_0^\infty \left| -te^{-st} g(t) \right| dt \le N \int_0^\infty te^{-(\sigma-\alpha)t} dt = \frac{N}{(\sigma - \alpha)^2} \; ; \; \sigma > \alpha.$$

In fact we get for n-th derivative

$$\begin{aligned}
\left| G^{(n)}(s) \right| &\le \int_0^\infty \left| (-t)^n e^{-st} g(t) \right| dt \le N \int_0^\infty t^n e^{-(\sigma-\alpha)t} dt \\
&= \frac{Nn!}{(\sigma - \alpha)^{n+1}} \; ; \; \sigma > \alpha.
\end{aligned}$$

This shows that not only $G(s)$ but also all its derivatives exist and are bounded for $\text{Re} \, s > \alpha$. Consequently $G(s)$ is an analytic function for $\text{Re} \, s > \alpha$. Therefore, in view of (4.15) and (4.16) $F(s)$ is also analytic for $\text{Re} \, s > \alpha$ with the possible exception of a simple pole at infinity. This shows that (4.8a) holds in the finite part of the complex plane. To prove that there is no pole at infinity, i.e., that in fact (4.8b) holds, we use the fact that the existence of the integral in (4.7) implies that given an $\varepsilon > 0$ there exists a δ such that $|g(t)| < \varepsilon$ for all $0 \le t \le \delta$. We now write the transform of $g(t)$ as the sum of two parts

$$G(s) = \int_0^\delta e^{-st} g(t) \, dt + \int_\delta^\infty e^{-st} g(t) \, dt \qquad (4.17)$$

and bound each integral separately. For the second integral we use the exponential bound on $g(t)$ in (4.12) so that

$$\left| \int_\delta^\infty e^{-st} g(t) \, dt \right| \le N \int_\delta^\infty e^{-(\sigma-\alpha)t} \, dt = N \frac{e^{-(\sigma-\alpha)\delta}}{\sigma - \alpha} \; ; \; \sigma > \alpha. \qquad (4.18)$$

In the first integral we first change variables and write $\int_0^\delta e^{-st} g(t) \, dt = \frac{1}{s} \int_0^{\delta s} e^{-x} g(x/s) \, dx$ and then setting $s = \sigma$ use the fact that in view of (4.7) $|g(t)|$ is bounded by ε. In this way we get

$$\left| \int_0^\delta e^{-\sigma t} g(t) \, dt \right| \le \frac{1}{\sigma} \left| \int_0^{\delta\sigma} e^{-x} g(x/\sigma) \, dx \right| \le \frac{\varepsilon}{\sigma} \left(1 - e^{-\delta\sigma} \right). \qquad (4.19)$$

Finally we add (4.18) and (4.19) and multiply the result by σ and in view of (4.17) obtain

$$F(\sigma) = \sigma G(\sigma) \le N \frac{\sigma e^{-(\sigma-\alpha)\delta}}{\sigma - \alpha} + \varepsilon \left(1 - e^{-\delta\sigma}\right) \ ; \ \sigma > \alpha. \qquad (4.20)$$

Since ε may be chosen as small as desired the preceding expression gives

$$\lim_{\sigma \to \infty} F(\sigma) = 0 \ \text{ for } \sigma > \alpha. \qquad (4.21)$$

Because $F(s)$ is analytic for $\operatorname{Re} s > \alpha$ (4.21) implies (4.8b).

The significance of (4.8) is that it forms the basis for the evaluation of the inverse transform, i.e., the determination of $f(t)$ from $F(s)$ using the techniques of the calculus of residues to be discussed in Sect. 4.1.4.

4.1.2 Singularity Functions

The analytic properties of $F(s)$ discussed in the preceding subsection are restricted to LT of piecewise smooth functions satisfying (4.5) and (4.7). We can also accommodate within the framework of LT theory the delta function as well as higher order singularity functions. As we shall see their transforms do not satisfy (4.8). On first thought the computation of the LT of a delta function appears to present a bit of a problem since $\delta(t)$, being defined in terms of limiting forms of kernels that are even in t, appears incompatible with the semi-infinite integration interval of the unilateral LT. This difficulty is easily removed by the artifice of permitting the lower limit of the LT integral to assume slightly negative vales, i.e.,

$$F(s) = \int_{-\epsilon}^{\infty} e^{-st} f(t) \, dt \qquad (4.21^*)$$

with ϵ an arbitrarily small positive quantity.[1] For ordinary functions the replacement of (4.3) by (4.21*) is inconsequential since such functions are defined as zero for negative t. However as long as $\epsilon \ne 0$ the limit for any of the kernels studied in Chapter 2 reads

$$\lim_{\Omega \to \infty} \int_{-\epsilon}^{\infty} K_\Omega(t) dt = 1$$

so that also

$$\lim_{\Omega \to \infty} \int_{-\epsilon}^{\infty} e^{-st} K_\Omega(t) dt = 1.$$

Thus we find

$$\mathfrak{L}\{\delta(t)\} = 1 \qquad (4.22)$$

---

[1] An alternative popular notation for 4.21* is $F(s) = \int_{0^-}^{\infty} e^{-st} f(t) \, dt$.

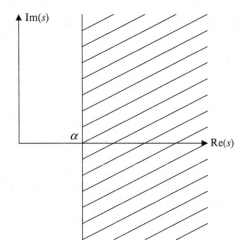

Figure 4.2: Region of analyticity of $F(s)$ defined by $\text{Re}(s) > \alpha$

i.e., the LT of a delta function is unity just like the corresponding FT. We see that in this case (4.8b) does not hold. Similarly for the higher order singularity functions we obtain for $n \geq 1$

$$\mathcal{L}\left\{\delta^{(n)}(t)\right\} = (s)^n \tag{4.23}$$

so that now both (4.8a) and (4.8b) are violated.

4.1.3   Some Examples

For piecewise smooth functions of exponential order the region of analyticity $\text{Re}\, s > \alpha$ in the complex s-plane may be identified by the cross-hatched region in Fig. 4.2.

For example, we easily find for $\text{Re}\, s > 0$ that

$$\mathcal{L}\left\{U(t)\right\} = 1/s. \tag{4.24}$$

The only singularity of this LT is a simple pole at $s = 0$ to the right of which $1/s$ is analytic. Thus in this case $\alpha = 0$. Similarly a simple integration yields the transform

$$\mathcal{L}\left\{e^{qt}\right\} = \frac{1}{s - q} \tag{4.25}$$

with q an arbitrary complex number. This transform has a simple pole at $s = q$ so that $\alpha = \text{Re}\, q$. As another example consider the function $\sqrt{t}$. Direct computation gives

$$\mathcal{L}\left\{\sqrt{t}\right\} = \int_0^\infty e^{-st}\sqrt{t}\,dt = 2\int_0^\infty x^2 e^{-sx^2}\,dx = \frac{\sqrt{\pi}}{2}s^{-3/2}. \tag{4.26}$$

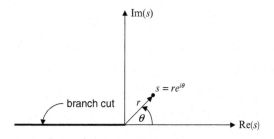

Figure 4.3: Branch cut to render $s^{-3/2}$ analytic in the right half-plane

Unlike (4.24) and (4.25) this transform is not a rational function. The singularity at $s = 0$ is not a pole but a branch point and the function is multivalued. Nevertheless we can still define an analytic function for $\operatorname{Re} s > \varepsilon$ where $\alpha = \varepsilon$ with ε an arbitrarily small positive quantity. We do this by choosing a branch cut along the negative real axis, as shown in Fig. 4.3.

This renders $s^{-3/2}$ analytic everywhere except along the branch cut. Another example of a function with a non-rational LT is $1/\sqrt{t}$. Unlike in examples (4.25) and (4.26) the derivative of this function does not have an LT since it fails to satisfy (4.7). On the other hand $\mathcal{L}\left\{1/\sqrt{t}\right\}$ exists and is given by

$$\mathcal{L}\left\{1/\sqrt{t}\right\} = \int_0^\infty e^{-st}\left(1/\sqrt{t}\right)dt = 2\int_0^\infty e^{-sx^2}dx = \sqrt{\frac{\pi}{s}}. \tag{4.27}$$

The same branch cut as in Fig. 4.3 can be used to ensure that $1/\sqrt{s}$ is analytic in the right half plane. More generally, we can obtain a formula for the LT of t^ν, where ν is an arbitrary real number greater than -1 by using the definition of the Gamma function

$$\Gamma(\nu) = \int_0^\infty x^{\nu-1}e^{-x}dx.$$

The result is

$$\mathcal{L}\left\{t^\nu\right\} = \frac{\Gamma(\nu+1)}{s^{\nu+1}}\ ;\ \nu > -1. \tag{4.27*}$$

4.1.4   Inversion Formula

Consider the function $w(t)$ defined by the integral

$$w(t) = \frac{1}{2\pi i}\int_{\gamma-i\infty}^{\gamma+i\infty}\frac{F(s)e^{st}}{s^2}ds, \tag{4.28}$$

where the integration is performed along the straight line path intercepting the axis of reals at $\operatorname{Re} s = \gamma$ with $\gamma > \alpha$, as shown in Fig. 4.4.

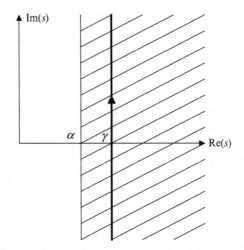

Figure 4.4: Integration path in the complex s-plane

Next we substitute for $F(s)$ in (4.28) the integral (4.3) and interchange the orders of integration. As a result we obtain

$$w(t) = \int_0^\infty f(\tau) \left\{ \frac{1}{2\pi i} \int_{\gamma - i\infty}^{\gamma + i\infty} \frac{e^{s(t-\tau)}}{s^2} ds \right\} d\tau. \tag{4.29}$$

We now evaluate the inner integral by residues. First consider the case $t - \tau < 0$. Since the exponential decays in the right half of the complex plane we form a closed contour by adding to the existing straight line path an infinite semicircular contour situated in the right half plane. Because $1/s^2$ approaches zero at infinity Jordan's lemma (see Appendix A) ensures that in the limit the contribution along the semicircular path vanishes. Therefore the integral along the straight line path in Fig. 4.4 may be equated to the negative sum of the residues at singularities located to the right of the integration path. Since, however, $1/s^2$ is analytic to the right of γ this integral equals zero. For $t - \tau > 0$ the exponential decays in the left half of the complex plane. We therefore close the contour with a large semicircle in the left half plane. This contribution again vanishes in accordance with Jordan's lemma. The resulting integration along the path in Fig. 4.4 is then equal to the residue at the second-order pole located at $s = 0$. A simple differentiation of the exponential shows that this residue equals $t - \tau$. Incorporating both results into a single expression we obtain

$$\frac{1}{2\pi i} \int_{\gamma - i\infty}^{\gamma + i\infty} \frac{e^{s(t-\tau)}}{s^2} ds = (t - \tau) U(t - \tau). \tag{4.30}$$

When the inner integral in (4.29) is replaced by the right side of (4.30) the upper limit of the integral may be truncated to t so that

$$w(t) = \int_0^t f(\tau)(t - \tau) d\tau. \tag{4.31}$$

Differentiating this we get $w'(t) = \int_0^t f(\tau) \, d\tau$ and with the aid of a second differentiation we find that $w''(t) = f(t)$. Thus to obtain $f(t)$ we merely need to differentiate both sides of (4.28) twice. The result is

$$f(t) = \frac{1}{2\pi i} \int_{\gamma - i\infty}^{\gamma + i\infty} F(s) \, e^{st} ds, \tag{4.32}$$

which is the LT inversion formula. Since the integral on the right represents the inverse operation of (4.3) it seems appropriate to use the notation

$$f(t) = \mathcal{L}^{-1}\{F(s)\}. \tag{4.33}$$

As a compact notation for both the direct transform (4.3) and its inverse (4.32) we shall also at times use the symbols

$$f(t) \overset{\mathcal{L}}{\iff} F(s). \tag{4.34}$$

4.1.5 Fundamental Theorems

Properties

Transformation of Derivatives and solution of DE. Because of the exponential nature of the kernel, the LT, just like the FT, can be employed to convert a differential operator with constant coefficients into a polynomial in the transform variable. In (2.161) in Sect. 2.2 we have demonstrated how this property can be exploited in conjunction with the FT to solve linear differential equations with constant coefficients. We can accomplish the same result by using the LT. In fact in this respect the LT has two advantages over the FT. The first of these is that given the initial conditions at say $t = t_0$ we can solve a DE for $t \geq t_0$ without the need of explicitly imposing the constraint that the output signal be identically zero for $t < t_0$. Instead we accomplish the same result by truncating the integral *defining* the transform to $t \geq t_0$. By convention in the (unilateral) LT t_0 is set equal to zero so that causality becomes an implicit attribute of the transform. The other advantage of the LT over the FT is that the given DE may not admit solutions which are both causal *and* possess an FT. On the other hand, as long as the excitation (input) is of exponential order as $t \sim \infty$ we are assured of the existence of an LT.

We now assume that the derivatives up to and including order N of the function satisfying an N-th order DE exist for $t > 0$. Thus we exclude functions with step discontinuities except at $t = 0$ and functions that may become unbounded at $t = 0$ as, e.g., the function in (4.27). Since the function will generally possess a step discontinuity at $t = 0$ we must distinguish between its right side value, which we denote by $f(0^+)$, and the value at $t = 0^-$ which by definition equals zero. We then define the LT of $f'(t)$ with the lower integration limit set to $t = 0^+$. This avoids differentiation at the discontinuity and implicitly identifies

$f'(t)$ as the derivative of the " smooth" part of the function. The Laplace integral is then transformed by an integration by parts as follows:

$$\mathcal{L}\{f'(t)\} = \int_{0+}^{\infty} e^{-st} f'(t)\, dt = e^{-st} f(t)\,|_{0+}^{\infty} + s \int_{0+}^{\infty} e^{-st} f(t)\, dt$$

$$= sF(s) - f(0^+), \tag{4.35}$$

where it is assumed that $\mathrm{Re}\, s > \alpha$. Alternatively, we can get the same result by using an explicit decomposition of $f(t)$ into the smooth (differentiable) part and a step, i.e.,

$$f(t) = f_s(t) + f(0^+)\, U(t).$$

Upon differentiation this becomes

$$f'(t) = f_s'(t) + f(0^+)\, \delta(t). \tag{4.36}$$

On the other hand differentiating both sides of the inversion formula (4.32) gives $\mathcal{L}\{f'(t)\} = sF(s)$. This derivative is to be interpreted as a generalized derivative that may include singularity functions. Taking the LT of both sides of (4.36) and identifying the LT of the left side with $sF(s)$ yields

$$\mathcal{L}\{f_s'(t)\} = sF(s) - f(0^+). \tag{4.37}$$

The preceding is identical to (4.35) provided $f'(t)$ is relabeled as $f_s'(t)$. The distinction between $f_s(t)$ and $f(t)$ is usually not made explicit with unilateral LT and we shall adhere to this custom by dropping the subscript s. Nevertheless, to avoid apparent inconsistencies this distinction should be kept in mind. For example, using (4.35) with $f(t) = U(t)$ we get on account of (4.24)

$$\mathcal{L}\left\{\frac{dU(t)}{dt}\right\} = s\frac{1}{s} - 1 = 0,$$

which in view of (4.22) and $dU(t)/dt = \delta(t)$ should equal unity. This contradiction is only apparent for formula (4.35) holds only for the smooth part of the function which according to the decomposition (4.36) is in this case identically zero.

The LT of higher order derivatives follows from a repetition of the steps in (4.35). For example, for the second derivative we obtain

$$\mathcal{L}\left\{\frac{d^2 f(t)}{dt^2}\right\} = s^2 F(s) - sf(0^+) - f^{(1)}(0^+), \tag{4.38}$$

where $f^{(1)}(0^+)$ is the first derivative at $t = 0^+$. Finally, for a derivative of any order, we get

$$\mathcal{L}\left\{\frac{d^n f(t)}{dt^n}\right\} = s^n F(s) - \sum_{\ell=1}^{n} s^{n-\ell} f^{(\ell-1)}(0^+). \tag{4.39}$$

We can use formula (4.39) to find the LT of the general solution of an N-th order linear differential equation with constant coefficients subject to initial conditions at $t = 0$. Thus we transform

$$\sum_{n=0}^{N} a_n \frac{d^n y(t)}{dt^n} = f(t) \tag{4.40}$$

as follows:

$$Y(s) \sum_{n=0}^{N} a_n s^n = \sum_{n=1}^{N} a_n \sum_{\ell=1}^{n} s^{n-\ell} f^{(\ell-1)}(0^+) + F(s), \tag{4.41}$$

where $Y(s) = \mathcal{L}\{y(t)\}$. Solving for $Y(s)$ we get

$$Y(s) = \frac{\sum_{n=1}^{N} a_n \sum_{\ell=1}^{n} s^{n-\ell} f^{(\ell-1)}(0^+)}{\sum_{n=0}^{N} a_n s^n} + \frac{F(s)}{\sum_{n=0}^{N} a_n s^n}. \tag{4.42}$$

Using the notation in (4.33) for the inverse transform the complete solution of (4.40) for $t \geq 0$ reads

$$y(t) = \mathcal{L}^{-1} \left\{ \frac{\sum_{n=1}^{N} a_n \sum_{\ell=1}^{n} s^{n-\ell} f^{(\ell-1)}(0^+)}{\sum_{n=0}^{N} a_n s^n} \right\} + \mathcal{L}^{-1} \left\{ \frac{F(s)}{\sum_{n=0}^{N} a_n s^n} \right\}. \tag{4.43}$$

We recognize the first term on the right as the zero input and the second term as the zero state response of the corresponding LTI system. Also, since the LT of a delta function is unity, the system impulse response is

$$h(t) = \mathcal{L}^{-1} \left\{ \frac{1}{\sum_{n=0}^{N} a_n s^n} \right\}. \tag{4.44}$$

The utility of this formula and (4.43) depends on the ease with which the inverse transforms can be evaluated. We shall discuss techniques of evaluating the inverse LT in 4.1.6.

Integration. In the analysis of linear systems one sometimes encounters integro-differential operators involving integrals of the form $\int_0^t f(\tau)d\tau$. We can readily find the corresponding LT using integration by parts. Thus assuming $\operatorname{Re} s > \alpha$ we compute

$$\mathcal{L} \left\{ \int_0^t f(\tau)d\tau. \right\} = \int_0^\infty e^{-st} \left\{ \int_0^t f(\tau)d\tau \right\} dt$$

$$= \left\{ -\frac{e^{-st}}{s} \int_0^t f(\tau)d\tau \right\} |_0^\infty + \frac{1}{s} \int_0^\infty f(t)e^{-st} dt.$$

Because $\left| \int_0^t f(\tau)d\tau \right| < Ne^{\alpha t}$ the first term to the right of the second equality vanishes at the upper limit and we obtain

$$\mathcal{L} \left\{ \int_0^t f(\tau)d\tau \right\} = \frac{F(s)}{s}. \tag{4.45}$$

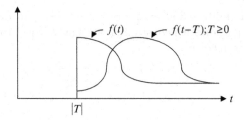

Figure 4.5: Truncation of $f(t)$ for $T < 0$

Initial and Final Value Theorems. An initial and final value theorem similar to the one we have obtained for the FT of causal functions (2.181) and (2.183) in Sect. 2.2 hold also for the LT. Thus if the LT of the derivative exists, it must satisfy (4.8b). Hence using (4.35) we have

$$\lim_{\substack{\mathrm{Re}\,s>\alpha,\ |s|\longrightarrow\infty}} sF(s) = f\left(0^+\right), \qquad (4.46)$$

which is the initial value theorem. For the final value theorem we can proceed as for FT. Assuming $\lim_{t\to\infty} f(t) = A$ exists we have

$$\lim_{s\longrightarrow 0} sF(s) - f\left(0^+\right) = \lim_{s\longrightarrow 0} \mathfrak{L}\left\{\frac{df\,(t)}{dt}\right\} = \lim_{s\longrightarrow 0}\int_{0^+}^{\infty} e^{-st}\frac{df\,(t)}{dt}dt = A - f\left(0^+\right)$$

and cancelling $f\left(0^+\right)$ from both sides we get

$$\lim_{s\longrightarrow 0} sF(s) = A. \qquad (4.47)$$

Differentiation with Respect to the Transform Variable. Differentiating (4.3) n times with respect to s we obtain

$$\mathfrak{L}\left\{t^n f(t)\right\} = (-1)^n\frac{d^n F(s)}{ds^n}. \qquad (4.48)$$

Time Shift. For a time-shifted signal $f(t - T)$ we have

$$\mathfrak{L}\left\{f(t - T)\right\} = \int_0^{\infty} f\left(t - T\right) e^{-st}dt = e^{-sT}\int_{-T}^{\infty} f\left(\tau\right) e^{-s\tau}d\tau.$$

Because $f(t) = 0$ for $t < 0$ the last integral equals $F(s)$ only if $T \geq 0$, i.e., when the signal is *delayed* in time. When $T < 0$ this integral equals the LT of a function that is equal to $f(t)$ only for $t \geq -T$ but identically zero in the interval $0 \leq t < -T$. This truncation is illustrated in Fig. 4.5.

We can write the final result as follows:

$$\mathfrak{L}\left\{f(t - T)\right\} = \begin{cases} e^{-sT}F(s) & ;T \geq 0, \\ e^{-sT}\int_{-T}^{\infty} f\left(\tau\right) e^{-s\tau}d\tau & ;T < 0. \end{cases} \qquad (4.49)$$

Multiplication by Exponential Functions. With q an arbitrary complex number and $f(t)$ of exponential order we form the function $y(t) = e^{qt} f(t)$. (Note that this modifies the growth rate at infinity to $O(e^{(\alpha + \operatorname{Re} q)t})$.) The corresponding LT is

$$Y(s) = \mathcal{L}\left\{e^{qt} f(t)\right\} = F(s - q) \tag{4.50}$$

so that the region of analyticity of $Y(s)$ is shifted to $\operatorname{Re}(s) > \alpha + \operatorname{Re} q$.

Convolution. For two causal functions $f_1(t)$ and $f_2(t)$ the convolution integral becomes

$$\int_{-\infty}^{\infty} f_1(\tau) f_2(t - \tau) d\tau = \int_0^t f_1(\tau) f_2(t - \tau) d\tau. \tag{4.51}$$

For the LT we obtain

$$
\begin{aligned}
\mathcal{L}\left\{\int_0^t f_1(\tau) f_2(t - \tau) d\tau\right\} &= \mathcal{L}\left\{\int_0^{\infty} f_1(\tau) f_2(t - \tau) U(t - \tau) d\tau\right\} \\
&= \int_0^{\infty} f_1(\tau) \mathcal{L}\{f_2(t - \tau) U(t - \tau)\} d\tau \\
&= \int_0^{\infty} f_1(\tau) \left\{\int_0^{\infty} e^{-st} f_2(t - \tau) U(t - \tau) dt\right\} d\tau \\
&= \int_0^{\infty} f_1(\tau) e^{-s\tau} d\tau \int_{-\tau}^{\infty} e^{-sx} f_2(x) U(x) dx \\
&= \int_0^{\infty} f_1(\tau) e^{-s\tau} d\tau \int_0^{\infty} e^{-sx} f_2(x) dx \\
&= F_1(s) F_2(s). \tag{4.52}
\end{aligned}
$$

Thus just like for the FT, the convolution in the time domain yields a product in the transform domain.

LT of Polynomial and Exponential Signals

The LT of signals comprised of polynomials and exponentials can be synthesized in terms of preceding results without evaluating the transform integral directly. For example, using (4.24) and (4.48) we get

$$\mathcal{L}\{t^n\} = \frac{n!}{s^{n+1}}. \tag{4.53}$$

Using (4.50) together with (4.53) we obtain

$$\mathcal{L}\{t^n e^{qt}\} = \frac{n!}{(s - q)^{n+1}}. \tag{4.54}$$

Also with $q = i\omega_0$ with ω_0 real and $f(t) = U(t)$ (4.50) yields

$$\mathcal{L}\{e^{i\omega_0 t}\} = \frac{1}{s - i\omega_0}$$

and using appropriate superpositions of $e^{i\omega_0 t}$ we readily obtain the transforms

$$\mathcal{L}\left\{\cos\omega_0 t\right\} = \frac{s}{s^2 + \omega_0^2}, \tag{4.55a}$$

$$\mathcal{L}\left\{\sin\omega_0 t\right\} = \frac{\omega_0}{s^2 + \omega_0^2}. \tag{4.55b}$$

With $\alpha > 0$ the LT of the damped sinusoid $e^{-\alpha t}\cos\omega_0 t$ can be found from (4.55a) in conjunction with (4.50). Thus

$$\mathcal{L}\left\{e^{-\alpha t}\cos\omega_0 t\right\} = \frac{s + \alpha}{(s + \alpha)^2 + \omega_0^2}. \tag{4.56}$$

Note that as long as $\alpha > 0$ the FT can be obtained from (4.56) by the substitution $s = i\omega$. Evidently this procedure fails when $\alpha = 0$ for it cannot account for the presence of the delta functions in the FT (see (2.138)). For $\alpha < 0$ the FT does not exist but (4.56) retains its validity. We shall discuss the relationship between the FT and the LT in detail in (2.3).

Periodic Signals

Let $f_0(t)$ be a signal of duration T, identically zero for $t \geq T$ and $t < 0$. In view of (4.5) $f_0(t)$ is of exponential order with $\alpha = 0$. The signal defined by the sum

$$f(t) = \sum_{n=0}^{\infty} f_0(t - nT) \tag{4.57}$$

is then periodic for $t \geq 0$ with period T. Setting

$$\mathcal{L}\left\{f_0(t)\right\} = F_0(s)$$

and applying (4.49) to (4.57) we obtain

$$\mathcal{L}\left\{f(t)\right\} = F(s) = F_0(s)\sum_{n=0}^{\infty} e^{-sTn}.$$

This geometric series converges for $\operatorname{Re} s > 0$ so that $f(t)$ is of exponential order with $\alpha > 0$. Summing the series we get

$$F(s) = \frac{F_0(s)}{1 - e^{-sT}}. \tag{4.58}$$

4.1.6   Evaluation of the Inverse LT

Rational Functions

The LT of the zero input response of an LTI system governed by ordinary DE is in general a ratio of two polynomials, i.e., a rational function. If the input is representable as a superposition of a combination of exponentials and polynomials, then the LT of the zero state response is also a rational function. In that case we can represent the LT of the total output by

$$\hat{Y}(s) = \frac{\hat{N}(s)}{D(s)}, \tag{4.59}$$

where $\hat{N}(s)$ and $D(s)$ are coprime polynomials (i.e., without common factors) and $degree\hat{N}(s) \leq degreeD(s)$. If the two polynomials have identical degrees, we can carry out a long division to obtain

$$\hat{Y}(s) = c + Y(s),$$

where c is a constant and

$$Y(s) = \frac{N(s)}{D(s)} = \frac{b_1 s^{N-1} + b_2 s^{N-2} + \cdots + b_N}{s^N + a_1 s^{N-1} + \cdots + a_N}. \tag{4.60}$$

The inverse transform of (4.59) is then

$$\hat{y}(t) = c\delta(t) + y(t)$$

with

$$y(t) = \mathcal{L}^{-1}\{Y(s)\} = \frac{1}{2\pi i} \int_{\gamma - i\infty}^{\gamma + i\infty} Y(s) e^{st} ds. \tag{4.61}$$

The region of analyticity of $Y(s)$ is defined by $\mathrm{Re}\, s > \alpha = \mathrm{Re}\{s_{\max}\}$ with $s_{\max}$ the zero of $D(s)$ with the largest real part. Consequently in (4.61) we must choose $\gamma > \mathrm{Re}\{s_{\max}\}$. Referring to Fig. 4.6, we now evaluate the integral of $Y(s) e^{st}$ along a closed path in the counterclockwise direction over the contour consisting of the straight line and the circular path in the left half plane. We choose the radius R of the circle large enough to enclose all the zeros of $D(s)$ and obtain

$$\frac{1}{2\pi i}\{\int_{\gamma - i\sqrt{R^2 - \gamma^2}}^{\gamma + i\sqrt{R^2 - \gamma^2}} Y(s) e^{st} ds + \int_{CR} Y(s) e^{st} ds\} = \sum_{n=1}^{K} r_n(t), \tag{4.62}$$

where the subscript CR refers to the circular path, $r_n(t)$ is the residue of $Y(s) e^{st}$ at the pole located at $s = s_n$, and $K \leq N$ is the number of poles. If we let R approach infinity (while keeping γ fixed) the first integral on the left of (4.62) reduces to (4.61). Because $Y(s) \to 0$ for $\mathrm{Re}\, s < 0$ as $R \to \infty$ the integral over the circular path approaches zero in accordance with Jordan's lemma (Appendix A). Hence

$$y(t) = \mathcal{L}^{-1}\{Y(s)\} = \frac{1}{2\pi i} \int_{\gamma - i\infty}^{\gamma + i\infty} Y(s) . e^{st} ds = \sum_{n=1}^{K} r_n(t). \tag{4.63}$$

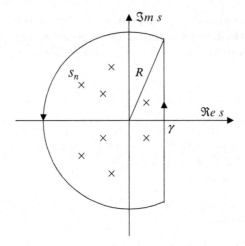

Figure 4.6: Evaluation of inverse LT of rational functions

For a pole of order m at $s = s_n$ the residue is (see Appendix A)

$$r_n(t) = \frac{1}{(m-1)!} \frac{d^{m-1}}{ds^{m-1}} \left[(s-s_n)^m \frac{N(s)e^{st}}{D(s)} \right]_{s=s_n}. \tag{4.64}$$

When the pole is simple ($m = 1$) the following alternative formula (Appendix A) may be used:

$$r_n(t) = \frac{N(s)e^{st}}{\frac{d}{ds}D(s)} |_{s=s_n}. \tag{4.65}$$

Example 1

$$F(s) = \frac{1}{s(s+2)} \tag{4.66}$$

The function has simple poles at $s = 0$ and $s = -2$. We choose $\gamma > 0$ and in accordance with (4.63) and (4.64) obtain

$$\mathcal{L}^{-1}\{F(s)\} = \frac{1}{s+2}e^{st} |_{s=0} + \frac{1}{s}e^{st} |_{s=-2}$$

$$= \frac{1}{2} - \frac{1}{2}e^{-2t}. \tag{4.67}$$

Example 2

$$F(s) = \frac{s}{(s+1)(s-3)^2} \tag{4.68}$$

This function has a simple pole at $s = -1$ and a pole of order 2 at $s = 3$. Here $\gamma > 3$ so that

$$\mathcal{L}^{-1}\{F(s)\} = \frac{se^{st}}{(s-3)^2}\Big|_{s=-1} + \frac{d}{ds}\frac{se^{st}}{(s+1)}\Big|_{s=3}$$

$$= -\frac{1}{16}e^{-t} + \frac{(s+1)(1+st)e^{st} - se^{st}}{(s+1)^2}\Big|_{s=3}$$

$$= -\frac{1}{16}e^{-t} + \frac{1+12t}{16}e^{3t}. \tag{4.69}$$

Example 3

$$F(s) = \frac{s^2}{(s+2)(s+3)} \tag{4.70}$$

In this case $F(s)$ approaches a finite value at infinity which must be subtracted before the residue formula can be applied. Since $F(\infty) = 1$ we write

$$F(s) = 1 - 1 + F(s) = 1 - \frac{5s+6}{(s+2)(s+3)}.$$

The inverse of the first term on the right is a delta function while the second term represents a transform that vanishes at infinity. To the second term we may apply the residue formula and obtain

$$\mathcal{L}^{-1}\{F(s)\} = \delta(t) - \frac{(5s+6)e^{st}}{s+2}\Big|_{s=-3} - \frac{(5s+6)e^{st}}{s+3}\Big|_{s=-2}$$

$$= \delta(t) - 9e^{-3t} + 4e^{-2t}. \tag{4.71}$$

Example 4

$$F(s) = \frac{4s}{[(s+1)^2 + 4](s+2)}. \tag{4.72}$$

This function has two simple poles at $s_{1,2} = -1 \pm 2i$ as well as a simple pole at $s = -2$. We obtain

$$\mathcal{L}^{-1}\{F(s)\} = \frac{4se^{st}}{[(s+1)^2 + 4]}\Big|_{s=-2}$$

$$+ \frac{4se^{st}}{2(s+1)(s+2)}\Big|_{s=-1+2i} + \frac{4se^{st}}{2(s+1)(s+2)}\Big|_{s=-1-2i},$$

where for the complex poles we have used the differentiation formula (4.65). Completing the algebra in the preceding expression we get

$$\mathcal{L}^{-1}\{F(s)\} = -\frac{8}{5}e^{-2t} + e^{-t}(\frac{8}{5}\cos 2t + \frac{6}{5}\sin 2t). \tag{4.73}$$

Transcendental Functions

Meromorphic Functions. In many physical problems one encounters LTs that are not rational functions. A particularly important example of nonrational functions is the class referred to as meromorphic functions. Their only singularities in the finite part of the complex plane are poles. Meromorphic functions differ from rational functions only in that the number of poles is allowed to approach infinity. For example, referring to (4.58), we note that the zeros of the denominator are infinite in number so that the LT of a periodic function will be meromorphic provided $F_0(s)$ is itself either meromorphic or rational. Let us examine a specific case:

$$F(s) = \frac{e^{-s} - e^{-2s}}{s(1 - e^{-2s})}. \tag{4.74}$$

Here $F_0(s) = \left(e^{-s} - e^{-2s}\right)/s$ so that $f_0(t)$ is the rectangular pulse $U(t-1) - U(t-2)$. Hence $f(t) = \mathcal{L}^{-1}\{F(s)\}$ is the periodic function displayed in Fig. 4.7.

Alternatively we can apply the inversion formula to (4.74). Since $F(s) \to 0$ as $|s| \to \infty$ the Jordan lemma applies so that for $t > 0$ $f(t)$ may be equated to sum of the residues corresponding to the infinite number of simple poles at $s = s_n = in\pi$, with $n = \pm 1, \pm 2, \ldots$ and the solitary double pole at $s = 0$. Setting $G = \left(e^{-s} - e^{-2s}\right)\left(1 - e^{-2s}\right)^{-1}$ the contribution from the latter is

$$\frac{d}{ds}\left(Gse^{st}\right)\big|_{s=0} = e^{st}\left[(1+st)G + s\frac{dG}{ds}\right]\big|_{s=0}.$$

As s approaches zero G approaches a finite value which we compute with the aid of Taylor expansions about $s = 0$ of the numerator and denominator as follows:

$$G \to \frac{1 - s + s^2/2 - \ldots - \left[1 - 2s + 2s^2 - \ldots\right]}{1 - \left[1 - 2s + 2s^2 - \ldots\right]} \to \frac{1}{2}.$$

Clearly $\frac{dG}{ds}\big|_{s=0}$ is also finite so that we obtain

$$\frac{d}{ds}\left(Gse^{st}\right)\big|_{s=0} = \frac{1}{2}. \tag{4.75}$$

For the residues at the infinite set of simple poles we get

$$\frac{\left(e^{-s} - e^{-2s}\right)e^{st}}{s\frac{d}{ds}\left(1 - e^{-2s}\right)}\big|_{s=in\pi} = \frac{[(-1)^n - 1]e^{in\pi t}}{2in\pi} = \begin{cases} 0 \,; \ n \text{ even}, \\ \frac{e^{in\pi t}}{in\pi}\,; \ n \text{ odd}. \end{cases} \tag{4.76}$$

Summing over all positive and negative n, setting $n = 2k+1$ and adding (4.75) we get the final result

$$f(t) = \frac{1}{2} + 2\sum_{k=0}^{\infty} \frac{\sin\left[\pi(2k+1)t\right]}{\pi(2k+1)}, \tag{4.77}$$

which is just the FS representation of the square wave in Fig. 4.7.

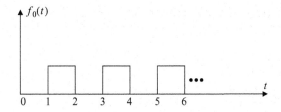

Figure 4.7: Inverse LT of (4.74)

Example 2
Next consider the LT

$$I(s) = \frac{1+\gamma}{2s} \frac{1+e^{-s\tau}}{1+\gamma e^{-e\tau}}, \tag{4.78}$$

where $0 \leq \gamma \leq 1$ and $\tau > 0$. This function represents the LT of the electric current $i(t)$ in the resistor R in the circuit shown in Fig. 4.8. The resistor is shunted by an open-circuited transmission line of length L and characteristic impedance Z_0, and the parallel combination is excited by the ideal current generator $U(t)$. The transmission line is assumed lossless with $\tau = L/v$, v the phase velocity, and $\gamma = (R - Z_0)(R + Z_0)^{-1}$.

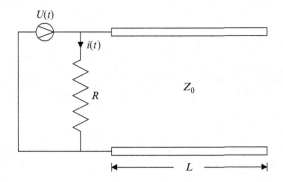

Figure 4.8: Electric circuit interpretation of (4.78)

As in Example 1 the LT (4.78) approaches zero as $|s| \to \infty$ so that $i(t)$ is obtained by summing the residues. Presently we have a simple pole at $s = 0$ and, unless $\gamma = 1$, an infinite set of simple poles corresponding to $1 + \gamma e^{-e\tau} = 0$. Denoting these by s_n we have

$$s_n = \ln \gamma / \tau + in\pi / \tau; \ n = \pm 1, \pm 3, \pm 5 \ldots \tag{4.79}$$

Note that the real parts of s_n are all identical and always negative and tend to $-\infty$ as γ approaches zero. Carrying out the residue calculation we get

$$i(t) = \frac{1+\gamma}{2} \frac{1+e^{-s\tau}}{1+\gamma e^{-e\tau}} e^{st} \mid_{s=0} + \sum_n \frac{1+\gamma}{2s} \frac{1+e^{-s\tau}}{\frac{d}{ds}(1+\gamma e^{-e\tau})} e^{st} \mid_{s=s_n}$$

$$= \frac{1}{2} - \frac{1-\gamma^2}{2\gamma}\gamma^{t/\tau} \sum_n \frac{e^{in\pi t}}{\ln\gamma + in\pi}. \tag{4.80}$$

Changing the summation index to k with $n = 2k+1$ the last series assumes the form of a Fourier series:

$$\sum_n \frac{e^{in\pi t}}{\ln\gamma + in\pi} = \sum_{k=0}^\infty \frac{\ln\gamma \cos\pi(2k+1)t + \pi(2k+1)\sin\pi(2k+1)t}{\ln^2\gamma + \pi^2(2k+1)^2}. \tag{4.80*}$$

Functions with Branch Point Singularities.   In a variety of physical problems one encounters multivalued. LT. Examples include lossy transmission lines, waveguides and, more generally, electromagnetic radiating structures (e.g., antennas). Unlike for rational and meromorphic functions analyticity of $F(s)$ for sufficiently large $|s|$ in the right half plane is for such functions not automatic but must be imposed by confining the range of s to a particular (usually top) Riemann sheet. This can be done through a judicious choice of branch cuts as illustrated in the following representative examples.

Example 1

$$F(s) = \frac{b\sqrt{s}}{\sqrt{s} - a}, \tag{4.81}$$

where a and b are real constants. Since $F(\infty) = b$, we subtract b from $F(s)$ yielding a function of s that tends to zero at infinity and then add it back again to maintain equality. Thus

$$F(s) = \frac{b\sqrt{s}}{\sqrt{s} - a} - b + b = b + \frac{ab}{\sqrt{s} - a}.$$

Hence

$$f(t) = \mathcal{L}^{-1}\{F(s)\} = b\delta(t) + ab\mathcal{L}^{-1}\{G(s)\}, \tag{4.82}$$

where

$$G(s) = \frac{1}{\sqrt{s} - a}. \tag{4.83}$$

Evidently $G(\infty) = 0$ which ensures the validity of the Jordan lemma for the inversion formula. We note that $G(s)$ has a branch point at $s = 0$ and an isolated singularity at $\sqrt{s} = a$. Since $(\sqrt{s} - a)^{-1} = (\sqrt{s} - a)(s - a^2)^{-1}$ we see that this singularity is a simple pole at $s = a^2$. To apply the LT inversion formula we must construct a $G(s)$ that is analytic for $\Re\sqrt{s} > \alpha$. We can accomplish this by defining the analytic function $\sqrt{s}$ by means of a branch cut along the negative real axis, with the result that $\Re\sqrt{s} > 0$ on the top Riemann

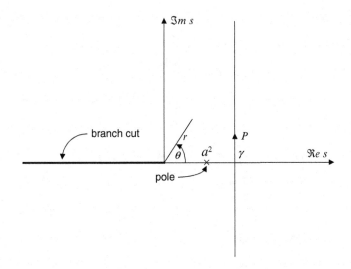

Figure 4.9: Integration path for the LT inversion formula in (4.84)

sheet. Consequently, the pole will exist on the top Riemann sheet only if $a > 0$. Assuming this to be the case $G(s)$ will be analytic for $\Re es > a^2$ $(\alpha > a^2)$ so that the inversion formula reads

$$g(t) = \mathcal{L}^{-1}\left\{\frac{1}{\sqrt{s}-a}\right\} = \frac{1}{2\pi i}\int_{\gamma-i\infty}^{\gamma+i\infty}\frac{e^{st}ds}{\sqrt{s}-a}, \tag{4.84}$$

where $\gamma > a^2$, and the disposition of the integration path P relative to the singularities is as shown in Fig. 4.9.

We close the path of integration in (4.84) with a semicircle in the left half plane and two integrals along the two sides of the branch cut, as shown in Fig. 4.10.

The resulting integral around a closed path equals $2\pi i$ times the residue at the enclosed pole. Thus

$$\frac{1}{2\pi i}\left\{\begin{array}{l}\int_{\gamma-iR\sin\psi}^{\gamma+iR\sin\psi}\frac{e^{st}ds}{\sqrt{s}-a} + \int_{\Gamma_{R+}}\frac{e^{st}ds}{\sqrt{s}-a} + \int_{BC+}\frac{e^{st}ds}{\sqrt{s}-a} \\ + \int_{BC-}\frac{e^{st}ds}{\sqrt{s}-a} + \int_{\Gamma_{R-}}\frac{e^{st}ds}{\sqrt{s}-a}\end{array}\right\}$$

$$= 2ae^{a^2t}. \tag{4.85}$$

As $R \to \infty$ the Jordan lemma ensures that for $t > 0$ the two integrals along the semicircular path approach zero. Because $\Re e \sqrt{s} > 0$ on the top Riemann sheet $\sqrt{s} = i\sqrt{r}$ along the upper side of the branch cut and $-i\sqrt{r}$ on the lower side. The integrations around the branch cut then yield the following:

$$\lim_{R\to\infty}\int_{BC+}\frac{e^{st}ds}{\sqrt{s}-a} = \int_\infty^0\frac{-e^{-rt}dr}{i\sqrt{r}-a} = -2\int_0^\infty\frac{e^{-tx^2}xdx}{a-ix},$$

$$\lim_{R\to\infty}\int_{BC-}\frac{e^{st}ds}{\sqrt{s}-a} = \int_0^\infty\frac{-e^{-rt}dr}{-i\sqrt{r}-a} = 2\int_0^\infty\frac{e^{-tx^2}xdx}{a+ix}.$$

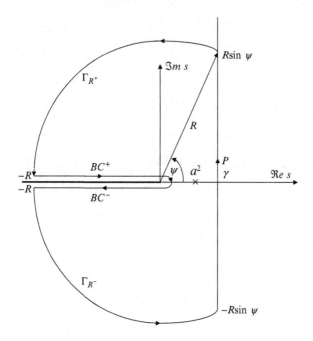

Figure 4.10: Deformation of the integration path in the evaluation of the inverse LT

Inserting these into (4.85) we get in the limit as $R \to \infty$

$$g(t) = \frac{1}{2\pi i} \int_{\gamma-i\infty}^{\gamma+i\infty} \frac{e^{st} ds}{\sqrt{s} - a} = 2ae^{a^2 t} + \frac{2}{\pi} Q(t), \qquad (4.86)$$

where

$$Q(t) = \int_0^\infty \frac{e^{-tx^2} x^2 dx}{a^2 + x^2}. \qquad (4.87)$$

The last integral can be expressed in terms of the error function by first setting $P(t) = \int_0^\infty e^{-tx^2} \left(a^2 + x^2\right)^{-1} dx$ and noting $Q(t) = -P'(t)$. Then since

$$\int_0^\infty \frac{e^{-tx^2}(x^2 + a^2) dx}{a^2 + x^2} = \int_0^\infty e^{-tx^2} dx = \frac{1}{2}\sqrt{\frac{\pi}{2}}$$

we get the differential equation

$$P'(t) - a^2 P(t) = -\frac{1}{2}\sqrt{\frac{\pi}{t}}.$$

The solution for $P(t)$ reads

$$P(t) = P(0)e^{a^2 t} - \frac{\sqrt{\pi}}{2} \int_0^t e^{a^2(t-\tau)} \tau^{-1/2} d\tau,$$

where $P(0) = \int_0^\infty \left(a^2 + x^2\right)^{-1} dx = \pi/2a$. Through a change of the integration variable to $x = a\tau^{1/2}$ the preceding becomes

$$P(t) = (\pi/2a)\, e^{a^2 t} - \frac{\sqrt{\pi}}{a}\, e^{a^2 t} \int_0^{a\sqrt{t}} e^{-x^2}\, dx,$$

so that

$$Q(t) = \frac{1}{2}\sqrt{\frac{\pi}{t}} - \frac{a\pi}{2}\, e^{a^2 t} + \frac{a\pi}{2}\, e^{a^2 t}\, \mathrm{erf}(a\sqrt{t}), \qquad (4.88)$$

where

$$\mathrm{erf}(a\sqrt{t}) = \frac{2}{\sqrt{\pi}} \int_0^{a\sqrt{t}} e^{-x^2}\, dx = 1 - \mathrm{erf}\, c(a\sqrt{t}) \qquad (4.89)$$

is the error function and its complement. Collecting our results we get the final expression

$$f(t) = b\delta\,(t) + 2a^2 b e^{a^2 t} - a^2 b e^{a^2 t}\, \mathrm{erf}\, c(a\sqrt{t}) + \frac{ab}{\sqrt{\pi t}}, \qquad (4.90)$$

which requires that $a > 0$. When $a < 0$ the pole exists on the bottom Riemann sheet so that the right side of (4.85) is zero. Proceeding as above we obtain

$$g(t) = \frac{1}{\sqrt{\pi t}} - |a|\, e^{a^2 t}\, \mathrm{erf}\, c(|a|\, \sqrt{t}), \qquad (4.91)$$

which then yields

$$f(t) = b\delta\,(t) - a^2 b e^{a^2 t}\, \mathrm{erf}\, c(|a|\, \sqrt{t}) + \frac{ab}{\sqrt{\pi t}}. \qquad (4.92)$$

Example 2

In this example we compute the response of a homogeneous electrically conducting medium to a unit step excitation. A plane wave field component initialized by a unit step at $x = 0$ and $t = 0$ that has propagated a distance x into the conducting medium can be represented at time t by

$$A(t, x) = \frac{1}{2\pi i} \int_{\gamma - i\infty}^{\gamma + i\infty} e^{st}\, \frac{e^{-(sx/v)\sqrt{1+(s\tau)^{-1}}}}{s}\, ds, \qquad (4.93)$$

where τ is the medium relaxation time and v the phase velocity. It will be convenient to introduce the nondimensional parameters $\hat{t} = t/\tau$, $\hat{x} = x/v\tau$ and change the complex variable s to $z = s\tau$. With these changes (4.93) transforms into

$$
\begin{aligned}
A(t, x) &\equiv \hat{A}(\hat{t}, \hat{x}) = \frac{1}{2\pi i} \int_{\hat{\gamma} - i\infty}^{\hat{\gamma} + i\infty} e^{z\hat{t}}\, \frac{e^{-z\hat{x}\sqrt{1+z^{-1}}}}{z}\, dz \\
&= \frac{1}{2\pi i} \int_{\hat{\gamma} - i\infty}^{\hat{\gamma} + i\infty} e^{z(\hat{t} - \hat{x})}\, \frac{e^{-\hat{x}\left(\sqrt{z(z+1)} - z\right)}}{z}\, dz. \qquad (4.94)
\end{aligned}
$$

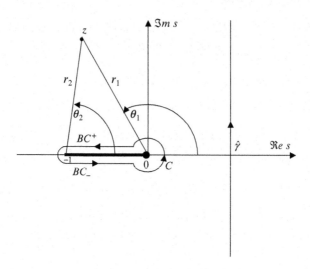

Figure 4.11: Disposition of singularities in the evaluation of the inverse LT (4.95)

Next we note that $\sqrt{z\,(z+1)} - z \to 0$ as $|z| \to \infty$ so that $p(z) = e^{-\hat{x}\left(\sqrt{z(z+1)}-z\right)}$ $/z \to 0$ as $|z| \to \infty$. Thus if we can demonstrate that $p(z)$ is analytic for $\Re$ $s > 0$ (which we do in the sequel by a particular choice of branch cut) then, because $\exp z\,(\hat{t} - \hat{x})$ decays in the left half plane whenever $\hat{t} - \hat{x} < 0$, it will follow from the Jordan lemma that $A(t, x) = 0$ for $\hat{t} - \hat{x} < 0$. This is equivalent to $t < x/v$, which is in agreement with our expectation that no signal can propagate faster than v. Since the signal evolution will commence at $\hat{t} = \hat{x}$ it is sensible to shift the time origin to coincide with the time of first signal arrival. This amounts to replacing (4.94) by

$$a\left(\hat{t}, \hat{x}\right) = \frac{1}{2\pi i} \int_{\hat{\gamma}-i\infty}^{\hat{\gamma}+i\infty} e^{z\hat{t}} \frac{e^{-\hat{x}\left(\sqrt{z(z+1)}-z\right)}}{z} dz. \qquad (4.95)$$

The singularities in the integrand are the two branch points of $\sqrt{z\,(z+1)}$: one at $z = 0$, where the integrand approaches ∞, and the other at $z = -1$ where the integrand is finite. We can render the function

$$p(z) = e^{-\hat{x}\left(\sqrt{z(z+1)}-z\right)}/z \qquad (4.96)$$

analytic for $\Re\ s > 0$ by connecting the branch points with the branch cut along the segment $(-1, 0)$ of the real axis as shown in Fig. 4.11.

This follows from the following definition of the square root:

$$\sqrt{z(z+1)} = \sqrt{r_1 r_2}\, e^{i(\theta_1+\theta_2)}, \qquad (4.97)$$

where the angles θ_1 and θ_2 shown in the figure are restricted to $-\pi \le \theta_{1,2} \le \pi$. Using (4.97) we observe that $p(z)$ in (4.96) is continuous everywhere except possibly on the branch cut. This implies that $p(z)$ is analytic outside every closed curve enclosing the branch cut. Turning our attention to the branch cut itself we note that on its upper side, BC^+, the angles approach $\theta_1 = \pi, \theta_2 = 0$ which results in $\sqrt{z(z+1)} = i\sqrt{r_1 r_2} = i\sqrt{r_1(1-r_1)}$. On the lower side of the branch cut, BC^-, the angles evaluate to $\theta_1 = -\pi, \theta_2 = 0$ which gives $\sqrt{z(z+1)} = -i\sqrt{r_1 r_2} = -ir_1(1-r_1)$. Thus in traversing the branch cut downward the square root undergoes a jump of $i2\sqrt{r_1(1-r_1)}$.

We have now all the necessary machinery to evaluate (4.95). Thus for $\hat{t} < 0$ the exponential decays in right half plane so that, invoking the Jordan lemma, $a\left(\hat{t}, \hat{x}\right) = 0$, in agreement with our previous conjecture. For $\hat{t} > 0$ the exponential decays in left half plane. Again taking account of the Jordan lemma and the analyticity of $p(z)$ for all points not on the branch cut we may deform the original path of integration into the path enclosing the branch cut shown in Fig. 5.15. The three contributions can then be represented as follows:

$$a\left(\hat{t}, \hat{x}\right) = \frac{1}{2\pi i}\left\{\int_C p(z)e^{z\hat{t}}dz + \int_{BC^+} p(z)e^{z\hat{t}}dz + \int_{BC^-} p(z)e^{z\hat{t}}dz\right\}. \qquad (4.98)$$

Evaluating each contribution in turn we get

$$\frac{1}{2\pi i}\int_C p(z)e^{z\hat{t}}dz = 1, \qquad (4.99a)$$

$$\frac{1}{2\pi i}\int_{BC^+} p(z)e^{z\hat{t}}dz = \frac{1}{2\pi i}\int_{0^-}^1 -dr_1 e^{-r_1\hat{t}}\frac{e^{-\hat{x}\left(i\sqrt{r_1(1-r_1)}+r_1\right)}}{-r_1}, \qquad (4.99b)$$

$$\frac{1}{2\pi i}\int_{BC^-} p(z)e^{z\hat{t}}dz = \frac{1}{2\pi i}\int_1^{0^-} -dr_1 e^{-r_1\hat{t}}\frac{e^{-\hat{x}\left(-i\sqrt{r_1(1-r_1)}+r_1\right)}}{-r_1}. \qquad (4.99c)$$

Upon summation and a change of the integration variable from r_1 to ξ via $r_1 = \xi^2$ we get

$$a\left(\hat{t}, \hat{x}\right) = \left\{1 - \frac{2}{\pi}\int_0^1 e^{-\xi^2(\hat{t}+\hat{x})}\frac{\sin\left(\hat{x}\xi\sqrt{1-\xi^2}\right)}{\xi}d\xi\right\}U\left(\hat{t}\right) \qquad (4.100)$$

or, reverting to the $\hat{A}(\hat{t}, \hat{x})$ in (4.94)

$$\hat{A}(\hat{t}, \hat{x}) = a\left(\hat{t}, \hat{x}\right) = \left\{1 - \frac{2}{\pi}\int_0^1 e^{-\xi^2\hat{t}}\frac{\sin\left(\hat{x}\xi\sqrt{1-\xi^2}\right)}{\xi}d\xi\right\}U\left(\hat{t}-\hat{x}\right). \qquad (4.101)$$

A plot of (4.101) is shown in Fig. 4.12 at three positions within the medium.

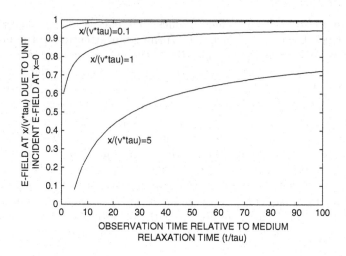

Figure 4.12: Response of conducting medium to unit step excitation

4.2 Double-Sided Laplace Transform

4.2.1 Definition and Analytic Properties

By analogy with functions of exponential order at $t \sim \infty$ we can define functions of exponential order at $t \sim -\infty$. Adapting our previous shorthand notation to this case we write

$$f(t) \sim \underset{t \sim -\infty}{O} (e^{\beta t}) \tag{4.102}$$

by which we shall mean that there exist constants t_0, β, and M such that $|f(t)| < Me^{\beta t}$ for all $t < t_0$. Again by analogy with (4.3) in Sect. 4.1 we now define the transform over the negative t-axis by

$$\hat{F}(s) = \int_{-\infty}^{0} f(t)e^{-st}dt. \tag{4.103}$$

Clearly with a trivial replacement of variables all the steps in the proof of (4.8a) and (4.8b) in Sect. 4.1 retain their validity so that

$$\hat{F}(s) \text{ is an analytic function of } s \text{ for } \operatorname{Re} s < \beta \tag{4.104a}$$

and

$$\lim_{|s| \to \infty} \hat{F}(s) \to 0 \text{ for } \operatorname{Re} s < \beta. \tag{4.104b}$$

We can combine the unilateral transform $F(s)$ and $\hat{F}(s)$ into a single function

$$F_{II}(s) = \hat{F}(s) + F(s), \tag{4.105}$$

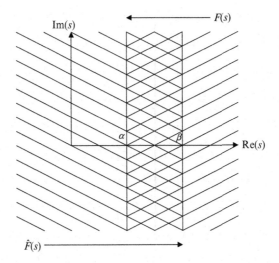

Figure 4.13: Strip of analyticity of a bilateral LT

where $\hat{F}(s)$ is analytic for $\operatorname{Re} s < \beta$ while $F(s)$ is analytic for $\operatorname{Re} s > \alpha$. If $\beta > \alpha$, then the two functions $\hat{F}(s)$ and $F(s)$ have a common region of analyticity, i.e., the vertical strip defined by

$$a < \operatorname{Re} s < \beta \tag{4.106}$$

as illustrated in Fig. 4.13.

Under these circumstances we *define* $F_{II}(s)$ as the double-sided (or bilateral) LT and write

$$F_{II}(s) = \int_{-\infty}^{\infty} f(t)e^{-st}dt, \tag{4.107}$$

which from the preceding discussion requires that $f(t)$ be of exponential order at $\pm\infty$, i.e.,

$$f(t) \sim \begin{cases} O\left(e^{\beta t}\right) & t \sim -\infty, \\ O\left(e^{\alpha t}\right) & t \sim \infty. \end{cases} \tag{4.108}$$

4.2.2 Inversion Formula

To derive the inversion formula for the bilateral LT we let $s = \gamma + i\omega$ with $a < \gamma < \beta$ and rewrite (4.107) as follows:

$$F_{II}(\gamma + i\omega) = \int_{-\infty}^{\infty} f(t)e^{-\gamma t}e^{-i\omega t}dt. \tag{4.109}$$

Evidently $F_{II}(\gamma + i\omega)$ is the FT of $f(t)e^{-\gamma t}$. Therefore we may employ the FT inversion formula and write

$$f(t)e^{-\gamma t} = \frac{1}{2\pi}\int_{-\infty}^{\infty} F_{II}(\gamma + i\omega)e^{i\omega t}d\omega. \tag{4.110}$$

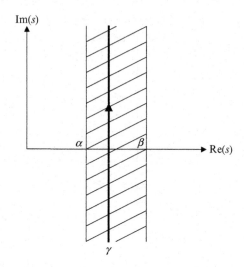

Figure 4.14: Integration path for the bilateral LT

If we now multiply both sides by $e^{\gamma t}$ and revert to the complex variable $s = \gamma + i\omega$, we get

$$f(t) = \frac{1}{2\pi} \int_{-\infty}^{\infty} F_{II}(s)e^{st}d\omega.$$

Since γ is fixed $ds = id\omega$. If we also express the limits of integration in terms of s, the preceding integral becomes

$$f(t) = \frac{1}{2\pi i} \int_{\gamma-i\infty}^{\gamma+i\infty} F_{II}(s)e^{st}ds, \tag{4.111}$$

which is the desired result. The integration path intercept γ may be located anywhere within the strip of analyticity as shown in Fig. 4.14.

Equation (4.111) is identical in appearance to the formula for the unilateral LT. The distinction between the two lies in the choice of the integration paths as defined by the intercept γ. Recall that for the unilateral transform we required analyticity for $\mathrm{Re}\, s > \alpha$ which in accordance with (4.106) implies that $\beta = \infty$. In general a given $F_{II}(s)$ may possess several strips of analyticity so that the corresponding signal is not unique but depends on the prescription of γ. We illustrate this in the following example.

Example 1

$$F_{II}(s) = \frac{1}{(s-2)(s-1)}.$$

The two poles define the following three strips of analyticity:

$$-\infty \;<\; \gamma < 1, \tag{i}$$
$$1 \;<\; \gamma < 2, \tag{ii}$$
$$2 \;<\; \gamma < \infty. \tag{iii}$$

Let us evaluate

$$f(t) = \frac{1}{2\pi i} \int_{\gamma-i\infty}^{\gamma+i\infty} \frac{1}{(s-2)(s-1)} e^{st} ds$$

for each of the three possible choices of integration paths.

case(i)

When $t < 0$ the exponential decays in the right half plane (Re $s > 0$). Since $F_{II}(s)$ decays at infinity, by Jordan's lemma an integral of $F_{II}(s)e^{st}$ taken over a circular contour in the right half plane vanishes as the radius of the circle is allowed to approach infinity. Therefore the line integral of $F_{II}(s)e^{st}$ taken over the path γ is equivalent to an integral over a closed contour enclosing all the singularities of $F_{II}(s)$ in the *clockwise direction*. Computing the (negative) residues at poles $s = 1$ and $s = 2$ we obtain

$$f(t) = -\frac{e^{st}}{(s-2)}\Big|_{s=1} - \frac{e^{st}}{(s-1)}\Big|_{s=2} = e^t - e^{2t}, \quad t < 0.$$

When $t > 0$ the exponential decays in the left half plane (Re $s < 0$). In this case, again applying Jordan's lemma, the integral over a circular path in the left half plane vanishes. The inversion integral is therefore equivalent to a closed contour integral enclosing all the singularities of $F_{II}(s)e^{st}$ in the *counterclockwise direction*. Since in the present case there are no singularities to the left of γ the result of the integration is identically zero. Hence

$$f(t) = 0, \quad t > 0.$$

case(ii)

Proceeding as above for $t < 0$ we close the contour with a circular path in the right half plane. The contributing pole is at $s = 2$ with a (negative) residue of

$$f(t) = -e^{2t}, \quad t < 0.$$

For $t > 0$ we close the contour in the left half plane. The contribution is a (positive) residue from the pole at $s = 1$ which is

$$f(t) = -e^t, \quad t > 0.$$

case(iii)

Now there are no singularities to the right of γ so that

$$f(t) = 0, \quad t < 0$$

while for $t > 0$ both poles contribute positive residues, i.e.,

$$f(t) = -e^t + e^{2t}, \quad t > 0.$$

Evidently case (iii) corresponds to the unilateral LT.

Note that in case(i) $\gamma = 0$ is a permissible choice of integration path. Consequently in this case with the substitution $s = i\omega$ the double-sided LT converts to the FT, viz.,

$$\left(e^t - e^{2t}\right) U(-t) \overset{\mathcal{F}}{\Longleftrightarrow} \frac{1}{(i\omega - 2)(i\omega - 1)}.$$

Example 2

The following transform has a pole of order 2 at $s = -2$ and a simple pole at $s = 3$:

$$F_{II}(s) = \frac{1}{(s+2)^2(s-3)}.$$

Again we have three strips of analyticity

$$-\infty \;<\; \gamma < -2, \tag{i}$$
$$-2 \;<\; \gamma < 3, \tag{ii}$$
$$3 \;<\; \gamma < \infty. \tag{iii}$$

Proceeding as in Example 1 we get the following three functions
case(i)

$$f(t) = \begin{cases} -\frac{d}{ds}\frac{e^{st}}{(s-3)}\big|_{s=-2} - \frac{e^{st}}{(s+2)^2}\big|_{s=3} = \frac{e^{-2t}}{5}\left(t + \frac{1}{5}\right) - \frac{e^{3t}}{25} \;;\; t < 0 \\ 0 \;;\; t > 0 \end{cases}$$

case(ii)

$$f(t) = \begin{cases} -\frac{e^{3t}}{25} \;;\; t < 0 \\ -\frac{e^{-2t}}{5}\left(t + \frac{1}{5}\right) \;;\; t > 0 \end{cases}$$

case(iii)

$$f(t) = \begin{cases} 0 \;;\; t < 0 \\ -\frac{e^{-2t}}{5}\left(t + \frac{1}{5}\right) + \frac{e^{3t}}{25} \;;\; t > 0 \end{cases}$$

Again case(iii) corresponds to the unilateral LT and case(ii) to the FT $F(\omega) = F_{II}(i\omega)$. Hence in the second case we may also write the transform relationship

$$-\frac{e^{3t}}{25}U(-t) - \frac{e^{-2t}}{5}\left(t + \frac{1}{5}\right)U(t) \overset{\mathcal{F}}{\Longleftrightarrow} \frac{1}{(i\omega + 2)^2(i\omega - 3)}.$$

Example 3

Next let us consider the transform

$$F_{II}(s) = \frac{1}{(s^2 + 4)(s + 3)},$$

which has two simple poles $\pm i2$ on the imaginary axis and a simple pole at $s = -3$. Again the inverse can be one of the three functions corresponding to the strips

$$-\infty < \gamma < -3, \tag{i}$$
$$-3 < \gamma < 0, \tag{ii}$$
$$0 < \gamma < \infty. \tag{iii}$$

Clearly case(iii) corresponds to the unilateral LT for which the inverse reads

$$
\begin{aligned}
f(t) &= \left\{ \frac{e^{st}}{(2s)(s+3)} \Big|_{s=i2} + \frac{e^{st}}{(2s)(s+3)} \Big|_{s=-i2} + \frac{e^{st}}{(s^2+4)} \Big|_{s=-3} \right\} U(t) \\
&= \left(-\frac{1}{13} \cos 2t + \frac{3}{26} \sin 2t + \frac{e^{-3t}}{13} \right) U(t)
\end{aligned}
$$

The FT of this function exists and we could compute it directly. Can we also determine it from $F_{II}(s)$? Since the strip of analyticity excludes the imaginary axis, i.e., $\gamma = 0$, the substitution $s = i\omega$ we employed in Examples 1 and 2 is no longer permissible. Our guidance here should be the inversion formula. Evidently the offending singularities are the two poles at $s = \pm i2$ which we can circumnavigate with two semicircular contours while keeping the remaining paths along the imaginary axis $\gamma = 0$, as shown in Fig. 4.15.
    We may then write the inversion formula as follows:

$$
\begin{aligned}
f(t) &= \frac{1}{2\pi} \lim_{\epsilon \to 0} \left(\int_{-\infty}^{-2-\epsilon} + \int_{-2+\epsilon}^{2-\epsilon} + \int_{2+\epsilon}^{\infty} \right) \frac{e^{i\omega t} d\omega}{(-\omega^2 + 4)(i\omega + 3)} \\
&\quad + \frac{1}{2\pi i} \lim_{\epsilon \to 0} \int_{C-} \frac{e^{st} ds}{(s^2+4)(s+3)} + \frac{1}{2\pi i} \lim_{\epsilon \to 0} \int_{C+} \frac{e^{st} ds}{(s^2+4)(s+3)}.
\end{aligned} \tag{4.112}
$$

The integral along the straight line path will be recognized as a sum of two CPV integrals while each of the two integrals along the circular paths equals one-half the residue at the respective pole. Therefore (4.112) is equivalent to

$$
\begin{aligned}
f(t) &= \frac{1}{2\pi} P \int_{-\infty}^{\infty} \frac{e^{i\omega t} d\omega}{(-\omega^2 + 4)(i\omega + 3)} \\
&\quad + \frac{e^{i2t}}{8(-2 + i3)} + \frac{e^{-i2t}}{8(-2 - i3)}.
\end{aligned} \tag{4.113}
$$

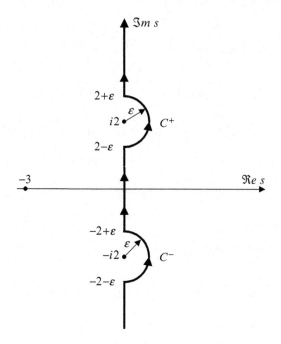

Figure 4.15: Deformation of LT integration path around simple poles on the imaginary axis

Note that consistency with the FT inversion formula is obtained by recognizing that the last two terms in (4.113) result from a pair of delta functions. The FT is then

$$F(\omega) = \frac{1}{(-\omega^2 + 4)(i\omega + 3)}$$

$$+\pi \frac{\delta(\omega - 2)}{4(-2 + i3)} + \pi \frac{\delta(\omega + 2)}{4(-2 - i3)}. \qquad (4.114)$$

To verify that the two terms involving delta functions yield the correct signal in the time domain we evaluate

$$\frac{1}{2\pi} \int_{-\infty}^{\infty} \left\{ \pi \frac{\delta(\omega - 2)}{4(-2 + i3)} + \pi \frac{\delta(\omega + 2)}{4(-2 - i3)} \right\} e^{i\omega t} d\omega,$$

which yields the last two terms in (4.113). It is worth noting that the noncausal signal corresponding to Case ii ($-3 < \gamma < 0$) also possesses a Fourier transform. In this case the path of integration in Fig. 5.12 will approach the imaginary axis from the left and the integrals taken over the two semi-circular indentations will each yield $-i\pi$ times the residue at each pole. As a result we get

$$F(\omega) = \frac{1}{(-\omega^2 + 4)(i\omega + 3)}$$
$$-\pi \frac{\delta(\omega - 2)}{4(-2 + i3)} - \pi \frac{\delta(\omega + 2)}{4(-2 - i3)}. \tag{4.115}$$

Example 4

Consider now the transform

$$F_{II}(s) = \frac{s^3}{(s^2 + 4)(s + 3)}. \tag{4.116}$$

Unlike in Example 3 this transform does not vanish at infinity so that the Jordan lemma does not apply. However, by subtracting its value at infinity we can represent $F_{II}(s)$ as the sum $F_{II}(s) = F_{II}(\infty) + \hat{F}_{II}(s)$ where $\hat{F}_{II}(\infty) = 0$. Thus

$$\begin{aligned} F_{II}(s) &= 1 + \left\{ \frac{s^3}{(s^2 + 4)(s + 3)} - 1 \right\} \\ &= 1 - \frac{3s^2 + 4s + 12}{(s^2 + 4)(s + 3)}. \end{aligned} \tag{4.117}$$

The inverse transform of the constant yields a delta function while the second term can represent three possible functions, depending on the choice of the intercept γ, as in Example 3. In particular, the Fourier transform of the causal signal ($\gamma > 0$) is

$$\begin{aligned} F(\omega) &= 1 - \frac{-3\omega^2 + i4\omega + 12}{(-\omega^2 + 4)(i\omega + 3)} \\ &+ 4\pi \left\{ \frac{\delta(\omega + 2)}{3 - i2} - \frac{\delta(\omega - 2)}{3 + i2} \right\}. \end{aligned} \tag{4.118}$$

4.2.3 Relationships Between the FT and the Unilateral LT

Determination of the FT from the Unilateral Laplace Transform

As evidenced by the preceding discussion the Fourier transform $F(\omega)$ of a signal with a unilateral LT $F_I(s)$ which is analytic for $\mathrm{Re}\, s \geq 0$ is $F_I(i\omega)$. On the other hand, as shown in Examples 3 and 4, if $F_I(s)$ is analytic for $\mathrm{Re}\, s > 0$ but fails to be analytic on the imaginary axis because of simple pole singularities at $s = i\omega_k$ the Fourier transform assumes the form

$$F(\omega) = F_I(i\omega) + \pi \sum_k res\left\{ F_I(i\omega_k) \right\} \delta(\omega - \omega_k), \tag{4.119}$$

where $res\left\{F_I(iw_k)\right\}$ is the residue of $F_I(s)$ at $s = iw_k$. For poles of higher
order (or for essential singularities) on the imaginary axis the Fourier Transform
does not exist since in that case the integration about a semicircular path (as,
e.g., in Fig. 5.12) enclosing the singularity tends to infinity as the radius of the
semicircle approaches zero. This is consistent with the observation that inverse
LTs of functions with poles on the imaginary axis of order higher than unity
yield signals that are unbounded at infinity (e.g., $1/s^2$). Cases for which $F_I(s)$
is analytic for $\operatorname{Re} s > 0$ with branch point singularities on the imaginary axis
generally do not yield delta functions. However, the replacement of s by iw in
such cases must be understood in the limit as the imaginary axis is approached
from the right half of the complex plane, i.e.,

$$F(\omega) = \lim_{\sigma \to 0^+} F_I(\sigma + iw). \tag{4.120}$$

Determination of the Unilateral Laplace Transform from the FT

It would appear that the solution to the converse problem, i.e., that of determining the unilateral LT from the FT of a causal signal amounts to the replacement
of variables $F(s/i)$ and, if necessary, making the adjustment for the presence of
simple poles as prescribed by (4.119). This approach always works for rational
functions. A formal approach that works in all cases can be constructed by
noting that with $s = iw + \sigma$

$$F_I(s) = \int_0^\infty f(t)e^{-(iw+\sigma)t}dt = \int_{-\infty}^\infty f(t)e^{-(iw+\sigma)t}U(t)dt \tag{4.121}$$

for $\sigma > 0$ and any $f(t)$ whose FT is $F(\omega)$. Since

$$\int_{-\infty}^\infty f(t)e^{-(iw+\sigma)t}U(t)dt = \int_{-\infty}^\infty \left[f(t)e^{-\sigma t}U(t)\right]e^{-iwt}dt \tag{4.122}$$

we are dealing with the FT of the product of two functions so that we can use
the frequency convolution theorem to obtain

$$\int_{-\infty}^\infty \left[f(t)e^{-\sigma t}U(t)\right]e^{-iwt}dt = \frac{1}{2\pi}\int_{-\infty}^\infty \frac{F(\eta)\,d\eta}{\sigma + i(\omega - \eta)}. \tag{4.123}$$

Hence with $s = iw + \sigma$ the preceding combined with (4.121) yields

$$F_I(s) = \frac{1}{2\pi}\int_{-\infty}^\infty \frac{F(\eta)\,d\eta}{s - i\eta}, \tag{4.124}$$

which is an analytic function for $\operatorname{Re}(s) > 0$. Note that (4.124) is valid, irrespective of whether $F(\omega)$ is the FT of a causal function. Thus in case the inverse of

$F(\omega)$ is not causal, the inverse of $F_I(s)$ is simply the inverse of $F(\omega)$ truncated to nonnegative time values. For example, for $F(\omega) = p_a(\omega)$ we obtain

$$F_I(s) = \frac{i}{2\pi} \ln\left(\frac{s - ia}{s + ia}\right), \tag{4.125}$$

This function has two logarithmic branch points at $s = \pm ia$. A branch cut represented by a straight line connecting these branch points can be used to ensure analyticity of $F_I(s)$ everywhere on the top Riemann sheet, and in particular for $\mathrm{Re}(s) > 0$. It can be shown by a direct application of the inversion formula that

$$\frac{\sin at}{\pi t} U(t) \overset{\mathcal{L}}{\Longleftrightarrow} \frac{i}{2\pi} \ln\left(\frac{s - ia}{s + ia}\right). \tag{4.126}$$

To find the FT of this causal signal requires the limiting form (4.120) to (4.125) which can be shown to yield the FT pair given in (3.35) in Sect. 3.2.

Problems

1. Solve the following differential equation using LTs:

$$\frac{d^2 y(t)}{dt^2} + 3\frac{dy(t)}{dt} + 2y(t) = f(t)$$

   when

   (a) $f(t) = 0$, $y(0^+) = 0$, $y'(0^+) = 1$; $t \geqslant 0$
   (b) $f(t) = e^{-t}U(t)$, $y(0^+) = y'(0^+) = 0$; $t \geqslant 0$
   (c) $f(t) = \cos 4t$ $U(t + 3)$; $y(-3^+) = y'(-3^+) = 0$; $t \geqslant -3$

2. Find the LT of the following functions:

   (a) $\frac{e^{-3(t-1)}}{\sqrt{2t}} U(t)$
   (b) $t^{3/2}e^{-5t}U(t)$
   (c) $\frac{e^{-2t}}{\sqrt{t-2}}U(t - 2)$
   (d) $e^{-4t}t^{5/2} \sin 2t U(t)$

3. Find the causal inverse LT of the following functions:

   (a) $\frac{s^2}{s^2+4}$
   (b) $\frac{e^{-s}}{\sqrt{s+2}}$
   (c) $\frac{1}{\sqrt{s}(1-e^{-2s})}$
   (d) $\frac{s^2+1}{(s+2)^2}$

4. For a linear system defined by the differential equation

$$\frac{d^2y(t)}{dt^2} - \frac{dy(t)}{dt} - 6y(t) = x(t)$$

where $y(t)$ is the output and $x(t)$ the input, find the impulse response $h(t)$ for each of the following cases:

(a) The system is causal.

(b) The system is stable.

(c) The system is neither causal nor stable.

5. The inverse of the LT of

$$F(s) = \frac{s-1}{(s+2)(s+3)(s^2+s+1)}$$

can represent several functions of time. Find the functions.

6. The inverse of the LT $F(s)$,

$$F(s) = \frac{s+3}{(s^2+9)(s-2)},$$

can represent several functions of time.

(a) Find the functions.

(b) Identify the function that has a single-sided LT.

(c) Identify the function that has a Fourier transform. Find the Fourier transform.

7. A causal LTI system is defined by the system function

$$H(s) = \frac{s+2}{s^2+2s+2}$$

Find the output when the input is $e^{-2|t|}$.

8. Find the causal inverse LT of

$$F(s) = \frac{e^{-a\sqrt{s-1}}}{s-1}$$

where $a > 0$.

Chapter 5

Bandlimited Functions Sampling and the Discrete Fourier Transform

5.1 Bandlimited Functions

5.1.1 Fundamental Properties

In the preceding chapters we have dealt exclusively with signals and their transformations defined on a continuous time interval. Such signals can be referred to as analogue signals. In this chapter we introduce the important topic of discrete signals, i.e., signals that are represented by sequences of numbers. Sometimes these numbers can be interpreted as samples of a continuous signal taken at discrete time intervals. Such an interpretation is, however, not always possible or for that matter even necessary. Nevertheless, the subject is best introduced by starting with the idea of sampling. For this purpose we shall initially restrict the class of analogue signals which we propose to sample to so-called bandlimited functions. We shall call a function (analogue signal) $f(t)$ bandlimited if its Fourier transform $F(\omega)$ vanishes outside a finite frequency interval, i.e.,

$$|F(\omega)| = 0; \quad |\omega| > \Omega. \tag{5.1}$$

No essential loss of generality is incurred if we simplify matters and also assume that[1]

$$\int_{-\Omega}^{\Omega} |F(\omega)| \, d\omega < \infty. \tag{5.2}$$

---

[1] In particular, this excludes functions of the type $(1/i\omega)\, p_\Omega(\omega)$.

W. Wasylkiwskyj, *Signals and Transforms in Linear Systems Analysis*,
DOI 10.1007/978-1-4614-3287-6_5, © Springer Science+Business Media, LLC 2013

In view of (5.1) $f(t)$ can be represented by

$$f(t) = \frac{1}{2\pi} \int_{-\Omega}^{\Omega} F(\omega) e^{i\omega t} d\omega \tag{5.3}$$

while the direct transform may be computed in the usual way from

$$F(\omega) = \int_{-\infty}^{\infty} f(t) e^{-i\omega t} dt. \tag{5.4}$$

Using (5.3) the n-th derivative of $f(t)$ is

$$\frac{d^n}{dt^n} f(t) = \frac{1}{2\pi} \int_{-\Omega}^{\Omega} (i\omega)^n F(\omega) e^{i\omega t} d\omega,$$

which in view of (5.2) converges for all n and all finite real and complex values of t. This shows that $f(t)$ possesses derivatives of all orders for all finite t. We recall that this property defines an analytic function. Consequently $f(t)$ is analytic for all finite t (entire function). A fundamental property of an analytic function is that it may not vanish over any finite segment of its independent variable without vanishing identically. In particular, this means that $f(t)$ cannot be identically zero on any finite segment of the real t axis. Thus a bandlimited function cannot be simultaneously timelimited (i.e., be truncated to a finite time segment, say $-T/2 < t < T/2$, while its Fourier transform satisfies (5.1) and (5.2)). Hence all bandlimited signals are necessarily of infinite duration. Since we know that all signals in practice must be of finite duration the concept of a bandlimited signal may initially strike one as hopelessly artificial. As will be discussed in the sequel, the practical utility of this concept arises largely from the fact that the actual theoretical duration of a bandlimited analogue signal is less important than the number of discrete samples in terms of which the signal may be represented. In practice the latter are always finite in number.

An important property of a bandlimited function is that the Fourier integral (5.4) may be replaced by a sum taken over discrete samples of the function. To obtain such a representation we expand the Fourier transform $F(\omega)$ in Fig. 5.1 in a Fourier series within the interval $(-\Omega, \Omega)$

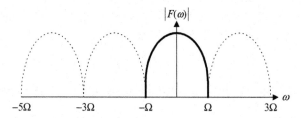

Figure 5.1: Fourier transform limited to a finite band and its periodic extension

$$F(\omega) = \sum_{n=-\infty}^{\infty} c_n e^{-in\pi\omega/\Omega}, \tag{5.5}$$

where the Fourier series expansion coefficients c_n may be computed in the usual way, i.e.,

$$c_n = \frac{1}{2\Omega} \int_{-\Omega}^{\Omega} F(\omega)\, e^{in\pi\omega/\Omega} d\omega. \tag{5.6}$$

The series (5.5) converges in the mean to the prescribed function within the closed interval $(-\Omega, \Omega)$ and to the periodic extension of $F(\omega)$ outside this interval, as indicated in Fig. 5.1. Presently we consider the latter an artifact of no direct interest to us. Comparing the last formula with (5.3) we readily make the identification

$$c_n = \Delta t f(n\Delta t), \tag{5.7}$$

where

$$\Delta t = \frac{\pi}{\Omega} \tag{5.8}$$

so that the *Fourier series* expansion coefficients of the *Fourier transform* are proportional to the samples $f(n\Delta t)$ of the function taken at uniform intervals Δt. We make this explicit by putting (5.5) into the form

$$F(\omega) = \sum_{n=-\infty}^{\infty} f(n\Delta t)\, e^{-i\omega n\Delta t} \Delta t. \tag{5.9}$$

Thus for a bandlimited function the Fourier transform (5.4) has an alternative representation in form of a series with coefficients proportional to the samples of the function spaced Δt apart. Upon closer examination we note that this series also has the form of as a Riemann Sum approximation to (5.4) which should converge to the integral only in the limit as Δt approaches zero. That for bandlimited functions convergence appears possible also for finite sampling intervals may appear initially as a startling result. It is, however, perfectly consistent with the fact that $\Delta t f(n\Delta t)$ is a Fourier series coefficient. Equation (5.9) can be used to represent $F(\omega)$ provided $\Delta t \leq \pi/\Omega$. This follows from the observation that when $\Delta t' \leq \Delta t$, any function bandlimited to $\pi/\Delta t$ is also bandlimited to $\pi/\Delta t'$.

From another perspective (5.9) as a Fourier series is subject to the same convergence constraints as the Fourier series in the time domain discussed in 2.1. The only difference lies in the interpretation. In Sect. 2.1 we dealt with the expansion of a waveform limited in time to the interval $(-T, T)$. Here we use the same representation for a Fourier Transform limited to the frequency band $(-\Omega, \Omega)$ and interpret the expansion coefficients as products of signal samples and the sampling interval. In fact one can conceive of a dual situation wherein the expansion coefficients of a time-limited waveform are interpreted as products of samples of the frequency spectrum and the frequency sampling interval.

Recall that improvements of resolution in the time domain generally require an increase in the number of Fourier series coefficients. Similarly, to obtain a better spectral resolution requires an increase in the number of signal samples. Since we are dealing with uniformly spaced samples any increase in the number of samples requires a longer set of data. This is illustrated in Fig. 5.2 which shows the magnitude of a spectral density tending to a pronounced peak as the signal duration is increased from 16 Δt to $64\Delta t$. It turns out that in this particular example nearly full resolution is reached with 64 samples so that further increases in signal duration will not alter the spectral density.

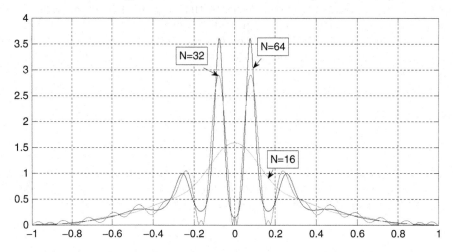

Figure 5.2: Spectral resolution and signal duration

A somewhat different situation arises when the spectral density changes significantly within a narrow band. An extreme example of this is the functional form

$$F(\omega) = \left\{ \begin{array}{l} 3 \; ; \; |\omega| < 0.5, \\ 1 \; ; \; 0.5 < |\omega| < 1. \end{array} \right.$$

A step change in this pure form is clearly not physically realizable. Nevertheless idealizations of this sort are frequently of value in preliminary assessments of the performance of bandpass and band suppression filters. Unlike the "smooth" spectral forms in Fig. 5.2 convergence of the FS at step discontinuities will be nonuniform. Figure 5.3 shows the spectra for signal durations of $16\Delta t$ and $64\Delta t$ As in the mathematically equivalent case in the time domain we see already for $N = 64$ the beginning of Gibbs characteristic oscillatory convergence.

5.1.2   The Sampling Theorem

Since (5.9) represents a Fourier Transform we are entitled to substitute it in (5.3). This gives us a representation of a signal on a continuum of t. With this substitution we obtain

$$f(t) \;\; = \;\; \frac{1}{2\pi} \int_{-\Omega}^{\Omega} \left\{ \sum_{n=-\infty}^{\infty} f(n\Delta t) e^{-in\Delta t \omega} \Delta t \right\} e^{i\omega t} d\omega$$

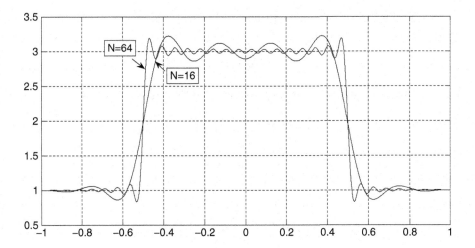

Figure 5.3: Spectrum with step discontinuities

Integrating the exponential yields

$$f(t) = \sum_{n=-\infty}^{\infty} f(n\Delta t) \frac{\sin(\Omega t - n\pi)}{(\Omega t - n\pi)}, \tag{5.10}$$

which is the famous Shannon sampling theorem. An alternative form wherein Ω is replaced by $\pi/\Delta t$ is

$$f(t) = \sum_{n=-\infty}^{\infty} f(n\Delta t) \frac{\sin[(t/\Delta t - n)\pi]}{(t/\Delta t - n)\pi} \tag{5.11}$$

which highlights the interpolation aspect of this representation. In fact the interpolation formula for sinusoids, 2.74, reduces to the Shannon sampling representation for sufficiently large M. The sampling interval Δt is generally referred to as the Nyquist interval and its reciprocal as the Nyquist sampling rate, usually expressed in Hz. If we denote the (single-sided) bandwidth of the signal by $B = \Omega/2\pi$, then $\Delta t = 1/2B$. As already noted in conjunction with (5.9) a signal bandlimited to B is also bandlimited to $B' > B$, so that (5.11) remains valid for any $1/2B' = \Delta t \leq 1/2B$. A signal sampled at a rate greater than the Nyquist rate is said to be oversampled. We will postpone the discussion of the consequences of sampling at a rate lower than the Nyquist rate (i.e., undersampling) to Sect. 5.3.

Observe that for any integer k substitution of $t = k\Delta t$ into the argument of the $\sin(x)/x$ function in (5.11) gives zero for $n \neq k$ and unity when $n = k$, thus verifying directly the convergence of the sum to $f(k\Delta t)$. Since the expansion coefficients of (5.11) are the signal samples and the sum reproduces the function at these sample values exactly, (5.11) may be considered an interpolation formula. It is actually more than an interpolation formula since unlike interpolation formulas in 2.1.8, it reproduces the given function *exactly* at all intermediate values of t. In fact (5.11) may also be viewed as an LMS representation of $f(t)$ in terms of the expansion functions

$$\phi_n(t/\Delta t) = \sin\left[(t/\Delta t - n)\pi\right] / \left[(t/\Delta t - n)\pi\right]. \tag{5.12}$$

It is not hard to show that these functions are orthogonal over $-\infty, \infty$:

$$\int_{-\infty}^{\infty} \phi_n(t/\Delta t)\,\phi_m(t/\Delta t)\,dt = \Delta t \delta_{nm}. \tag{5.13}$$

Since the series converges pointwise, the LMS error must, of course, be zero. If in (5.11) we set $f(n\Delta t) = \hat{f}_n$ we can apply Parseval's formula (1.126) to obtain

$$\int_{-\infty}^{\infty} |f(t)|^2\,dt = \sum_{n=-\infty}^{\infty} |f(n\Delta t)|^2\,\Delta t, \tag{5.14}$$

This result could also have been obtained by first applying Parseval's theorem to the Fourier series (5.9) and then to the Fourier transform, i.e., using

$$\frac{1}{2\pi}\int_{-\Omega}^{\Omega} |F(\omega)|^2\,d\omega = \int_{-\infty}^{\infty} |f(t)|^2\,dt.$$

Eq. (5.14) states that the energy in the bandlimited analogue signal may be viewed as if it were distributed among its sampled values. The right side of (5.14), just as the Fourier series (5.9), looks like a Riemann sum approximation but is in fact an exact representation of the integral to its left. As in (5.9) we have the option of choosing for Δt any value less or equal to $1/2B$.

5.1.3 Sampling Theorem for Stationary Random Processes*

Frequently it is more appropriate to model signals as random processes rather than as definite deterministic functions. A particularly important class of random processes are stationary processes one of whose attributes is that they extend over an infinite time interval. Here we examine the meaning of bandlimiting for such processes as well as their representation in terms of sampled values. Unlike deterministic signals, sample functions of stationary random processes do not possess Fourier transforms so that the definition of a bandlimited random function as in (5.1) and (5.2) cannot be used directly. On way out of this difficulty is not to focus on the properties of individual sample functions but define bandlimiting in terms a statistical (ensemble) average. A suitable average is the autocorrelation function the FT of which is the power spectrum of the process. Thus a bandlimited stationary random process may be defined as a process whose power spectrum $S(\omega)$ vanishes identically for $|\omega| > \Omega$, i.e.,

$$S(\omega) = \int_{-\infty}^{\infty} R(\tau)\,e^{-i\omega\tau}d\tau = 0 \; ; \; |\omega| > \Omega, \tag{5.15}$$

where

$$R(\tau) = \langle \underline{x}(t+\tau)\underline{x}^*(t) \rangle \tag{5.16}$$

is the autocorrelation function and $\underline{x}(t)$ is a sample function of the process. We will show that for such a sample function

$$\lim_{K \to \infty} \left\langle \left| \underline{x}(t) - \sum_{\ell=-K}^{\ell=K} \underline{x}(\ell\Delta t)\phi_\ell(t/\Delta t) \right|^2 \right\rangle = 0, \tag{5.17}$$

where $\phi_\ell(t/\Delta t)$ is given by (5.12) and $\Delta t \leq \pi/\Omega$. Equation (5.17) states that the Shannon sampling theorem is also valid for stationary random processes provided one interprets the convergence in the (statistical) mean squared sense.

To prove (5.16) we first expand squared magnitude in (5.16) and perform the statistical average on each term. Thus

$$\left\langle \left| \underline{x}(t) - \sum_{\ell=-K}^{\ell=K} \underline{x}(\ell\Delta t)\phi_\ell(t/\Delta t) \right|^2 \right\rangle$$

$$= R(0) - \sum_{\ell=-K}^{\ell=K} R(\ell\Delta t - t)\phi_\ell(t/\Delta t) - \sum_{\ell=-K}^{\ell=K} R(t - \ell\Delta t)\phi_\ell(t/\Delta t)$$

$$+ \sum_{n=-K}^{n=K} \sum_{\ell=-K}^{\ell=K} R\left[(\ell-n)\Delta t \right] \phi_\ell(t/\Delta t) \, \phi_n(t/\Delta t). \tag{5.18}$$

Next we substitute for the correlation function in (5.18) its spectral representation (5.15) to obtain

$$\left\langle \left| \underline{x}(t) - \sum_{\ell=-K}^{\ell=K} \underline{x}(\ell\Delta t)\phi_\ell(t/\Delta t) \right|^2 \right\rangle$$

$$= \frac{1}{2\pi} \int_{-\Omega}^{\Omega} d\omega \, S(\omega)$$

$$\times \left[\begin{array}{l} 1 - \sum_{\ell=-K}^{\ell=K} e^{i\omega(\ell\Delta t - t)}\phi_\ell(t/\Delta t) - \sum_{\ell=-K}^{\ell=K} e^{-i\omega(\ell\Delta t - t)}\phi_\ell(t/\Delta t) \\ + \sum_{n=-K}^{n=K} \sum_{\ell=-K}^{\ell=K} e^{i[(\ell-n)\Delta t]}\phi_\ell(t/\Delta t) \, \phi_n(t/\Delta t) \end{array} \right]$$

$$= \frac{1}{2\pi} \int_{-\Omega}^{\Omega} d\omega \, S(\omega) \left| e^{i\omega t} - \sum_{\ell=-K}^{\ell=K} e^{i\omega\ell\Delta t}\phi_\ell(t/\Delta t) \right|^2. \tag{5.19}$$

On the other hand, a direct development of $e^{i\omega t}$ in a Fourier series in ω in the interval $|\omega| \leq \pi/\Delta t$ with t as a parameter gives

$$e^{i\omega t} = \sum_{\ell=-\infty}^{\infty} e^{i\omega\ell\Delta t}\phi_\ell(t/\Delta t). \tag{5.20}$$

Hence in the limit as $K \to \infty$ the squared magnitude in the second integrand of (5.19) vanishes, thus proving (5.17).

5.2 Signals Defined by a Finite Number of Samples

As we have seen, one fundamental attribute of bandlimiting is that the corresponding signal is necessarily of infinite duration. No incompatibility with the bandlimited nature of the signal is implied if we assume that the number of nonzero samples is finite. For definiteness, assume that these correspond to indices $n = 0, 1, 2, \ldots N - 1$. Then (5.11) becomes

$$f(t) = \sum_{n=0}^{N-1} f(n\Delta t) \, \frac{\sin\left[(t/\Delta t - n)\,\pi\right]}{(t/\Delta t - n)\,\pi} \tag{5.21}$$

and the corresponding FT reads

$$F(\omega) = \sum_{n=0}^{N-1} f(n\Delta t) \, e^{-in\Delta t\omega} \Delta t \; ; |\omega| \le \Omega. \tag{5.22}$$

Note that the signal itself is of *infinite* duration as required by the bandlimited nature of the spectrum. For example, a bandlimited signal of infinite duration that can be represented by only one nonzero Nyquist sample is

$$f(t) = f(0) \, \frac{\sin\left[(t/\Delta t)\,\pi\right]}{(t/\Delta t)\,\pi}$$

for which the spectrum is the rectangle $\Delta t f(0)\, p_{\pi/\Delta t}(\omega)$. A bandlimited signal comprised of the seven nonzero samples $6, 4, 7, 6, 6, 4, 5$ is plotted in Fig. 5.4.

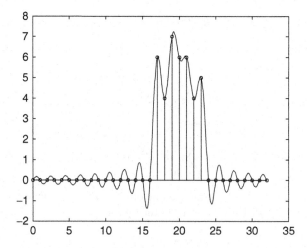

Figure 5.4: Bandlimited signal comprised of seven nonzero samples

The sampling interval equals 1 s so that these samples are numerically equal to the expansion coefficients of the Fourier series expansion of the signal spectrum which is

$$F(\omega) = \begin{cases} \sum_{n=17}^{23} f(n) e^{-i\omega n} & ; |\omega| \le \pi, \\ 0 & ; |\omega| > \pi. \end{cases}$$

A plot of $|F(\omega)|$ is shown in Fig. 5.5.

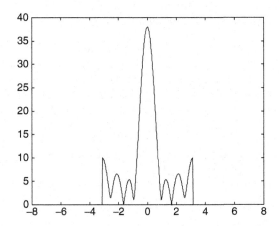

Figure 5.5: Magnitude of the FT of the function in Fig. 5.4

In accordance with (5.14) the total energy in this signal is simply the sum of the squares of the seven samples. Thus even though the analogue signal itself occupies an infinite time interval it is reasonable to assign it an effective duration of only 7 s. In general when the nonzero samples span the time interval $0 \le t \le T$ with $T = N\Delta t$ as in (5.22) the signal energy is

$$E = \int_{-\infty}^{\infty} |f(t)|^2 \, dt = \sum_{n=0}^{N-1} |f(n\Delta t)|^2 \, \Delta t. \tag{5.23}$$

If $f(0)$ and $f[(N-1)\Delta t]$ are not zero T may be taken as the effective signal duration. Of course this definition does not preclude that one or more samples in the interior of the interval equal zero but only that all samples falling outside $0 \le t \le T$ are zero.

The product of the (double sided) signal bandwidth $2B \; Hz$ and the (effective) signal duration T, i.e., $(2\Omega/2\pi) T = 2BT \equiv N$ is an important dimensionless parameter in signal analysis. It represents the number of sinusoids comprising the signal spectrum or in view of the Shannon representation (5.21), the number of sinc functions comprising the signal itself. In the spirit of our discussion of alternative representations in Chap. 2 we may think of these expansion functions as independent coordinates and regard N as the geometrical dimensionality of the signal space, or, to borrow a term from Mechanics, as the *number of degrees of freedom* of the signal. In the general case the dimensionality of a bandlimited signal can be infinite and truncation to a finite number of Nyquist samples must lead to errors. The nature of these errors can be assessed by recalling the one-to-one correspondence between the Nyquist samples and the coefficients in the Fourier series representation of the spectrum of the signal. As will be recalled from the examples discussed in Chap. 2, truncation of a Fourier series representing a signal in the time domain results in a loss of detail (resolution). Because (5.9) is a Fourier expansion in the frequency domain its truncation results in a reduction in resolution in the frequency domain. For example, consider the signal comprised of two pure tones

$$f(t) = \cos\left[(\Omega/4)\,t\right] + \cos\left[(5\Omega/16)\,t\right]. \tag{5.24}$$

Because the higher of the two frequencies is $5\Omega/16$ rps, the maximum permissible sampling interval may not exceed $16\pi/5\Omega$ s. Choosing $\Delta t = \pi/\Omega$ gives

$$f\left(n\Delta t\right) = \cos(n\pi/4) + \cos(n5\pi/16) \tag{5.25}$$

with $n = 0, 1, \ldots N - 1$. Employing (5.22), the magnitude of the Fourier Transform based on 16 samples is plotted in Fig. 5.6.

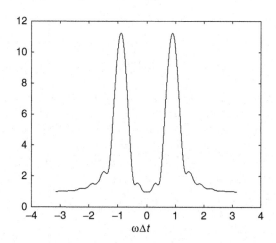

Figure 5.6: Spectrum of signal in (5.24) using 16 samples

Only two peaks appear, symmetrically disposed with respect to the zero frequency, which is consistent with the presence of a single tone. Evidently the two frequencies present in (5.24) are not resolved. This is not surprising for based on the effective signal duration of $T = 16\pi/\Omega$ s the Rayleigh resolution limit is $2\pi/T = \Omega/8$ rps whereas the frequency separation of the two tones is only $\Omega/16$ rps. Increasing the number of samples to 32 doubles the signal duration so that the Rayleigh limit just matches the frequency separation. As confirmed by the plot in Fig. 5.7 the two tones are just beginning to be resolved.

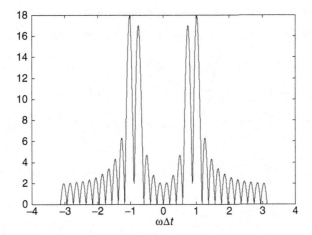

Figure 5.7: Spectrum of signal in (5.24) using 32 samples

Use of more samples would further reduce the width of the spikes representing the individual frequency components eventually approaching two pairs of delta functions as the number of samples tends to infinity.

5.2.1   Spectral Concentration of Bandlimited Signals

Suppose we have the freedom to assign arbitrary weights to N signal samples. We would like to find a set of weights that will maximize the concentration of energy within a prescribed sub-band $(-\Omega_0, \Omega_0)$ of a signal with fixed energy bandlimited to $(-\Omega, \Omega)$. To find such an optimum sequence we again represent the Fourier transform of the signal in terms of its N samples, i.e.,

$$F_N(\omega) = \sum_{n=0}^{N-1} f(n\Delta t)\, e^{-i\omega n\Delta t}\Delta t. \tag{5.26}$$

The corresponding total energy is then

$$
\begin{aligned}
E_\Omega &= \frac{1}{2\pi} \int_{-\Omega}^{\Omega} \left| \sum_{n=0}^{N-1} f(n\Delta t)\, e^{-i\omega n\Delta t}\Delta t \right|^2 d\omega \\
&= \frac{\Delta t^2}{2\pi} \sum_{n=0}^{N-1}\sum_{m=0}^{N-1} f(n\Delta t)\, f^*(m\Delta t) \int_{-\Omega}^{\Omega} e^{-i\omega(n-m)\Delta t} d\omega \\
&= \sum_{n=0}^{N-1} |f(n\Delta t)|^2 \, \Delta t,
\end{aligned}
\tag{5.27}
$$

where in the final step we made use of (5.8). Next we carry out a similar calculation for the total energy within $(-\Omega_0, \Omega_0)$. Again using (5.26) we get

$$
\begin{aligned}
E_{\Omega_0} &= \frac{1}{2\pi} \int_{-\Omega_0}^{\Omega_0} \left| \sum_{n=0}^{N-1} f(n\Delta t) e^{-i\omega n \Delta t} \Delta t \right|^2 d\omega \\
&= \frac{\Delta t^2}{2\pi} \sum_{n=0}^{N-1} \sum_{m=0}^{N-1} f(n\Delta t) f^*(m\Delta t) \int_{-\Omega_0}^{\Omega_0} e^{-i\omega(n-m)\Delta t} d\omega \\
&= \sum_{n=0}^{N-1} \sum_{m=0}^{N-1} \frac{\sin\left[\pi \frac{\Omega_0}{\Omega}(n-m)\right]}{\pi \frac{\Omega_0}{\Omega}(n-m)} f(n\Delta t) f^*(m\Delta t) \Delta t.
\end{aligned}
\tag{5.28}
$$

Our objective is to maximize the ratio

$$
\rho = \frac{E_{\Omega_0}}{E_\Omega}
\tag{5.29}
$$

for a constant E_Ω. The necessary manipulations are simplified if we rewrite (5.29) in matrix form. Accordingly we define the column vector

$$
\mathbf{f} = [f(0) \; f(\Delta t) \; f(2\Delta t) \; f(3\Delta t) \; \dots \; f((N-1)\Delta t)]^T
\tag{5.30}
$$

and the matrix $\mathbf{A}$ with elements

$$
A_{nm} = \frac{\sin\left[\pi \frac{\Omega_0}{\Omega}(n-m)\right]}{\pi \frac{\Omega_0}{\Omega}(n-m)}.
\tag{5.31}
$$

As a result we get

$$
\rho = \frac{\mathbf{f}^T \mathbf{A} \mathbf{f}}{\mathbf{f}^T \mathbf{f}}.
\tag{5.32}
$$

The deviation of ρ from its maximum must vanish which we express by

$$
\delta \left(\frac{\mathbf{f}^T \mathbf{A} \mathbf{f}}{\mathbf{f}^T \mathbf{f}} \right) = 0.
\tag{5.33}
$$

Since $\mathbf{f}^T \mathbf{f}$ is constant the preceding is equivalent to

$$
\delta \left(\mathbf{f}^T \mathbf{A} \mathbf{f} \right) = 0.
\tag{5.34}
$$

But the $\mathbf{f}$ that satisfies (5.34) maximizes ρ and therefore the energy E_{Ω_0}. Since according to (5.34) $\mathbf{f}^T \mathbf{A} \mathbf{f}$ is constant, this constant must necessarily equal E_{Ω_0} so that we have

$$
\mathbf{f}^T \mathbf{A} \mathbf{f} = E_{\Omega_0}.
\tag{5.35}
$$

Multiplying this from the left by $\mathbf{f}^T$ (5.35) yields $\mathbf{f}^T \mathbf{A} \mathbf{f}^T \mathbf{f} = \mathbf{f}^T E_{\Omega_0}$ but since $\mathbf{f}^T \mathbf{f} = E_\Omega$ we get $\mathbf{f}^T \mathbf{A} = \mathbf{f}^T E_{\Omega_0}/E_\Omega$. Transposing and taking account of the fact that $\mathbf{A}$ is a symmetric matrix give us the final result

$$
\mathbf{A} \mathbf{f} = \left(\frac{E_{\Omega_0}}{E_\Omega} \right) \mathbf{f}
\tag{5.36}
$$

Thus the maximum relative concentration of power within the prescribed fractional band Ω_0/Ω is in given by $E_{\Omega_0}/E_{\Omega} = \lambda_{\max}$, the maximum eigenvalue of $\mathbf{A}$. The samples of the signal whose spectrum provides this maximum are elements of the corresponding eigenvector (5.30). It can be shown that the N eigenvalues of $\mathbf{A}$ are positive. They are usually arranged to form a monotonically decreasing sequence which makes $\lambda_{\max}$ the first eigenvalue. The eigenvectors of $\mathbf{A}$ form an orthogonal set and represent a special case of discrete Prolate Spheroidal Wave Functions that have been studied extensively and have a variety of applications.[2]

A useful normalization of the fractional bandwidth is obtained when it is expressed in terms of the time-bandwidth product of the output signal.

Setting $\Delta t = 1/2B$ we get $\Omega_0/\Omega = 2\pi B_0/2\pi B = 2B_0\Delta t = 2B_0T/N$ so that $2B_0T = N(B_0/B)$. In Fig. 5.8 $\lambda_{\max}$ is plotted as a function of $2B_0T$ for $N = 1024$. Figure 5.9 shows plots of components of three eigenvectors as function of the normalized variable k/N (with $N = 1024$). The corresponding eigenvalues are shown on the plot. The fractional bandwidths for these cases (not shown on the plot) are .004, .0015, and .7938. The corresponding values of $2B_0T$ can be found from Fig. 5.9 except for the first value for which $2B_0T$ is 4.096.

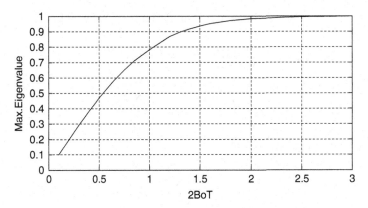

Figure 5.8: Spectral concentration

5.2.2 Aliasing

We now suppose that $f(t)$ is not bandlimited. Let us pick, quite arbitrarily, a sampling interval Δt and inquire what happens when we attempt to compute the FT using the sum (5.9). In anticipation that presently this sum may not converge to $F(\omega)$ we denote it by $\hat{F}(\omega)$ and write

$$\hat{F}(\omega) = \sum_{n=-\infty}^{\infty} f(n\Delta t)\, e^{-in\Delta t\omega}\, \Delta t \ . \qquad (5.37)$$

---

[2]One of the early works on this subject is D. Slepian, H.O. Pollack, and H.T. Landow, "Prolate Spheroidal Wave Functions, Fourier Analysis and Uncertainty Principle," *Bell System Technical J.*, 40, no.1 pp. 43–84. January, 1961.

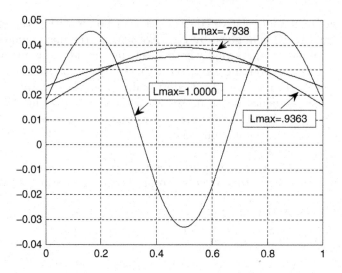

Figure 5.9: Selected eigenvectors (prolate spheroidal wave functions) of **A**

Upon substituting the FT inversion formula for $f(n\Delta t)$ we obtain

$$
\begin{aligned}
\hat{F}(\omega) &= \frac{\Delta t}{2\pi} \sum_{n=-\infty}^{\infty} \int_{-\infty}^{\infty} F\left(\omega'\right) e^{in\Delta t\left(\omega'-\omega\right)} d\omega' \\
&= \frac{\Delta t}{2\pi} \int_{-\infty}^{\infty} F\left(\omega'\right) d\omega' \sum_{n=-\infty}^{\infty} e^{in\Delta t\left(\omega'-\omega\right)}.
\end{aligned}
\tag{5.38}
$$

Taking account of (2.31) we have

$$
\begin{aligned}
\sum_{n=-\infty}^{\infty} e^{in\Delta t\left(\omega'-\omega\right)} &= 2\pi \sum_{k=-\infty}^{\infty} \delta\left[\Delta t\left(\omega'-\omega\right)+2\pi k\right] \\
&= \frac{2\pi}{\Delta t} \sum_{k=-\infty}^{\infty} \delta\left[\omega'-\omega+\frac{2\pi k}{\Delta t}\right].
\end{aligned}
$$

Substituting the last expression into (5.22) and integrating with respect to ω' we obtain

$$
\hat{F}(\omega) = \sum_{k=-\infty}^{\infty} F\left(\omega - \frac{2\pi k}{\Delta t}\right).
\tag{5.39}
$$

Thus while the term corresponding to $k = 0$ in this series gives the correct FT it is embedded in an infinite set of frequency shifted transforms which we shall refer to as the *images* of the spectrum. When the function is bandlimited to $|\omega| < \pi/\Delta t = \Omega$ the images do not overlap and we are back to the picture in Fig. 5.1: the images coincide with what we had previously referred to as periodic extensions of the spectrum. When the signal spectrum extends outside the bandwidth defined by $\pi/\Delta t$ the images overlap and we have the situation depicted in Fig. 5.10. The individual image spectra together with $F(\omega)$ are shown in Fig. 5.10b where for the sake of simplicity we suppose that the signal

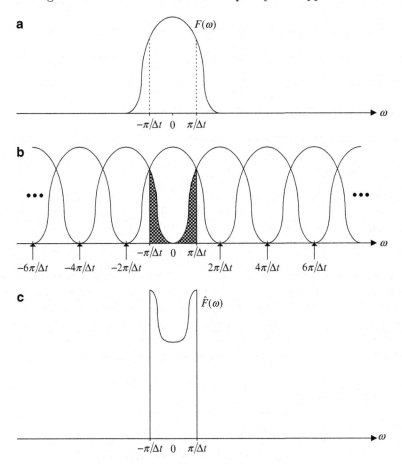

Figure 5.10: Signal spectrum corrupted by aliasing

spectrum is purely real. Note that the overlap (shown shaded in the diagram) with $F(\omega)$ of the two image spectra corresponding to $k = \pm 1$ can be viewed geometrically as a folding of $F(\omega)$ along the two vertical lines defined by $\omega = \pm\pi/\Delta t$ whereby the high frequencies $|\omega| > \pi/\Delta t$ are mapped or "folded" into the *baseband* $|\omega| < \pi/\Delta t$. For this reason $\pi/\Delta t$ is sometimes referred to as the *folding frequency*. Such fold lines correspond to all frequencies that are multiples of $\pm\pi/\Delta t$, as illustrated in more detail in Fig. 5.11. In effect the frequencies $2\pi/\Delta t - \omega, 4\pi/\Delta t - \omega, \ldots \omega - 3\pi/\Delta t, \omega - 5\pi/\Delta t, \ldots$ all appear as

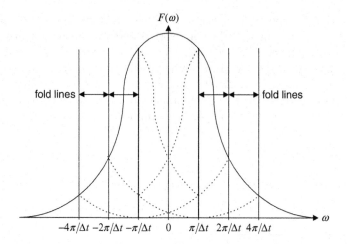

$F(\omega)$

fold lines

fold lines

$-4\pi/\Delta t$ $-2\pi/\Delta t$ $-\pi/\Delta t$ 0 $\pi/\Delta t$ $2\pi/\Delta t$ $4\pi/\Delta t$

ω

Figure 5.11: Aliasing in terms of spectral folding

ω within the band $0 \leq \omega \leq \pi/\Delta t$ and similarly for negative ω. This mapping of the high frequencies of the image spectra into the lower frequency baseband regime is referred to as *aliasing*. When the signal spectrum extends to infinity (e.g., as in the case of a Gaussian function) the number of aliased images is theoretically infinite. In the example in Fig. 5.10 only the two image spectra on either side of $F(\omega)$ contribute so that the signal spectrum with the superposed aliases appears as in Fig. 5.10c. In practice alias-induced spectral distortion can be reduced by increasing the sampling rate or (and) bandlimiting the signal to $|\omega| < \pi/\Delta t$. For this purpose a low-pass filter (generally referred to as an *anti aliasing* filter) is employed. Since perfect bandlimiting can only be achieved with filters having infinitely sharp cutoff aliasing effects can never be completely eliminated but only reduced below some tolerable level.

5.3 Sampling

5.3.1 Impulse Sampling

The process of sampling may be visualized as a multiplication of the analogue signal by a sequence of uniformly spaced pulses sufficiently narrow so that the signal remains practically constant during each pulse period. As an extreme idealization of this process we replace these pulses by delta functions and refer to the sampling as *impulse sampling*. The impulse train

$$P(t) \equiv \Delta t \sum_{n=-\infty}^{\infty} \delta(t - n\Delta t) \tag{5.40}$$

when multiplied by the analogue signal $f(t)$ will be denote $f_p(t)$. The properties of delta functions permit us to write

$$f_p(t) = \sum_{n=-\infty}^{\infty} f(n\Delta t)\, \Delta t \delta\, (t - n\Delta t). \tag{5.41}$$

With

$$f_p(t) \overset{\mathcal{F}}{\Longleftrightarrow} F_p(\omega) \tag{5.42}$$

we have

$$F_p(\omega) = \sum_{n=-\infty}^{\infty} f(n\Delta t)\, e^{-in\Delta t\omega}\, \Delta t . \tag{5.43}$$

We can mechanize this idealized sampling operation as shown in Fig. 5.12 where the block labeled I/D represents the conversion of the impulse sampled function

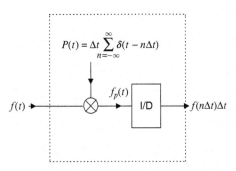

Figure 5.12: Impulse sampling

$f_p(t)$ to the sequence of samples $f(n\Delta t)\, \Delta t$. The reverse process that of converting $f_p(t)$ to the original analogue signal $f(t)$ can be accomplished by passing $f_p(t)$ through an ideal low-pass filter with transfer function $H(\omega) = p_{\pi/\Delta t}(\omega)$. This follows directly by convolving the impulse response $\sin(\pi t/\Delta t)/\pi t$ with (5.22):

$$\sin(\pi t/\Delta t)/\pi t * f_p(t) = \sum_{n=-\infty}^{\infty} f(n\Delta t)\, \frac{\sin\left[(t/\Delta t - n)\,\pi\right]}{(t/\Delta t - n)\,\pi}$$

which is just the Shannon sampling representation (5.11). Thus, as expected, we get a perfect reconstruction only if $f(t)$ is perfectly bandlimited to $|\omega| < \pi/\Delta t$. This reconstruction is summarized in Fig. 5.13 where the block labeled D/I represents the conversion of the discrete sequence $f(n\Delta t)\, \Delta t$ to impulse samples.

5.3.2 Zero-Order Hold Sampling and Reconstruction

One difficulty with the implementation of the reconstruction scheme in Fig. 5.13 is that it requires a train of delta functions. However, this is not a fundamental flaw since impulses can always be approximated by pulses of sufficiently

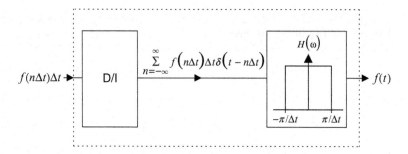

Figure 5.13: Reconstruction of analogue signal from sampled values

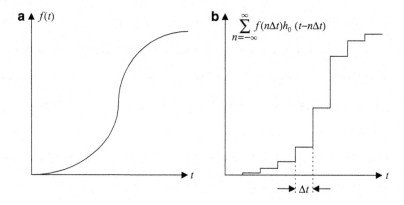

Figure 5.14: Zero-order hold reconstruction

short duration. A more fundamental difficulty is that the process is noncausal. Practically this means that before reconstruction can be attempted one has to observe the entire data record (past and future). In other words a real-time implementation of impulse sampling is not possible. One technique which does not suffer from this drawback is the so-called zero-order hold method. This scheme employs a circuit whose response to an impulse is constant for the duration of the sampling interval and subsequently resets to zero. The resulting transformation of a typical analogue signal is illustrated in Fig. 5.14. The net effect is to replace the signal in Fig. 5.14a by the staircase approximation in Fig. 5.14b. Mathematically the output is represented by

$$y\left(t\right) = \sum_{n=-\infty}^{\infty} f\left(n\Delta t\right) \Delta t \, h_0\left(t - n\Delta t\right), \tag{5.44}$$

where

$$h_0\left(t\right) = \begin{cases} 1/\Delta t & ; 0 \leq t \leq \Delta t, \\ 0 \; otherwise \end{cases} \tag{5.45}$$

is the impulse response of the zero-order hold circuit. The zero-order hold sampling representation (5.44) can be mechanized as shown in Fig. 5.15. To examine the faithfulness of such a reconstruction we take the FT of (5.44) to obtain

$$Y(\omega) = \frac{2\sin(\omega\Delta t/2)}{\omega\Delta t}e^{-i\omega\Delta t/2}\sum_{n=-\infty}^{\infty} f(n\Delta t)e^{-in\Delta t\omega}\Delta t$$

$$= \frac{2\sin(\omega\Delta t/2)}{\omega\Delta t}e^{-i\omega\Delta t/2}\sum_{k=-\infty}^{\infty} F\left(\omega - \frac{2\pi k}{\Delta t}\right). \tag{5.46}$$

Thus the reconstructed spectrum differs from the ideal case obtained with impulse sampling by the factor

$$H(\omega) = \frac{2\sin(\omega\Delta t/2)}{\omega\Delta t}e^{-i\omega\Delta t/2}, \tag{5.47}$$

i.e., the transfer function of the zero-order hold circuit. In addition to the phase shift causing a signal delay of $\Delta t/2\,\mathrm{s}$, which in itself is relatively unimportant, the magnitude of the transfer function decreases from unity at $\omega = 0$ to $2/\pi \approx .6366$ at the folding frequency, representing an amplitude reduction of about $4\,\mathrm{dB}$, i.e., $20\log(.636)$. Amplitude tapers of this magnitude are generally not acceptable so that compensating techniques must be employed.

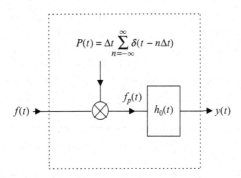

Figure 5.15: Zero-hold sampling

One approach is to increase the sampling rate. For example, if the actual signal bandwidth is $2\pi/\Delta t$ we choose the sampling interval equal to $\Delta' t = \Delta t/r$ where $r > 1$ may be defined as the order of oversampling. The magnitude of the transfer function at the actual band edges of the signal spectrum then becomes $|H(\pi/\Delta t)| = 2r\sin(\pi/2r)/\pi$. For example, if we oversample by a factor of 2 the total amplitude taper is seen to be $2\sqrt{2}/\pi$ which is only about to $0.9\,\mathrm{dB}$. The geometrical relationship between the signal spectrum and the transfer functions for the Nyquist and double Nyquist sampling rates is displayed in Fig. 5.16.

Alternatively one can attempt to compensate for the amplitude taper of the zero hold filter with an additional filter with transfer function magnitude proportional to the reciprocal of $|H(\omega)|$. Unfortunately such filters can be realized only approximately so that a variety of compromises are generally employed in practice usually involving both filtering and oversampling.

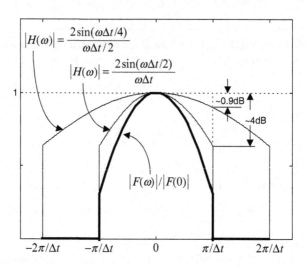

Figure 5.16: Reduction in the spectral amplitude taper by oversampling

5.3.3   BandPass Sampling

In many applications an analogue signal that is to be sampled will be modulated by a carrier (as, e.g., in a wireless communications link) or for other reasons may contain significant energy only within a relatively narrow band around a center frequency. Such a signal can of course be treated formally as being bandlimited to its maximum frequency, say $\omega_{\max}$, and, in principle, could be sampled without alias-induced degradations at $\omega_{\max}/\pi$ Hz. We say in principle since such an approach would generally lead to unacceptably high sampling rates. An acceptable solution would be to be able to sample at or near the sampling rate dictated by the much narrower bandwidth of the baseband signal. On first thought it might appear that this cannot be done without serious signal degradation due to aliasing. Surprisingly, it turns out that we can actually take advantage of this aliasing and reduce the sampling rate substantially below $\omega_{\max}/\pi$. We accomplish this by forcing the ratio of the carrier frequency to the modulation bandwidth to be an integer. As a result an aliased band can be made to correspond exactly to the low-pass equivalent of the modulation spectrum. To see how this comes about consider a bandpass signal whose FT is represented by

$$Z\left(\omega\right) = Z_{-}\left(\omega\right) + Z_{+}\left(\omega\right), \tag{5.48}$$

where $Z_{-}\left(\omega\right)$ and $Z_{+}\left(\omega\right)$ are bandpass spectra, corresponding to the negative and positive frequency ranges, respectively, as illustrated in Fig. 5.17. The spectrum of the signal sampled at intervals Δt is the periodic function

$$\hat{Z}(\omega) = \sum_{n=-\infty}^{\infty} z\left(n\Delta t\right) e^{-i\omega n\Delta t}\Delta t = \sum_{n=-\infty}^{\infty} Z\left(\omega - \frac{2\pi n}{\Delta t}\right). \tag{5.49}$$

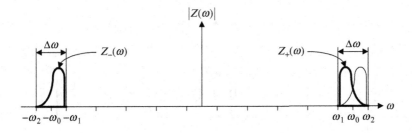

Figure 5.17: Bandpass spectrum

Assuming the carrier frequency ω_0 is fixed and not at our disposal we adjust the nominal modulation bandwidth $\Delta\omega$ (by simply adding extra guard bands around the given modulation band) such that

$$\omega_0/\Delta\omega = K = \text{integer}. \tag{5.50}$$

We choose K to be the largest possible integer compatible with the given modulation band (i.e., $\Delta\omega$ is nearly identical with the "true" modulation band). Setting

$$\Delta t = \frac{2\pi}{\Delta\omega} = \frac{2\pi K}{\omega_0} \tag{5.51}$$

the $\hat{Z}(\omega)$ becomes the periodic repetition of the inphase component of the base-band spectrum (c.f. 2.225a)

$$Z_- (\omega - \omega_0) + Z_+ (\omega + \omega_0) = X(\omega) \tag{5.52}$$

illustrated in Fig. 5.18. Hence we can unambiguously extract $X(\omega)$ from any frequency segment having a total bandwidth $2\pi/\Delta t$. For example, we could use the interval symmetrically disposed with respect to the zero frequency highlighted in Fig. 5.18. Note that this spectrum is actually an alias of the original (undersampled) waveform. An alternative (and perhaps more direct) interpretation of this result can be had if we recall that $Z_- (\omega - \omega_0) + Z_+ (\omega + \omega_0)$ is just the FT of the inphase component of the signal, i.e., $x(t) \cos(\omega_0 t)$, so that its sampled value is $x(n\Delta t) \cos(\omega_0 \Delta t n)$. Evidently choosing the sampling interval such that $\omega_0 \Delta t = 2\pi K$ reduces this sampled value to $x(n\Delta t)$. Thus we are able to demodulate the inphase component of the bandpass signal by simply sampling it at an appropriate rate. There are, however, significant practical limitations in implementing such a demodulation scheme, one of which is the requirement that the time-aperture (duration of the sampling pulse) of the A/D converter must be much less than the period of the carrier. To assess the effect of a finite time-aperture let us assume a sampling pulse duration $\delta \ll \Delta t$ so that the estimated sampled value of $x(t) \cos(\omega_0 t)$ is

$$\hat{x}(n\Delta t) = \int_{-\infty}^{\infty} \frac{1}{\delta} p_{\delta/2}(t - n\Delta t) x(t) \cos(\omega_0 t) dt$$

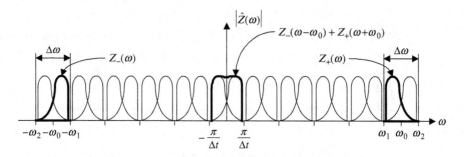

Figure 5.18: Spectrum of sampled bandpass signal

$$= \frac{1}{\delta} \int_{-\delta/2}^{\delta/2} x(\xi + n\Delta t) \cos\left[\omega_0\left(\xi + n\Delta t\right)\right] d\xi$$

$$\approx x(n\Delta t)\frac{1}{\delta} \int_{-\delta/2}^{\delta/2} \cos\left[\omega_0\left(\xi + n\Delta t\right)\right] d\xi$$

$$= x(n\Delta t)\frac{1}{\delta} \int_{-\delta/2}^{\delta/2} \cos(\omega_0\xi)d\xi = x(n\Delta t)\frac{\sin\left(\omega_0\delta/2\right)}{\omega_0\delta/2}$$

$$= x(n\Delta t)\frac{\sin\left(\pi\delta/T\right)}{\pi\delta/T}, \tag{5.53}$$

where T is the carrier period. Thus the error incurred as a result of a finite pulse width is

$$x(n\Delta t) - \hat{x}(n\Delta t) = x(n\Delta t)\left[1 - \frac{\sin\left(\pi\delta/T\right)}{\pi\delta/T}\right].$$

From this we infer that for a given δ/T the maximum possible resolution that can be achieved with an A/D converter is approximately

$$N = -\log_2\left[1 - \frac{\sin\left(\pi\delta/T\right)}{\pi\delta/T}\right] \text{ bits}, \tag{5.54}$$

which is plotted in Fig. 5.19. We observe that to achieve a 12-bit resolution requires pulse widths approximately 1% of the carrier period.

To achieve digital demodulation of the complete bandpass signal $x(t)$ $\cos(\omega_0 t) - y(t)\sin(\omega_0 t)$ requires that the carrier of the quadrature component be phase shifted by 90°, as illustrated in Fig. 5.20. Note that this scheme assumes that the phase shift has no effect on the signal envelope. This is not difficult to achieve with communication signals at microwave frequencies where the bandwidth of the envelope is usually much smaller than the carrier frequency. (A simple way of implementing such a phase shift is to use a quadrature hybrid.)

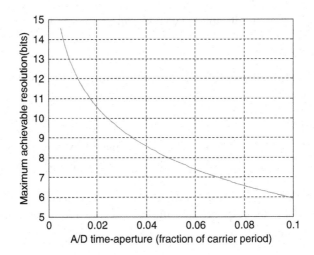

Figure 5.19: Time-aperture requirements for digital demodulation of bandpass signals

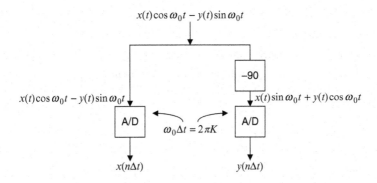

Figure 5.20: Demodulation using bandpass sampling

5.3.4 Sampling of Periodic Signals

In the following we discuss aliasing effects engendered by sampling of periodic signals. The interesting cases arise when we assume that the periodic function can be represented by a truncated FS. This is nearly always true in practice. The significance of this assumption is that a signal comprised of a finite number of FS terms is bandlimited and therefore representable in terms of its samples exactly. Let us first consider a simple sinusoid such as $\cos(\omega_0 t)$ and parameterize the sampling interval as follows:

$$\Delta t = \frac{\pi}{\omega_0} + \xi.$$

As long as $-\frac{\pi}{\omega_0} < \xi < 0$ aliasing is avoided and the series defining the sampled spectrum gives

$$\sum_{n=-\infty}^{\infty} \cos(\omega_0 n\Delta t)\,\Delta t e^{-i\omega n \Delta t} = \pi\delta(\omega - \omega_0) + \pi\delta(\omega + \omega_0),\qquad (5.55)$$

wherein $\omega_0 < \pi/\Delta t$. Moreover the Shannon sampling theorem gives

$$\cos(\omega_0 t) = \sum_{n=-\infty}^{\infty} \cos(\omega_0 n\Delta t)\,\frac{\sin[\pi(t/\Delta t - n)]}{\pi(t/\Delta t - n)}.\qquad (5.56)$$

A plot of (5.56) using 13 samples with $\omega_0\xi = -\pi/2$ (twice the Nyquist rate) is shown in Fig. 5.21. When $0 < \xi$ the sampled spectrum is aliased since then $\omega_0 > \pi/\Delta t$. Restricting ξ to $\xi < \frac{\pi}{\omega_0}$ we get for $|\omega| < \pi/\Delta t = \omega_0/(1 + \omega_0\xi/\pi)$

$$\sum_{n=-\infty}^{\infty} \cos(\omega_0 n\Delta t)\,\Delta t e^{-i\omega n \Delta t} = \pi\delta\left\{\omega + \omega_0\left[\frac{\pi/\omega_0 - \xi}{\pi/\omega_0 + \xi}\right]\right\}$$

$$+\pi\delta\left\{\omega - \omega_0\left[\frac{\pi/\omega_0 - \xi}{\pi/\omega_0 + \xi}\right]\right\}.\qquad (5.57)$$

The inverse FT of the right side is $\cos\left\{\omega_0\left[\frac{\pi/\omega_0 - \xi}{\pi/\omega_0 + \xi}\right]t\right\}$. Thus undersampling of a pure tone produces a pure tone of lower frequency which can be adjusted from ω_0 down to zero by varying ξ. In fact (5.56) can in this case be replaced by

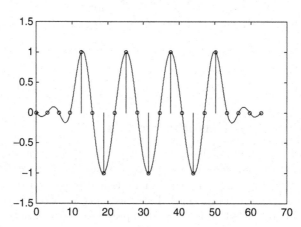

Figure 5.21: Representation of $\cos(\omega_0 t)$ using 13 samples (sampled at twice the Nyquist rate)

$$\cos\left\{\omega_0\left[\frac{\pi/\omega_0 - \xi}{\pi/\omega_0 + \xi}\right]t\right\} = \sum_{n=-\infty}^{\infty} \cos(\omega_0 n\Delta t)\,\frac{\sin[\pi(t/\Delta t - n)]}{\pi(t/\Delta t - n)}.\qquad (5.58)$$

Plots of (5.58) together with stem diagrams indicating the sampled values of the signal are shown in Fig. 5.22 for $\omega_0 = 1\pi \times 100\,\text{rps}$, $\xi = 0$ (Nyquist sampling, $\Delta t = 5\,\text{ms}$) and $\xi = .001667$ ($\Delta t = 6.667\,\text{ms}$). As is evident from the figure the frequency of the undersampled and reconstructed signal is exactly one half of the original analogue signal.

The reduction of the signal frequency through aliasing finds application in the sampling oscilloscope. The idea is to increase the timescale (reduce the frequency) of a rapidly varying waveform to facilitate its observation. Here the periodicity of the signal plays an essential role. Thus suppose $x(t)$ is the rapidly varying waveform whose FS representation reads

$$x(t) = \sum_{m=-M}^{M} \hat{x}_m e^{i2\pi mt/T}. \tag{5.59}$$

Let us sample this signal at intervals $\Delta t = T + \tau$ with τ a positive time increment to be specified. The sampled (and possibly aliased) spectrum

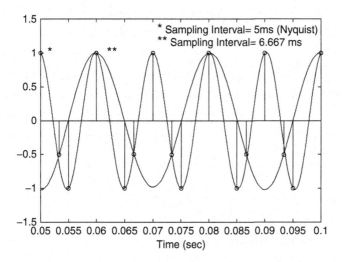

Figure 5.22: Reduction in frequency by undersampling is, by definition

$$
\begin{aligned}
\hat{X}(\omega) &= (T+\tau) \sum_{n=-\infty}^{\infty} x\left[n\left(T+\tau\right)\right] e^{-in\omega(T+\tau)} \\
&= (T+\tau) \sum_{n=-\infty}^{\infty} x\left[n\tau\right] e^{-in\omega(T+\tau)},
\end{aligned} \tag{5.60}
$$

where in the last expression we have taken advantage of the periodicity in T and set $x\left[n\left(T+\tau\right)\right] = x\left[n\tau\right]$. Next we evaluate (5.59) at $t = n\tau$ and substitute in (5.60) to obtain

$$\hat{X}(\omega) = (T+\tau) \sum_{m=-M}^{M} \hat{x}_m \sum_{n=-\infty}^{\infty} e^{i2\pi mn\tau/T} e^{-in\omega(T+\tau)}$$

$$= 2\pi (T+\tau) \sum_{m=-M}^{M} \hat{x}_m \sum_{\ell=-\infty}^{\infty} \delta\left[\frac{2\pi m\tau}{T} - \omega(T+\tau) + 2\pi\ell\right]$$

$$= 2\pi \sum_{m=-M}^{M} \hat{x}_m \sum_{\ell=-\infty}^{\infty} \delta\left[\omega - \frac{2\pi m\tau}{T(T+\tau)} - \frac{2\pi\ell}{T+\tau}\right], \qquad (5.61)$$

where we have employed the delta function representation of the infinite sum of sinusoids. Alternatively, we may rewrite the preceding as follows:

$$\hat{X}(\omega) = \sum_{\ell=-\infty}^{\infty} X(\omega - \frac{2\pi\ell}{T+\tau}), \qquad (5.62)$$

where

$$X(\omega) = 2\pi \sum_{m=-M}^{M} \hat{x}_m \delta\left[\omega - \frac{2\pi m\tau}{T(T+\tau)}\right]. \qquad (5.63)$$

Since the bandwidth of $X(\omega)$ is limited to $|\omega| \le 2\pi M\tau/T(T+\tau)$, and the sampling is done at intervals $T+\tau$ the frequency components $2\pi\ell/(T+\tau)$ in (5.62) for $\ell \ne 0$ represent aliases. To avoid aliasing (see Fig. 5.10) we choose τ such that $2\pi M\tau/T(T+\tau) < \pi/(T+\tau)$ or, equivalently,

$$\tau < T/2M. \qquad (5.64)$$

In that case the expression for the spectrum (5.61) simplifies to

$$\hat{X}(\omega) = 2\pi \sum_{m=-M}^{M} \hat{x}_m \delta\left[\omega - \frac{2\pi m\tau}{T(T+\tau)}\right] \qquad (5.65)$$

and taking the inverse transform we get

$$\hat{x}(t) = \sum_{m=-M}^{M} \hat{x}_m e^{i\frac{2\pi m\tau}{T(T+\tau)}t}.$$

Comparing this with (5.59) yields

$$\hat{x}(t) = x\left(\frac{t\tau}{T+\tau}\right) \qquad (5.66)$$

so that the sampled and reconstructed signal is just the original periodic signal scaled in time by the factor $\tau/(T+\tau)$. This technique is illustrated in Fig. 5.23 which shows 10 cycles of a periodic signal with period of 4 s comprised of a fundamental frequency and a third harmonic. Superposed is a period of the time-scaled signal reconstructed from samples spaced 4.4 s apart. Here $M = 3$ so that (5.64) is satisfied since $\tau = 0.4 < 2/3$.

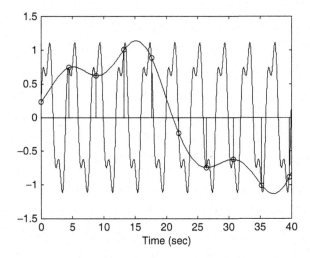

Figure 5.23: Stretching the timescale of a periodic waveform by undersampling

5.4   The Discrete Fourier Transform

5.4.1   Fundamental Definitions

Let us now revisit the model of a bandlimited function whose spectrum can
be represented by a FS comprised of a finite number of terms. The FT is then
given by (5.22) and the periodic extensions outside the baseband may be ignored
since they exist where the actual transform is defined as identically zero. On
the other hand, it turns out that for reasons of mathematical symmetry it is
advantageous to incorporate a periodic extension of the FT into the definition of
the Discrete Fourier Transform (DFT). The relationship between this extension
and the actual transform is best illustrated with the following simple example.
Consider a bandlimited function defined by the two non-zero samples $f(0) = f(\Delta t) = 1$. Its FT is

$$F(\omega) = \left\{ \begin{array}{c} \left(1 + e^{-i\omega\Delta t}\right)\Delta t \;; |\omega| \leq \pi/\Delta t \\ 0 \;; |\omega| > \pi/\Delta t. \end{array} \right. \tag{5.67}$$

A plot of the magnitude and phase of $F(\omega)/\Delta t$ within the baseband is shown
in Fig. 5.24. On the other hand, the identical information can be extracted from
Fig. 5.25 where the magnitude and phase of $1 + e^{-i\omega\Delta t}$ are plotted in the range
$0 \leq \omega < 2\pi/\Delta t$ provided of course one recognizes that the folding frequency now
falls in the center of the plot, while the negative frequencies have been mapped
into the region $\pi/\Delta t < \omega < 2\pi/\Delta t$. The impetus for this convention was apparently driven by the insistence that the time interval (which is normally taken
as the positive segment of the time axis with zero as the reference) occupied
by the signal map into a corresponding positive segment of the frequency axis.

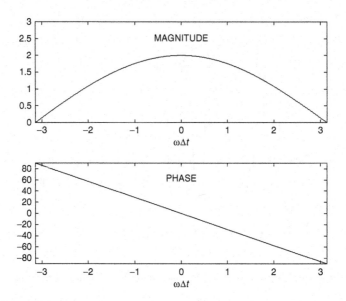

Figure 5.24: Magnitude and phase of the FT in (5.67)

Despite the somewhat anomalous situation of finding negative frequencies on the second half of the positive frequency axis this convention has found almost universal acceptance and is incorporated in the standard definition of the DFT. To derive it we first rewrite (5.22) with a small notational change:

$$F\left(\omega\right) = \sum_{n=0}^{N-1} f\left[n\right] e^{-in\Delta t\omega} \tag{5.68}$$

$$f\left[n\right] = f\left(n\Delta t\right) \Delta t. \tag{5.69}$$

An inversion formula for the recovery of the sample values $f\left[n\right]$ in terms of $F\left(\omega\right)$ is, of course, available in the form of the FS integral (5.6). On the other hand, (5.68) also states that $F\left(\omega\right)$ is represented *exactly* by N sinusoids. We recognize this as an interpolation problem analogous to the one already dealt with in 2.1.8 in the time domain. This suggests that an alternative to evaluating the integral (5.6) is to specify $F\left(\omega\right)$ at N *discrete* frequencies and solve the resulting NXN system of algebraic equations for $f\left[n\right]$. Although, in principle, these sample points need not be uniformly distributed only a uniform distribution leads to an algebraically simple inversion formula. Accordingly we divide the frequency interval $0, 2\Omega$ into N bins each of width $\Delta\omega = 2\Omega/N = 2\pi/N\Delta t$, and set $\omega = m\Delta\omega$.[3] We introduce these into (5.68) to obtain

$$F\left[m\right] = \sum_{n=0}^{N-1} f\left[n\right] e^{-i2\pi nm/N} \quad ; m = 0, 1, 2, \ldots N, \tag{5.70}$$

---

[3]Note that with the choice $0, 2\Omega$ for the baseband rather than $-\Omega, \Omega$ we are adopting the periodic extension definition of Fig. 5.18.

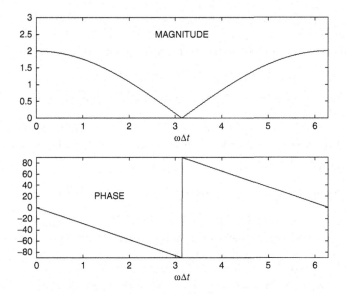

Figure 5.25: Plots of periodic extension of FT in (5.67)

where for the sake of notational consistency with (5.69) we denote the frequency samples by

$$F[m] = F(m\Delta\omega).$$ (5.71)

The final step is to use the orthogonality relationship (2.70) to solve for $f[n]$. The result is

$$f[n] = \frac{1}{N} \sum_{m=0}^{N-1} F[m] e^{i2\pi nm/N} \quad ; n = 0, 1, 2, \ldots N.$$ (5.72)

The transform pairs (5.70) and (5.72) are, respectively, the standard forms of the direct and the inverse DFT. In addition it is customary to set

$$W_N \equiv e^{-i2\pi/N}$$ (5.73)

which recasts (5.70) and (5.72) into the following form:

$$F[m] = \sum_{n=0}^{N-1} W_N^{mn} f[n] \quad ; m = 0, 1, 2, \ldots N$$ (5.74a)

$$f[n] = \frac{1}{N} \sum_{m=0}^{N-1} W_N^{-nm} F[m] \quad ; n = 0, 1, 2, \ldots N.$$ (5.74b)

Although both transformations involve only discrete samples the underlying bandlimited time waveform and the frequency spectrum can be reconstructed with suitable interpolation formulas. We already know that in the time domain

$f(t)$ can be recovered via the Shannon sampling formula (5.11). An interpolation formula for $F(\omega)$ can be obtained by substituting $f[n]$ from (5.72) into (5.68) and summing the geometric series with respect to n. The result is

$$F(\omega) = \sum_{k=0}^{N-1} F[k]\, e^{i\pi\left(\frac{N-1}{N}\right)(k-\omega/\Delta\omega)} \frac{\sin\left[\pi\left(k-\omega/\Delta\omega\right)\right]}{N\sin\left[\pi\frac{1}{N}\left(k-\omega/\Delta\omega\right)\right]}, \tag{5.75}$$

which is very similar to the time domain interpolation formula for Fourier series in (2.74). An alternative interpolation technique is the so-called "zero padding" described on page 308.

5.4.2   Properties of the DFT

Even though we have derived the DFT with reference to bandlimited functions and the continuum this transform may be viewed in its own right as a linear operation on discrete sequences. In fact it constitutes a significant processing tool in diverse DSP (Discrete Signal Processing) applications quite independent from any relationship to analogue waveforms. In the following we investigate some of its fundamental properties.

Periodicity

Since $W_N^{Nk} = 1$ for any integer k, every sequence $f[n]$ of length N transforms into a periodic sequence $F[m]$ of period N i.e.,

$$F[m+Nk] = F[m] \quad ; k = \pm 1, \pm 2, \ldots$$

In this sense the situation is very similar to Fourier series so that we may regard $F[m]$ for $|m| > N$ as a periodic extension. Similarly every sequence $F[m]$ of length N transforms into a periodic sequence $f[n]$ of period N. Even though normally the specification of a data sequence $f[n]$ is restricted to the index range $0 \le n \le N-1$ negative indexes can be accommodated through periodic extensions. For example, the sequence defined by

$$f[n] = \begin{cases} 1 - \frac{|n|}{10}; |n| \le 10 \\ 0 \text{ otherwise} \end{cases}$$

shown plotted in Fig. 5.26a with respect to the index range $-32 \le n < 32$ appears as in Fig. 5.26b when plotted with respect to $0 \le n \le 63$

Orthogonality

According to (2.70) the DFT matrix W_N^{mn} in (5.74) is comprised of orthogonal columns and rows. Because of periodicity of W_N^{mn} the orthogonality generalizes to

$$\sum_{n=0}^{N-1} W_N^{mn} W_N^{-nk} = \sum_{n=0}^{N-1} W_N^{mn} W_N^{*nk} = N\delta_{m,k+N\ell} \quad ; \ell = 0, \pm 1, \pm 2\ldots \tag{5.76}$$

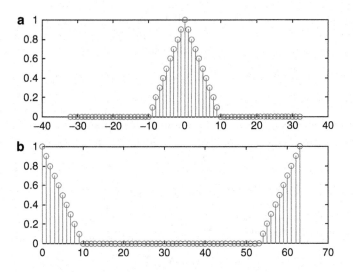

Figure 5.26: Periodic extensions for the DFT

Matrix Properties

To emphasize the matrix properties of the DFT we rewrite (5.74a) as follows:

$$\mathbf{F} = \mathbf{W}_N \mathbf{f} \tag{5.77}$$

where $\mathbf{W}_N$ is the $N \times N$ matrix with elements W_N^{mn} and

$$\mathbf{f} = \begin{bmatrix} f[0] & f[1] & f[2] & \ldots & f[N-1] \end{bmatrix}^T, \tag{5.78a}$$

$$\mathbf{F} = \begin{bmatrix} F[0] & F[1] & F[2] & \ldots & F[N-1] \end{bmatrix}^T. \tag{5.78b}$$

The DFT matrix is symmetric which we may write formally as

$$\mathbf{W}_N = \mathbf{W}_N^T. \tag{5.79}$$

Its structure is best appreciated with the aid of the display

$$\mathbf{W}_N = \begin{bmatrix} 1 & 1 & 1 & . & 1 \\ 1 & W_N^1 & W_N^2 & . & W_N^{N-1} \\ 1 & W_N^2 & W_N^4 & . & W_N^{2(N-1)} \\ . & . & . & . & . \\ 1 & W_N^{N-1} & W_N^{2(N-1)} & . & W_N^{(N-1)(N-1)} \end{bmatrix}. \tag{5.80}$$

In view of (5.73) the elements of $\mathbf{W}_N$ in (5.80) are powers of the N-th roots of unity. Hence the elements of a typical column of $\mathbf{W}_N$, say $\mathbf{w}_k$,

$$\mathbf{w}_k = \begin{bmatrix} 1 & W_N^k & W_N^{k2} & \ldots & W_N^{k(N-1)} \end{bmatrix}^T$$

when plotted in the complex plane will be uniformly spaced on a unit circle. Moreover the next vector $\mathbf{w}_{k+1}$ is generated by successive rotations of the complex numbers representing the elements of $\mathbf{w}_k$ through the angles

$$0, 2\pi/N, 2\,(2\pi/N)\,, 4\,(2\pi/N)\,,\dots(N-1)\,(2\pi/N)$$

radians. The direct DFT (5.77) may be viewed as an expansion of $\mathbf{F}$ in terms of the orthogonal basis vectors represented by the columns of $\mathbf{W}_N$. In terms of these basis vectors the orthogonality (5.76) may be expressed as

$$\mathbf{w}_k^H \mathbf{w}_n = N\delta_{kn} \tag{5.81}$$

or equivalently in block matrix form

$$\mathbf{W}_N \mathbf{W}_N^H = \mathbf{W}_N \mathbf{W}_N^* = N\mathbf{I}. \tag{5.82}$$

With the aid (5.81) the inverse DFT follows at once as

$$\mathbf{f} = \frac{1}{N} \mathbf{W}_N^* \mathbf{F}. \tag{5.83}$$

Recall that a matrix $\mathbf{U}$ with the property $\mathbf{U}\mathbf{U}^H = \mathbf{I}$ is said to be unitary. Thus in virtue of (5.92) the matrix

$$\mathbf{U}_N = \frac{1}{\sqrt{N}} \mathbf{W}_N \tag{5.84}$$

is unitary and according to (5.79) it is also symmetric. Thus except for a normalization factor the DFT is a unitary symmetric transformation.[4]

Consider now the eigenvalue problem of the normalized DFT matrix (5.83):

$$\mathbf{U}_N \mathbf{e}_k = \lambda_k \mathbf{e}_k; \;\; k = 1, 2, \dots, N. \tag{5.85}$$

We recall that a unitary and symmetric matrix can be diagonalized a real orthogonal transformation. Hence all the $\mathbf{e}_k$ are real and

$$\mathbf{e}_k^T \mathbf{e}_m = \delta_{km}. \tag{5.86}$$

Therefore an arbitrary N-dimensional sequence can be represented by the superposition

$$\mathbf{f} = \sum_{k=0}^{N-1} \hat{f}_k \mathbf{e}_k, \tag{5.87}$$

where $\hat{f}_k = \mathbf{e}_k^T \mathbf{f}$. An interesting feature of this basis is that each $\mathbf{e}_k$ is proportional to its own DFT. Thus if we take the DFT of (5.87) we find

$$\mathbf{F} = \sqrt{N} \sum_{k=0}^{N-1} \hat{f}_k \lambda_k \mathbf{e}_k \tag{5.88}$$

from which the DFT of $\mathbf{e}_k$ follows as $\sqrt{N}\lambda_k \mathbf{e}_k$.

---

[4]Recall that an analogous result holds for FT which, except for the factor $1/2\pi$, is also a unitary transformation.

By virtue of the special symmetry of the DFT matrix we can find the eigen-values λ_k without having to solve directly for the roots of the characteristic polynomial. To see this let us first evaluate the elements of the matrix $\mathbf{Q} \equiv \mathbf{U}^2$. We have

$$Q_{km} = \frac{1}{N} \sum_{n=0}^{N-1} W_N^{kn} W_N^{nm} = \delta_{k,-m+\ell N} \quad ; \ell = 0, \pm 1, \pm 2 \dots \tag{5.89}$$

Since the indices k, m are restricted to run from 0 to $N - 1$ the right side of (5.89) is zero unless $\ell = 0, m = 0$ or $\ell = 1$ and $m = \ell N - k$, in which case it equals unity. Therefore we get for $\mathbf{Q}$ the elementary matrix

$$\mathbf{Q} = \begin{bmatrix} 1 & 0 & 0 & . & 0 & 0 & 0 \\ 0 & 0 & 0 & . & 0 & 0 & 1 \\ 0 & 0 & 0 & . & 0 & 1 & 0 \\ . & . & . & . & . & . & . \\ 0 & 0 & 0 & . & 0 & 0 & 0 \\ 0 & 0 & 1 & . & 0 & 0 & 0 \\ 0 & 1 & 0 & . & 0 & 0 & 0 \end{bmatrix} .$$

It is not hard to see that $\mathbf{Q}^2 = \mathbf{U}^4 = \mathbf{I}$ so that in view of (5.85) for any $\mathbf{e}_k$

$$\mathbf{U}^4 \mathbf{e}_k = \lambda_k^4 \mathbf{e}_k = \mathbf{e}_k$$

and $\lambda_k^4 = 1$. From this we see that the N eigenvalues λ_k are restricted to the four roots of unity which are ± 1 and $\pm i$. Consequently for large N the eigenvalue problem for the DFT matrix is highly degenerate (many eigenvectors correspond to the same eigenvalue). Let us tag each of the four groups of eigenvectors with the corresponding eigenvalue and rewrite the expansion (5.88) as follows:

$$\begin{aligned} \mathbf{F} = &\sqrt{N} \sum_{N^{(1)}} \hat{f}_k^{(1)} \mathbf{e}_k^{(1)} - \sqrt{N} \sum_{N^{(-1)}} \hat{f}_k^{(-1)} \mathbf{e}_k^{(-1)} \\ &+ i\sqrt{N} \sum_{N^{(i)}} \hat{f}_k^{(i)} \mathbf{e}_k^{(i)} - i\sqrt{N} \sum_{N^{(i)}} \hat{f}_k^{(-i)} \mathbf{e}_k^{(-i)}, \end{aligned} \tag{5.90}$$

where $N = N^{(1)} + N^{(-1)} + N^{(i)} + N^{(-i)}$. From this we see that a general sequence represented by the vector $\mathbf{f}$ projects into four orthogonal subspaces spanned by the vectors $\mathbf{e}_k^{(1)}, \mathbf{e}_k^{(-1)}, \mathbf{e}_k^{(i)} \ \mathbf{e}_k^{(-i)}$. Except for a constant factor, each of these projections is seen to be its own transform. The content of (5.90) may therefore be restated as follows. Any sequence $f[n]$ can be written uniquely as a sum of four orthogonal sequences each of which is proportional to its DFT. When N is chosen such that $\log_2 N$ is an integer one can show that each of the four subspaces has the following dimension: $N^{(1)} = \frac{N}{4} + 1, N^{(-1)} = \frac{N}{4}, N^{(i)} = \frac{N}{4} - 1, N^{(-i)} = N/4$. Thus for large N the number of linearly independent eigenvectors in each subspace is essentially identical.

Parseval Identities

Consider two sequences represented by the vectors $\mathbf{f}$ and $\mathbf{g}$ and their respective DFTs by $\mathbf{F}$ and $\mathbf{G}$. We compute the inner product

$$\mathbf{f}^H\mathbf{g} = \frac{1}{N}\left(\mathbf{W}_N^H\mathbf{F}\right)^H\frac{1}{N}\mathbf{W}_N^H\mathbf{G} = \frac{1}{N^2}\mathbf{F}^H\mathbf{W}_N\mathbf{W}_N^H\mathbf{G}$$

and using (5.82) in the last term obtain

$$\mathbf{f}^H\mathbf{g} = \frac{1}{N}\mathbf{F}^H\mathbf{G}. \tag{5.91}$$

This is one form of the Parseval identity for the DFT. If one again ignores the asymmetry caused by N, (5.91) is a statement of the preservation of the inner product under a unitary transformation. In the special case $\mathbf{f} = \mathbf{g}$ this reduces to

$$\|\mathbf{f}\|^2 = \frac{1}{N}\|\mathbf{F}\|^2. \tag{5.92}$$

If we are dealing with bandlimited analogue signals the preceding can also be written in terms of integrals as follows:

$$\|\mathbf{f}\|^2 = \Delta t\int_{-\infty}^{\infty}|f(t)|^2\,dt = \frac{\Delta t}{2\pi}\int_{-\Omega}^{\Omega}|F(\omega)|^2\,d\omega = \frac{1}{N}\|\mathbf{F}\|^2 \tag{5.93}$$

Alternatively from last equality we get

$$\frac{1}{2\pi}\int_{-\Omega}^{\Omega}|F(\omega)|^2\,d\omega = \frac{1}{2\pi}\sum_{n=0}^{N-1}|F(n\Delta\omega)|^2\,\Delta\omega, \tag{5.94}$$

which highlights the fact that for bandlimited signals a finite Riemann sum represents the energy integral exactly also in the spectral domain.

Time and Frequency Shift

As we know, the special attribute of the FT that renders it particularly useful in the analysis of LTI systems is the transformation of a time-delayed signal $f(t-T)$ into $F(\omega)\,e^{-i\omega t}$. This relationship is also maintained between the samples and the FT of the time-delayed version of a bandlimited function. This follows from

$$f(t-T) \overset{\mathcal{F}}{\Longleftrightarrow} \sum_{n=-\infty}^{\infty}f(n\Delta t - K\Delta t)\,e^{-i\omega n\Delta t}\Delta t$$

$$= e^{-i\omega K\Delta t}\sum_{n=-\infty}^{\infty}f(\ell\Delta t)\,e^{-i\omega n\Delta t}\Delta t = e^{-i\omega K\Delta t}F(\omega), \tag{5.95}$$

where $T = K\Delta t$, and K an integer. Using the notation (5.69) we get

$$\sum_{n=-\infty}^{\infty} f[n-K] e^{-i\omega n \Delta t} = e^{-i\omega K \Delta t} \sum_{\ell=-\infty}^{\infty} f[\ell] e^{-i\omega \ell \Delta t}. \qquad (5.96)$$

Unfortunately an analogous relationship cannot hold without restriction for the DFT since it is defined with reference to the fixed data window of length N. To see this explicitly let us introduce the notation for the DFT pair

$$f[n] \overset{\mathcal{DF}}{\Longleftrightarrow} F[m] \qquad (5.97)$$

and then compute

$$f[n-K] \overset{\mathcal{DF}}{\Longleftrightarrow} \sum_{n=0}^{N-1} W_N^{mn} f[n-K] = W_N^{mK} \sum_{\ell=-K}^{N-K-1} W_N^{m\ell} f[\ell]. \qquad (5.98)$$

Clearly the preceding sum is in general not equal to $F[m]$. There are two conditions when this will be the case. First, if the sequence $f[n]$ is identically zero outside the data window and is of length $L \le N$ provided $N - K \ge L$. Second, if $f[n]$ is either periodic with principal period N or is *defined* outside the data window by its periodic extension. If one of these conditions holds, we may write

$$f[n-K] \overset{\mathcal{DF}}{\Longleftrightarrow} F[m] W_N^{mK}. \qquad (5.99)$$

By symmetry, if the same constraints are imposed on $F[m]$ we have

$$f[n] W_N^{-nK} \overset{\mathcal{DF}}{\Longleftrightarrow} F[m-K]. \qquad (5.100)$$

Convolution

The convolution operation for discrete sequences is defined in a manner analogous to analogue signals. Thus

$$y[n] = \sum_{m=-\infty}^{\infty} f[n-m] h[m] = \sum_{m=-\infty}^{\infty} h[n-m] f[m]. \qquad (5.101)$$

When each of the sequences is of finite length, say $f[n]$ is of length N_f and $h[n]$ of length N_h, (5.101) can be modified by taking explicit account of the truncations of the sequences. For example, the first of the sums in (5.101) transforms into

$$y[n] = \begin{cases} \sum_{m=0}^{n} f[n-m] h[m] & ; 0 \le n \le N_h - 1, \\ \sum_{m=0}^{N_h-1} f[n-m] h[m] & ; N_h - 1 < n \le N_h + N_f - 2. \end{cases} \qquad (5.102)$$

The sequence $y[n]$ is of length $N_f + N_h - 1$, as may also be seen from the following matrix representation of (5.102) (assuming $N_f > N_h$):

$$
\begin{bmatrix}
y[0] \\
y[1] \\
\cdot \\
y[N_h] \\
\cdot \\
y[N_h + N_f - 2]
\end{bmatrix}
=
\begin{bmatrix}
f[0] & 0 & \cdot & 0 \\
f[1] & f[0] & \cdot & 0 \\
\cdot & \cdot & \cdot & \cdot \\
f[N_h - 1] & f[N_h - 2] & \cdot & f[0] \\
\cdot & \cdot & & \cdot \\
0 & 0 & \cdot & f[N_f - 1]
\end{bmatrix}
$$

$$
\begin{bmatrix}
h[0] \\
h[1] \\
\cdot \\
h[N_h - 1]
\end{bmatrix}.
\tag{5.103}
$$

We shall abbreviate the convolution by writing

$$
\mathbf{y} = conv(\mathbf{f}, \mathbf{h}) \equiv conv(\mathbf{h}, \mathbf{f}).
\tag{5.104}
$$

It is worth noting at this juncture that the elements of the vector $\mathbf{y}$ can be interpreted as the coefficients of a polynomial that results from the multiplication of two polynomials whose coefficients are given by the vectors $\mathbf{f}$ and $\mathbf{h}$. To see this let $P_{N_f}(z) = \sum_{\ell=0}^{N_f - 1} f[n] z^\ell$ and $P_{N_h}(z) = \sum_{k=0}^{N_h - 1} h[k] z^k$. Multiplication gives

$$
\begin{aligned}
P_{N_f}(z) P_{N_h}(z) &= \sum_{\ell=0}^{N_f-1} \sum_{k=0}^{N_h-1} f[\ell] h[k] z^{\ell+k} \\
&= \sum_{n=0}^{N_h+N_f-2} \left(\sum_{k=0}^{N_h-1} f[n-k] h[k] \right) z^n.
\end{aligned}
\tag{5.105}
$$

The sum

$$
y[n] = \sum_{k=0}^{N_h-1} f[n-k] h[k]
\tag{5.106}
$$

with the upper limit set to n whenever $n < N_h - 1$ is just the convolution (5.102). It is also the n-th coefficient of an $N_h + N_f - 1$th order polynomial. Clearly (5.105) is independent of the nature of z. In fact if we replace z by W_N^m we can identify the two polynomials with DFTs provided the data window N is chosen sufficiently long to accommodate both sequences, i.e., $N \geq N_h + N_f - 1$, and we extend each of the sequences to length N by appending $N - N_f$ and $N - N_h$ zeros to $f[n]$ and $h[n]$, respectively. For such zero-padded sequences (5.105) with $z = W_N^m$ is equivalent to the transform pair

$$
conv(\mathbf{f}, \mathbf{h}) \overset{\mathcal{DF}}{\Longleftrightarrow} F[m] H[m] \; ; \; N \geq N_h + N_f - 1.
\tag{5.107}
$$

Similarly, by replacing the sequences in (5.105) with their zero-padded transforms $\mathbf{F}$ and $\mathbf{H}$, we deduce in almost identical fashion the frequency domain convolution formula[5]

$$f[n]\,h[n] \overset{\mathcal{DF}}{\Longleftrightarrow} \frac{1}{N}conv\,(\mathbf{F},\mathbf{H}) \; ; \; N \geq N_H + N_F - 1, \qquad (5.108)$$

where the vectors $\mathbf{F}$ and $\mathbf{H}$ have lengths N_F and N_H.

The zero-padding of the sequences in (5.107) may be viewed as a device to avoid the cumbersome notation in (5.102) resulting from the truncation of the sequences. A systematic approach to this is to introduce the so-called *circular convolution* that employs the periodic extension of one of the sequences to fill the convolution matrix. This type of convolution is defined by

$$circonv\,(\mathbf{f},\mathbf{h}) = \sum_{k=0}^{N-1} f\,[(n-k) \bmod N]\,h\,[k]\,, \qquad (5.109)$$

where N is the data window as well as the nominal length of each of the sequences. The symmetry of the matrix in (5.109) may be clarified by the display

$$\begin{bmatrix} y\,[0] \\ y\,[1] \\ y\,[2] \\ \\ y\,[N-2] \\ y\,[N-1] \end{bmatrix} = \begin{bmatrix} f\,[0] & f\,[N-1] & . & f\,[1] \\ f\,[1] & f\,[0] & . & f\,[2] \\ .f\,[2] & f\,[1] & . & .f\,[3] \\ . & . & . & . \\ f\,[N-2] & f\,[N-3] & . & f\,[N-1] \\ f\,[N-1] & f\,[N-2] & . & f\,[0] \end{bmatrix} \begin{bmatrix} h\,[0] \\ h\,[1] \\ \\ h\,[N-1] \end{bmatrix}.$$

$$(5.110)$$

The relationship between the elements in adjacent columns shows that the entire matrix can be generated starting with a uniform arrangement of the elements comprising the first column along the circumference of a circle as shown in Fig. 5.27.

Each succeeding column vector is then generated by a clockwise rotation through one element position (hence the name circular convolution). By substituting the DFT representation for each sequence on the right of (5.109) it is not hard to show that

$$circonv\,(\mathbf{f},\mathbf{h}) \overset{\mathcal{DF}}{\Longleftrightarrow} F\,[m]\,H\,[m] \qquad (5.111)$$

and similarly for the inverse transform

$$f[n]\,h[n] \overset{\mathcal{DF}}{\Longleftrightarrow} \frac{1}{N}circonv\,(\mathbf{F},\mathbf{H})\,. \qquad (5.112)$$

---

[5]It is important to note an essential difference between zero-padding of the (time-domain) sequences and their transforms. While zero-padding a sequence can be done by simply appending any number of zero without changing the form of the signal, zero-padding in the frequency domain requires that the symmetry relation between the spectral components on either side of the folding frequency be maintained.

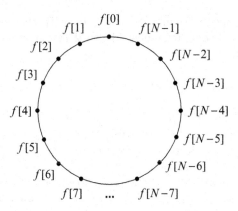

Figure 5.27: Matrix elements for circular convolution

This shows that for circular convolution the direct and inverse DFT factorization formulas (5.111) and (5.112) hold without any restrictions on the sequence lengths. This is, of course, entirely due to our use of the periodic extension of one of the sequences. By comparing (5.110) with (5.103) it is not hard to see that a circular convolution of two sequences which have been zero-padded to $N \geq N_h + N_f - 1$ reduces to an ordinary convolution. One utility of the circular convolution is that it provides, albeit indirectly, a numerically more efficient for the evaluation of ordinary convolutions than does the direct summation formula (5.102). The procedure is to zero-pad the sequences to $N = N_h + N_f - 1$ and compute their DFT. The inverse DFT of the product of these two DFTs then computed which, by virtue of (5.111), is formally equal to the circular convolution but because the initial zero-padding is the same as the ordinary convolution.

Zero-padding can also be used as an interpolation technique. Recall that given a DFT at N frequencies we can interpolate to any frequency using (5.75). An alternative scheme is to append K zeros to the original sequence of length N thereby creating a DFT of length $N + K$. Since this new DFT extends over the entire frequency band (appending zeros does not change the spectrum of the signal) we obtain a denser sampling of the signal spectrum by the factor $(N + K)/K$. As an illustration consider the sequence

$$g(n) = \cos\left[(n - 1 - N/2)\frac{\pi}{N}\right] + \sin\left[(n - 1 - N/2)\frac{\pi}{N}\right] \; ; \; n = 1, 2, \ldots N$$

(5.113)

and its DFT

$$F(m) = \sum_{n=0}^{M} W_M^{nm} \, g(n),$$

(5.114)

where $M \geq N$.

Figure 5.28 shows stem diagrams of the sequence (5.113) on the left and the magnitude of its DFT for $M = N = 64$ on the right.

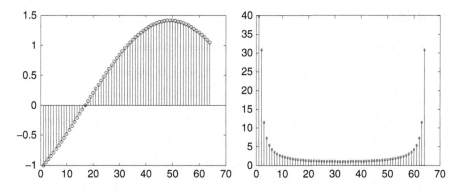

Figure 5.28: Stem diagrams of sequence (5.113) and the magnitude of its DFT

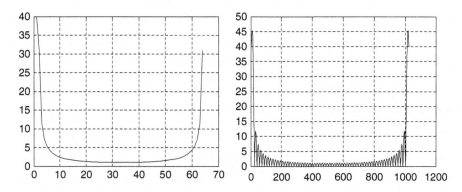

Figure 5.29: Resolution improvement by zero-padding from $N = 64$ to $N = 1024$

When the sequence is zero-padded to $M = 1024$ the resolution improves significantly. This is demonstrated by the plots in Fig. 5.29 which compare the DFT magnitudes before and after zero-padding.

In our discussion of the DFT we have not addressed the problem of its actual numerical evaluation. It is not hard to see that a direct computation requires N^2 multiplications plus additions. This placed a severe damper on the use of DFTs in signal processing until the invention of numerically efficient algorithms. The first of these was the Cooley and Tukey (1965) FFT (Fast Fourier Transform) which reduced the number of operations from N^2 to $O(N \log_2 N)$ thereby revolutionizing digital signal processing. Today there is a large number of "fast algorithms" for DFT evaluation. The presentation of their structure and their relative merits belongs in courses on digital signal processing and numerical analysis. A discussion of the classic Cooley and Tukey radix 2 algorithm can be found in virtually every introductory text on digital signal processing and we will not repeat it. For our purposes the MATLAB implementations of the FFT are sufficient.

Problems

1. The Fourier transform $F(\omega)$ of a bandlimited signal $f(t)$ is given by

$$F(\omega) = \begin{cases} \cos^2\left(\frac{\pi\omega}{2\Omega}\right) e^{-i\frac{\pi\omega}{\Omega}} & ; |\omega| \leq \Omega, \\ 0 & ; |\omega| > \Omega. \end{cases}$$

   (a) The signal is sampled at intervals $\Delta t = \pi/\Omega$. Find the samples.

   (b) Using the sampled values find $f(t)$.

   (c) Suppose the signal is sampled at intervals $\Delta t' = 2\pi/\Omega$ and its Fourier transform approximated by

$$\hat{F}(\omega) = \Delta t' \sum_{n=-\infty}^{\infty} f(n\Delta t') e^{-i\omega n\Delta t'}.$$

   Compute and sketch $\hat{F}(\omega)$.

2. A bandlimited signal $f(t)$ is represented by the finite number of samples $f(0), f(\Delta t), f(2\Delta t), \ldots f((N-1)\Delta t)$. Show that the ratio of the signal energy within the time interval $0 < t < N\Delta t$ to the total energy is given by the largest eigenvalue of the $N \times N$ matrix $\mathbf{A}$ with elements

$$A_{mn} = \int_0^N \frac{\sin[\pi(s-n)]\sin[\pi(s-m)]}{\pi(s-n)\pi(s-m)} ds.$$

3. Derive (5.75).

4. Use (5.79) with $N = 64$ to interpolate the sequence (5.113). Interpolate this sequence using zero padding with $M = 256$. Compare the resolution improvement obtained with the two techniques.

Chapter 6

The Z-Transform
and Discrete Signals

6.1 The Z-Transform

6.1.1 From FS to the Z-Transform

We return to the FT of a bandlimited function as given by (5.9) in Sect. 5.1. Setting $\omega \Delta t = \theta$ we have the FS

$$F(\omega) = F(e^{i\theta}) = \sum_{n=-\infty}^{\infty} f[n] e^{-in\theta} \tag{6.1}$$

with the coefficients $f[n] = f(n\Delta t)\Delta t$ computed in the usual way, viz.,

$$f[n] = \frac{1}{2\pi} \int_{-\pi}^{\pi} F(e^{i\theta}) e^{in\theta} d\theta. \tag{6.2}$$

As we have seen in Chap. 5, for discrete systems (6.1) can also be interpreted as DFT of the sequence $f[n]$ without any requirement that this sequence corresponds in any way to samples of a bandlimited analogue signal. However regardless whether we are dealing with samples of an analogue signal or simply with the DFT of a sequence of numbers mathematically formulas (6.1) and (6.2) represent, respectively, nothing more than the classic FS and its inverse and hence subject to all the restrictions and convergence properties of FS discussed in Chap. 2. In particular, for infinite sequences (6.1) can converge only if $f[n]$ decays sufficiently rapidly as $n \to \pm\infty$. The convergence issues here are similar to those we encountered in connection with the theory of the FT of analogue signals where we had to impose restrictions on the growth of the time function as $t \to \pm\infty$ for its FT to exist. In Chap. 4 we also found that by appending a convergence factor to the imaginary exponential of the FT we could enlarge the class of admissible signals and obtain convergent integral transforms

W. Wasylkiwskyj, *Signals and Transforms in Linear Systems Analysis*, DOI 10.1007/978-1-4614-3287-6_6, © Springer Science+Business Media, LLC 2013

for function that may grow at infinity. This suggests that with the aid of a suitable convergence factor in (6.1) we should be able to obtain DFTs of divergent infinite sequences. Thus suppose $f[n]$ grows as ρ_{min}^n for $n \to \infty$. By analogy with the notation employed in connection with the Laplace transform theory we shall denote this by

$$f[n] \underset{n \sim \infty}{\sim} O(\rho_{min}^n). \tag{6.3}$$

To simplify matters let us first suppose that $f[n] = 0$ for $n < 0$. We shall refer to such a sequence as a causal sequence. If we introduce the real positive number ρ, the series

$$\sum_{n=0}^{\infty} f[n]\rho^{-n}e^{-in\theta}$$

will converge provided $\rho > \rho_{min}$. Since the left side of (6.1) is a function of $e^{i\theta}$ the introduction of the convergence factor permits us to write the preceding as follows:

$$F^+(\rho e^{i\theta}) = \sum_{n=0}^{\infty} f[n](\rho e^{i\theta})^{-n}, \tag{6.4}$$

where we have added the superscript to draw attention to the causal nature of the corresponding sequence. Of course, (6.4) is still an FS with coefficients $f[n]\rho^{-n}$ so that the usual inversion formula applies which gives

$$f[n]\rho^{-n} = \frac{1}{2\pi} \int_{-\pi}^{\pi} F^+(\rho e^{i\theta})e^{in\theta}d\theta.$$

Clearly no substantive change will result if we simply multiply both sides by ρ^n and write

$$f[n] = \frac{1}{2\pi} \int_{-\pi}^{\pi} F^+(\rho e^{i\theta})(\rho e^{i\theta})^n d\theta. \tag{6.5}$$

As a notational convenience we now introduce the complex variable $z = \rho e^{i\theta}$ in (6.4) in which case the series assumes the form

$$F^+(z) = \sum_{n=0}^{\infty} f[n]z^{-n}. \tag{6.6}$$

In view of the convergence factor just introduced this series converges for all complex z such that $|z| = \rho > \rho_{min}$, i.e., outside a circle with radius ρ_{min}. Equation (6.6) will be recognized as the principal part of the Laurent series expansion of an analytic function (see Appendix) in the annular region $\rho_{min} < |z| < \infty$. The formula for the FS coefficients (6.5) can now be interpreted as a contour integral carried out in the counterclockwise direction along a circular path of radius ρ. We see this directly by changing the variable of integration from θ to $z = \rho e^{i\theta}$. Computing $dz = \rho i e^{i\theta}d\theta = izd\theta$ and substituting in (6.5) give

$$f[n] = \frac{1}{2\pi i} \oint_{|z|=\rho} F^+(z)z^{n-1}dz, \tag{6.7}$$

where $\rho > \rho_{min}$. The integration contour and the region of analyticity of $F^+(z)$ are shown in Fig. 6.1. Equations (6.6) and (6.7) represent, respectively, the

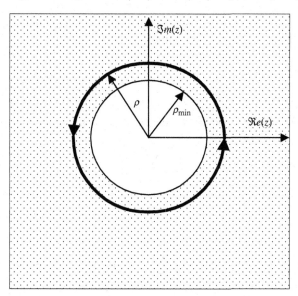

Figure 6.1: Region of analyticity and inversion contour for the unilateral Z-transform $F^+(z)$

(single-sided or unilateral) Z-transform and its inverse. We shall denote it by

$$f[n] \overset{\mathcal{Z}}{\Longleftrightarrow} F^+(z). \tag{6.8}$$

Just like the analogous inversion formula for the unilateral Laplace transform, (6.8) yields zero for negative times, i.e., for $n < 0$. This is automatically guaranteed by the analyticity of $F(z)$ outside on the integration contour. We can see this directly by applying the Cauchy residue theorem to a closed contour formed by adding to the circular contour in Fig. 6.1 a circle at infinity. The contribution from the integral along the latter vanishes as we see from the following bounding argument:

$$\frac{1}{2\pi} \left| \oint_{|z|=\rho} F^+(z)z^{n-1}dz \right| \leq \frac{1}{2\pi} \int_{-\pi}^{\pi} |F^+(\rho e^{i\theta})| \, \rho^n d\theta$$

$$\leq |F^+(\rho e^{i\theta})|_{max} \rho^n. \tag{6.9}$$

Because $F^+(z)$ is finite as $\rho \to \infty$ the last term tends to zero for negative n.

When $\rho_{min} < 1$ the integration path in Fig. 6.1 can be chosen to coincide with the unit circle. In that case the inversion formula (6.7) may be replaced by

$$f[n] = \frac{1}{2\pi} \int_{-\pi}^{\pi} F^+(e^{i\theta})e^{in\theta} d\theta \tag{6.10}$$

and we may interpret $F^+(e^{i\theta})$ either as the DFT of a causal sequence or as the FT of a bandlimited analogue signal whose sample values on the negative time axis are zero.

The restriction to causal sequences is not always convenient. To encompass negative indexes we suppose, by analogy with (6.3), that

$$f[n] \underset{n \sim -\infty}{\sim} O(\rho_{\max}^n). \tag{6.11}$$

Let us suppose that this sequence is zero for nonnegative indexes. The transform for such an "anti-causal" sequence may be defined by

$$F^-(z) = \sum_{n=-\infty}^{-1} f[n]z^{-n}. \tag{6.12}$$

With $z = \rho e^{i\theta}$ we see that this series converges for $0 \leq \rho < \rho_{\max}$ and may be interpreted as an FS, i.e.,

$$F^-(\rho e^{i\theta}) = \sum_{n=-\infty}^{-1} f[n]\rho^{-n}e^{-in\theta}, \tag{6.13}$$

wherein the coefficients are given by

$$f[n]\rho^{-n} = \frac{1}{2\pi} \int_{-\pi}^{\pi} F^-(\rho e^{i\theta})e^{in\theta} d\theta. \tag{6.14}$$

Changing the integration variable to z transforms (6.14) into the contour integral

$$f[n] = \frac{1}{2\pi i} \oint_{|z|=\rho} F^-(z)z^{n-1} dz, \tag{6.15}$$

where the radius of the integration contour $\rho < \rho_{\max}$ lies in the shaded region of Fig. 6.2, corresponding to the region of analyticity of $F^-(z)$. Note that the transforms $F^+(z)$ and $F^-(z)$ are analytic functions in their respective regions of analyticity. In general these regions would be disjoint. If, however, $\rho_{\max} > \rho_{\min}$, the regions of analyticity of $F^+(z)$ and $F^-(z)$ overlap, the overlap being the annular region

$$\rho_{\min} < |z| < \rho_{\max} \tag{6.16}$$

shown in Fig. 6.3. In this case $F(z)$,

$$F(z) = F^+(z) + F^-(z) \tag{6.17}$$

is an analytic function within this annulus (6.16) and defines the *bilateral* Z-transform of the sequence $f[n]$ represented by the series

$$F(z) = \sum_{n=-\infty}^{\infty} f[n]z^{-n}. \tag{6.18}$$

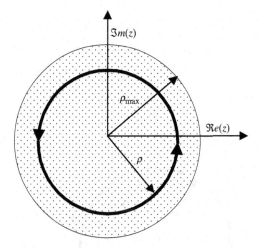

Figure 6.2: Region of analyticity and inversion contour for $F^-(z)$

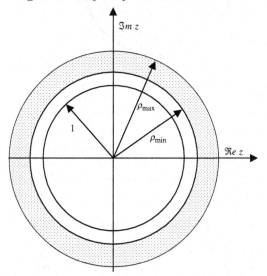

Figure 6.3: Typical annular region of analyticity of the doublesided Z-transform $F(z)$

The annular analyticity region is also the region of convergence (ROC) of (6.18). This convergence is guaranteed as long as the sequence $f[n]$ exhibits the asymptotic behavior specified by (6.3) *and* (6.11). Equation (6.18) will be recognized as the Laurent series (see Appendix) expansion of $F(z)$ about $z = 0$. The inversion formula is now given by the sum of (6.7) and (6.15), i.e.,

$$f[n] = \frac{1}{2\pi i} \oint_{|z|=\rho} F(z) z^{n-1} dz \tag{6.19}$$

with ρ lying within the shaded annulus in Fig. 6.3. Of course due to analyticity of the integrand we may use any closed path lying within the annular region of analyticity. If the ROC includes the unit circle as, e.g., in Fig. 6.4, the inversion formula (6.19) may be replaced by (6.2) in which case $F(z) = F(e^{i\theta})$ reduces to the FT.

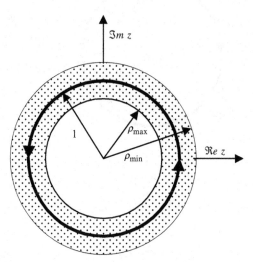

Figure 6.4: Evaluation of the inversion integral of the bilateral Z-transform along the unit circle

In summary (6.18) and (6.19) constitute, respectively, the direct and inverse bilateral Z-transform (ZT) of a sequence. As is customary, we shall use in the sequel the common symbol to denote both the bilateral and the unilateral Z-transform. Thus our compact notation for both transforms will be

$$f[n] \overset{\mathcal{Z}}{\Longleftrightarrow} F(z). \tag{6.20}$$

Other notations that will be employed in various contexts are : $F(z) = \mathcal{Z}\{f[n]\}$ and $f[n] = \mathcal{Z}^{-1}\{F(z)\}$.

6.1.2   Direct ZT of Some Sequences

The principal tool in the evaluation of the direct ZT is the geometric series summation formula. For example, for the exponential sequence $a^n u[n]$, we obtain

$$F(z) = \sum_{n=0}^{\infty} a^n z^{-n} = \sum_{n=0}^{\infty} (\frac{a}{z})^n = \frac{z}{z-a} \tag{6.21}$$

provided $|z| > |a| = \rho_{min}$. This ZT has only one singularity, viz., a simple pole at $z = a$. Using the notation (6.20) we write

$$a^n u[n] \overset{\mathcal{Z}}{\Longleftrightarrow} \frac{z}{z-a}. \tag{6.22}$$

For the special case of the unit step we get

$$u[n] \overset{\mathcal{Z}}{\Longleftrightarrow} \frac{z}{z-1}. \tag{6.23}$$

Differentiating both sides of (6.21) k times with respect to a we get

$$\sum_{n=0}^{\infty} [n(n-1)\dots(n-k+1)] a^{n-k}z^{-n} = \frac{zk!}{(z-a)^{k+1}}.$$

Dividing both sides by $k!$ and using the binomial symbol results in the transform pair

$$\binom{n}{k} a^{n-k} \overset{\mathcal{Z}}{\Longleftrightarrow} \frac{z}{(z-a)^{k+1}} \tag{6.24}$$

a ZT with a $k+1$-th order pole at $z = a$. Another canonical signal of interest in discrete analysis is the unit impulse, denoted either by the Kronecker symbol δ_{mn} or by its equivalent $\delta[n-m]$. Formula (6.18) gives the corresponding ZT as

$$\delta[n-m] \overset{\mathcal{Z}}{\Longleftrightarrow} z^{-m}. \tag{6.25}$$

For $m > 0$ this ZT is analytic everywhere except for a pole of order m at $z = 0$ which for $m < 0$ becomes a pole of the same order at $z = \infty$.

An example of a sequence that leads to a nonrational ZT is the sequence $f[n] = u[n-1]/n$

$$u[n-1]/n \overset{\mathcal{Z}}{\Longleftrightarrow} -\ln \frac{z-1}{z}, \tag{6.26}$$

which may be obtained from a Taylor expansion of $\ln(1-w)$ for $|w| < 1$. This ZT may be defined as analytic function for $|z| > 1$ by connecting the two logarithmic singularities at $z = 1$ and $z = 0$ with the branch cut shown in Fig. 6.5.

As in case of the LT derivation of more complex transform pairs is facilitated by the application of several fundamental properties of the ZT which we discuss in the sequel.

6.1.3   Properties

Time Shift

Forward.   We are given $F(z) = \mathcal{Z}\{f[n]\}$ and wish to compute $\mathcal{Z}\{f[n+k]\}$ where $k \geq 0$. Here we must distinguish between the unilateral and the bilateral transforms. In the latter case we have

$$\begin{aligned} \mathcal{Z}\{f[n+k]\} &= \sum_{n=-\infty}^{\infty} f[n+k]z^{-n} = z^k \sum_{m=-\infty}^{\infty} f[m]z^{-m} \\ &= z^k F(z). \end{aligned} \tag{6.27}$$

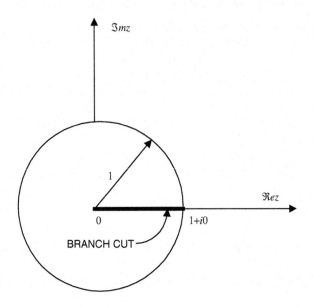

Figure 6.5: Branch cut that renders $-\ln\left[(z-1)/z\right]$ analytic outside the unit circle

On the other hand for the unilateral transform we obtain

$$\mathcal{Z}\{f[n+k]\} = \sum_{n=0}^{\infty} f[n+k]z^{-n} = z^k \sum_{m=k}^{\infty} f[m]z^{-m}. \tag{6.28}$$

Since $f[m] = 0$ for $m < 0$ then whenever $k \leq 0$ we may replace the lower limit of the last sum by zero in which case we again obtain (6.27). A different situation arises when $k > 0$ for then the last sum in (6.28) omits the first $k-1$ values in the sequence. Note that this is similar to the signal truncation for the unilateral LT in (4.49) in 4.1. We can still express (6.28) in terms of $F(z)$ by adding and subtracting the series $\sum_{m=0}^{k-1} f[m]z^{-m}$. We then obtain

$$\mathcal{Z}\{f[n+k]\} = z^k[F(z) - \sum_{m=0}^{k-1} f[m]z^{-m}]. \tag{6.29a}$$

The last result is particularly useful in the solution of finite difference equations with constant coefficients with specified initial conditions.

Backward Shift. For $k \geq 0$ we also compute $\mathcal{Z}\{f[n-k]\}$, referred to as a backward shift. Instead of (6.28) we get

$$\mathcal{Z}\{f[n-k]\} = \sum_{n=0}^{\infty} f[n-k]z^{-n} = z^{-k} \sum_{m=-k}^{\infty} f[m]z^{-m}. \tag{6.29b}$$

Note that in this case the initial conditions are specified for negative indices. When the sequence is causal[1] these vanish and we get

$$\mathcal{Z}\{f[n-k]\} = z^{-k}F(z). \tag{6.29c}$$

Time Convolution

By direct calculation we get

$$
\begin{aligned}
\mathcal{Z}\{\sum_{k=-\infty}^{\infty} f[n-k]g[k]\} &= \sum_{k=-\infty}^{\infty} \mathcal{Z}\{f[n-k]\}g[k] \\
&= \sum_{k=-\infty}^{\infty} F(z)z^{-k}g[k] \\
&= F(z)G(z).
\end{aligned}
\tag{6.30}
$$

Since $\sum_{k=-\infty}^{\infty} u[n-k]f[k] \equiv \sum_{k=-\infty}^{n} u[n-k]f[k] \equiv \sum_{k=-\infty}^{n} f[k]$ we get as a by-product the formula

$$\sum_{k=-\infty}^{n} f[k] \overset{\mathcal{Z}}{\Longleftrightarrow} \frac{zF(z)}{z-1}, \tag{6.31}$$

which is useful in evaluating the ZT of sequences defined by sums. For example, applying to (6.26) results in

$$\sum_{k=1}^{n} \frac{1}{k} \overset{\mathcal{Z}}{\Longleftrightarrow} -\frac{z \ln \frac{z-1}{z}}{z-1}. \tag{6.32}$$

Frequency Convolution

The ZT of the product of two sequences is

$$
\begin{aligned}
\mathcal{Z}\{f[n]g[n]\} &= \sum_{n=-\infty}^{\infty} f[n]g[n]z^{-n} \\
&= \sum_{n=-\infty}^{\infty} \frac{1}{2\pi i} \oint_{|z|=\rho'} F(z')z'^{n-1}dz'\,g[n]z^{-n} \\
&= \frac{1}{2\pi i} \oint_{|z|=\rho'} F(z')G(z/z')z'^{-1}dz'.
\end{aligned}
\tag{6.33}
$$

When the inversion contour is the unit circle the preceding becomes

$$\mathcal{Z}\{f[n]g[n]\} = \frac{1}{2\pi} \int_{-\pi}^{\pi} F(\theta')G(\theta-\theta')d\theta'. \tag{6.34}$$

---

[1]This does not imply that the underlying system is not causal. The shift of initial conditions to negative time is just a convenient way to handle forward differences.

Initial Value Theorem

For a causal sequence we have

$$\lim_{|z|\to\infty} F(z) = f[0], \tag{6.35}$$

which follows directly from (6.18) and analyticity of $F(z)$ for $|z| > \rho_{\min}$.

Differentiation

Differentiating (6.18) we get the transform pair

$$\mathcal{Z}\{nf[n]\} = -z\frac{dF(z)}{dz}. \tag{6.36}$$

For example, starting with $u[n] = f\{n\}$ and applying formula (6.36) twice we get

$$n^2 u[n] \overset{\mathcal{Z}}{\Longleftrightarrow} \frac{z(z+1)}{(z-1)^3}. \tag{6.37}$$

6.2 Analytical Techniques in the Evaluation of the Inverse ZT

From the preceding discussion we see that the ZT bears the same relationship to the FS as the LT to the FT. Just like the LT inversion formula, the ZT inversion formula (6.54) yields signals whose characteristics depend on the choice of the integration contour. Whereas in case of the LT this nonuniqueness is due to the possibility of the existence of several strips of analyticity within each of which we may place a linear integration path, in case of the ZT we may have several annular regions of analyticity wherein we may locate the circular integration contour. We illustrate this with several examples.

Example 1   Let us return to the simple case of a ZT whose only singularity is a simple pole at $z = a$, i.e.,

$$F(z) = \frac{z}{z-a}. \tag{6.38}$$

There are two annular regions of analyticity (i) $|a| < \rho < \infty$ ($\rho_{\min} = |a|$ and $\rho_{\max} = \infty$) and (ii) $0 \le \rho < |a|$ ($\rho_{\min} = 0$ and $\rho_{\max} = |a|$). In accordance with (6.19) we have to compute

$$f[n] = \frac{1}{2\pi i} \oint_{|z|=\rho} \frac{z^n}{z-a} dz. \tag{6.39}$$

Consider first case (i). For $n < 0$ the integrand decays as $|z| \to \infty$. In fact (6.39) taken over a circle of infinite radius vanishes. To see this let $z = R\,e^{i\theta}$ so that $dz = R\,e^{i\theta}id\theta$ and the integral may be bounded as follows

$$\left| \oint_{|z|=R} \frac{z^n}{z-a} dz \right| \leq \int_{-\pi}^{\pi} \frac{R}{|R\, e^{i\theta} - a|} R^n d\theta. \tag{6.40}$$

Since $n < 0$ the right side of the preceding expression approaches zero as $R \to \infty$. (It is not hard to see that the limit will also be zero for any $F(z)$ that approaches a constant as $|z| \to \infty$ a result we shall rely on repeatedly in the sequel.) Because there are no intervening singularities between the circle $|z| = \rho$ and ∞ this integral vanishes. For $n \geq 0$ the integrand in (6.39) is analytic except at $z = a$ and a residue evaluation gives a^n. In summary, for case (i) we have the causal sequence

$$f[n] = \begin{cases} 0 \; ; \; n < 0, \\ a^n \; ; \; n \geq 0. \end{cases} \tag{6.41}$$

so that (6.38) together with the choice of contour represents the unilateral transform. For case (ii) the pole at $z = a$ lies between the inversion contour and ∞. Because the integral over the infinite circular contour vanishes for $n < 0$ equation (6.39) gives the negative residue at the pole, i.e., the integration around the pole is, in effect, being carried out in the clockwise direction. This may be visualized in terms of the composite closed contour shown in Fig. 6.6 comprised of the circle with radius ρ, the circle at infinity, and two linear segments along which the individual contributions mutually cancel. As we proceed along the composite contour we see that the pole is being enclosed in the clockwise direction. For $n \geq 0$ the integrand is analytic within the integration contour and we get zero. Thus, in summary, for case (ii) we obtain the anti-causal sequence

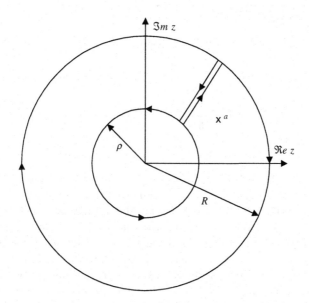

Figure 6.6: Residue evaluation using an auxiliary contour

$$f[n] = \begin{cases} -a^n \; ; \; n < 0, \\ 0 \; ; \; n \geq 0. \end{cases} \tag{6.42}$$

From the preceding example we see that if $|a| > 1$, i.e., the pole is outside the unit circle, the unilateral ZT (case (i)) corresponds to a sequence that grows with increasing n. Clearly for this sequence the FT does not exist. On the other hand for case (ii) $|a| > 1$ results in a sequence that decays with large negative n. Since in this case the inversion contour can be chosen to coincide with the unit circle the FT exists and equals

$$F(\omega) = \frac{e^{i\omega\Delta t}}{e^{i\omega\Delta t} - a}. \tag{6.43}$$

For $|a| < 1$ one obtains the reverse situation: the unilateral ZT corresponds to a decaying sequence whereas the anti-causal sequence in case (ii) grows. The four possibilities are illustrated in Fig. 6.7.

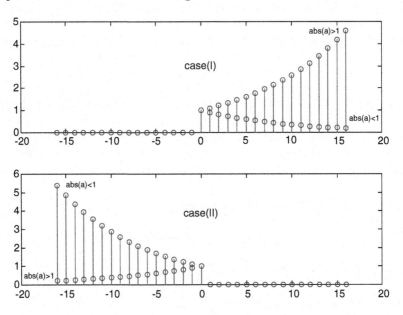

Figure 6.7: Sequences corresponding to a ZT with a simple pole

Note, however, as long as $|a| \neq 1$ formula (6.43) remains valid and represents the FT of either the causal or the anti-causal decaying sequence for $|a| < 1$ and $|a| > 1$, respectively. What happens when $|a| \to 1$? To answer this question we cannot simply substitute the limiting form of a in (6.43) for (6.43) is not valid for a on the unit circle. Rather we must approach the limit by evaluating (6.19) along a contour just inside or just outside the unit circle. The answer, not surprisingly, depends on which of the two options we choose. Suppose we start with a contour just outside the unit circle. Since we are looking for the FT we should like our integration contour to follow the path along the unit circle as much as possible. Our only obstacle is the pole at $z = e^{i\theta_0}$. To remain outside the circle we circumnavigate the pole with a small semicircle of radius ϵ resulting in the composite integration path shown in Fig. 6.8. The contribution to the integral along the circular contour is represented by a CPV integral. For nonnegative n, the sum of the two contributions equals the residue at the enclosed pole, and vanishes for negative n. This is, of course, identical to (6.41)

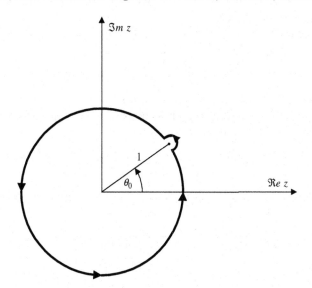

Figure 6.8: Integration path along unit circle in presence of a simple pole

with $a = e^{i\theta_0}$. Thus summing the two contributions along the closed contour we have

$$\frac{1}{2}e^{in\theta_0} + \frac{1}{2\pi}CPV \int_{-\pi}^{\pi} \frac{e^{in\theta}}{1 - e^{-i(\theta-\theta_0)}}\,d\theta \equiv u\,[n]\,e^{in\theta_0} = f\,[n]. \qquad (6.44)$$

Using the identity

$$\frac{1}{1 - e^{-i(\theta-\theta_0)}} = \frac{1}{2} - \frac{i}{2}\cot\left[(\theta - \theta_0)/2\right]$$

we observe that (6.44) implies the transform pair

$$u\,[n]\,e^{in\theta_0} \overset{\mathcal{F}}{\Longleftrightarrow} \pi\delta\,(\theta - \theta_0) + \frac{1}{2} - \frac{i}{2}\cot\left[(\theta - \theta_0)/2\right]. \qquad (6.45)$$

We leave it as an exercise to find the FT when the unit circle is approached from the interior.

Example 2   As another example consider the ZT with two simple poles

$$F(z) = \frac{z^2}{(z - 1/2)(z - 2)}. \tag{6.46}$$

In this case there are three annular regions of analyticity: (i) $2 < \rho_1$, (ii) $1/2 < \rho_2 < 2$, and (iii) $0 \le \rho_3 < 1/2$. We distinguish the three sequences by the superscript k and evaluate

$$f^{(k)}[n] = \frac{1}{2\pi i} \oint_{|z|=\rho_k} \frac{z^{n+1}}{(z - 1/2)(z - 2)} dz, \quad k = 1, 2, 3.$$

As in the preceding example $F(\infty)$ is finite so that the contribution to the inversion integral for $n < 0$ from a circle at ∞ is zero. The three sequences are then found by a residue evaluation as follows.

case (i)

When $n \ge 0$ we have

$$f^{(1)}[n] = \frac{z^{n+1}}{(z - 1/2)} \Big|_{z=2} + \frac{z^{n+1}}{(z - 2)} \Big|_{z=1/2} = \frac{4}{3} 2^n - \frac{1}{3} 2^{-n} \tag{6.47}$$

so that

$$f^{(1)}[n] = \begin{cases} 0 \; ; \; n < 0, \\ \frac{4}{3} 2^n - \frac{1}{3} 2^{-n} \; ; \; n \ge 0. \end{cases} \tag{6.48}$$

case (ii)

For $n < 0$ the negative of the residue at $z = 2$ contributes and for $n \ge 0$ the positive residue as in (6.47). Thus we obtain

$$f^{(2)}[n] = \begin{cases} -\frac{4}{3} 2^n \; ; \; n < 0, \\ -\frac{1}{3} 2^{-n} \; ; \; n \ge 0. \end{cases} \tag{6.49}$$

case (iii)

No singularities are enclosed for $n \ge 0$ so that $f^{(3)}[n] = 0$. For $n < 0$ the negatives of the two residues in (6.48) contribute, so that the final result reads

$$f^{(3)}[n] = \begin{cases} -\frac{4}{3} 2^n + \frac{1}{3} 2^{-n} \; ; \; n < 0, \\ 0 \; ; \; n \ge 0. \end{cases} \tag{6.50}$$

We note that only the sequence $f^{(2)}[n]$ possesses an FT, which is

$$F(\omega) = \frac{e^{2i\omega \Delta t}}{(e^{i\omega \Delta t} - 1/2)(e^{i\omega \Delta t} - 2)}. \tag{6.51}$$

This sequence is, however, not causal. Note that this is due to the presence of a pole outside the unit circle. This same pole gives rise to the exponential growth with n of the sequence in case (i). This sequence is causal and corresponds to the single-sided ZT. Clearly to obtain a sequence which is both causal and stable requires that all the poles of the ZT lie within the unit circle.

Example 3   In this next example

$$F(z) = \frac{z}{(z + 1/3)(z + 3)^2} \tag{6.52}$$

we again have two poles one of which is a double pole. We have again three annular regions of analyticity and we evaluate

$$f^{(k)}[n] = \frac{1}{2\pi i} \oint_{|z|=\rho_k} \frac{z^n}{(z + 1/3)(z + 3)^2} dz, \quad k = 1, 2, 3 \tag{6.53}$$

on each of three circular contours defined by $0 \le \rho_1 < 1/3, 1/3 < \rho_2 < 3, 3 < \rho_3$. The residue evaluation gives

$$f^{(1)}[n] = \begin{cases} 0 \; ; n \ge 0, \\ -\frac{9}{64}(-1/3)^n + \frac{(9-8n)(-3)^n}{64} \; ; n < 0, \end{cases} \tag{6.54a}$$

$$f^{(2)}[n] = \begin{cases} \frac{(-1/3)^n}{(-1/3+3)^2} = \frac{9}{64}(-1/3)^n \; ; n \ge 0, \\ -\frac{d}{dz}\left\{\frac{z^n}{z+1/3}\right\}|_{z=-3} = \frac{(9-8n)(-3)^n}{64} \; ; n < 0, \end{cases} \tag{6.54b}$$

$$f^{(3)}[n] = \begin{cases} \frac{9}{64}(-1/3)^n - \frac{(9-8n)(-3)^n}{64} \; ; n \ge 0, \\ 0 \; ; n < 0. \end{cases} \tag{6.54c}$$

Of the three sequences only $f^{(2)}[n]$ possesses an FT which is

$$F(\omega) = \frac{e^{i\omega\Delta t}}{(e^{i\omega\Delta t} + 1/3)(e^{i\omega\Delta t} + 3)^2}. \tag{6.55}$$

Recall that the sample values of the corresponding analogue signal are $f^{(2)}[n]/\Delta t$ so that the reconstruction of this signal via the Shannon sampling theorem reads

$$f(t) = \frac{1}{\Delta t}\sum_{n=-\infty}^{-1} \frac{(9-8n)(-3)^n}{64}\sin\left[\pi(t/\Delta t - n)\right] / \left[\pi(t/\Delta t - n)\right]$$

$$+ \frac{1}{\Delta t}\sum_{n=0}^{\infty} \frac{9}{64}(-1/3)^n \; \sin\left[\pi(t/\Delta t - n)\right] / \left[\pi(t/\Delta t - n)\right]. \tag{6.56}$$

Example 4   In the preceding examples the ZT was a proper rational function. If this is not the case, a causal inverse does not exist. Consider, for example,

$$F(z) = \frac{z^4}{z^2 - 1/4}. \tag{6.57}$$

In addition to the two simple poles at $z = \pm 1/2$ there is a second order pole at infinity. Accordingly two analyticity regions are $1/2 < \rho_1 < \infty$ and $0 \le \rho_2 < 1/2$. The sequence corresponding to the first region is

$$f^{(1)}[n] = \frac{1}{2\pi i} \oint_{|z|=\rho_1} \frac{z^{n+3}}{z^2 - 1/4} dz. \tag{6.58}$$

The integral over the circle at infinity vanishes provided $n + 3 \leq 0$ so that $f^{(1)}[n] = 0$ for $n \leq -3$. For $n > -3$ the only singularities within $|z| < \rho_1$ are the two simple poles of $F(z)$. Summing the two residues gives

$$f^{(1)}[n] = \frac{1}{8}\left[\left(\frac{1}{2}\right)^n + \left(-\frac{1}{2}\right)^n\right] u[n+2]. \tag{6.59}$$

For the second sequence

$$f^{(2)}[n] = \frac{1}{2\pi i} \oint_{|z|=\rho_2} \frac{z^{n+3}}{z^2 - 1/4} dz \tag{6.60}$$

the integration yields a null result for $n + 3 \geq 0$. For $n + 3 < 0$ the integral over the infinite circle again vanishes so that we can sum the residues of the poles lying outside the integration contour resulting in

$$f^{(2)}[n] = -\frac{1}{8}\left[\left(\frac{1}{2}\right)^n + \left(-\frac{1}{2}\right)^n\right] u[-n-4]. \tag{6.61}$$

An alternative approach to dealing with an improper rational function is to employ long division and reduce the given function to a sum of a polynomial and a proper rational function. The inverse ZT of the polynomial is then just a sum of Kronecker deltas while the inverse ZT of the proper rational function is evaluated by residues. Using this approach in the present example we have

$$\frac{z^4}{z^2 - 1/4} = z^2 + \frac{1}{4}\frac{z^2}{z^2 - 1/4}. \tag{6.62}$$

The inverse of z^2 is $\delta[n+2]$ while the residue evaluation involving the second term yields either a causal or anticausal sequence. In the former case we get for the final result

$$f^{(1)}[n] = \delta[n+2] + \frac{1}{8}\left[\left(\frac{1}{2}\right)^n + \left(-\frac{1}{2}\right)^n\right] u[n], \tag{6.63}$$

which is easily seen as just an alternative way of writing (6.59).

Example 5   Let us find the FT of the causal sequence whose ZT is

$$F(z) = \frac{z - 2}{(z - 1/2)(z - e^{i\theta_0})(z - e^{-i\theta_0})}. \tag{6.64}$$

This function has two simple poles on the unit circle and one interior pole. Consequently the inversion contour for the FT along the unit circle will have to be modified by two small semi-circles surrounding the poles (instead of one as in Fig. 6.8). The integration along each of these semi-circles will contribute one-half the residue at the respective pole while the integral along the unit circle must be defined as a CPV integral. As a result we obtain

$$
f[n] = \frac{1}{2\pi} CPV \int_{-\pi}^{\pi} F\left(e^{i\theta}\right) e^{in\theta} d\theta +
$$

$$
+ \frac{1}{2} \frac{e^{i\theta_0(n-1)}\left(e^{i\theta_0} - 2\right)}{(e^{i\theta_0} - 1/2)(e^{i\theta_0} - e^{-i\theta_0})} + \frac{1}{2} \frac{e^{-i\theta_0(n-1)}\left(e^{-i\theta_0} - 2\right)}{(e^{-i\theta_0} - 1/2)(e^{-i\theta_0} - e^{i\theta_0})}. \tag{6.65}
$$

By absorbing the two residue contributions as multipliers of delta functions we can write the complete FT as follows:

$$
F(\theta) = F\left(e^{i\theta}\right) + \pi \frac{e^{-i\theta_0}\left(e^{i\theta_0} - 2\right)}{(e^{i\theta_0} - 1/2)(e^{i\theta_0} - e^{-i\theta_0})} \delta(\theta - \theta_0)
$$

$$
+ \pi \frac{e^{i\theta_0}\left(e^{-i\theta_0} - 2\right)}{(e^{-i\theta_0} - 1/2)(e^{-i\theta_0} - e^{i\theta_0})} \delta(\theta + \theta_0). \tag{6.66}
$$

6.3 Finite Difference Equations and Their Use in IIR and FIR Filter Design

The "method of finite differences" generally refers to the approximation of derivatives in a differential equation using finite increments of the independent variable. The approximate solution for the dependent variable is then found by algebraic means. The finite difference approximation can be of the forward or of the backward type. Thus if the finite difference is defined as

$$
\frac{y(t + \Delta t) - y(t)}{\Delta t} \tag{6.67}
$$

it is of the forward type. If it is defined by

$$
\frac{y(t) - y(t - \Delta t)}{\Delta t} \tag{6.68}
$$

it is referred to as a backward difference. Whereas (6.67) is more common when dealing directly with numerical solutions of differential equations (6.68) is generally preferred in digital signal processing mainly because $y(t - \Delta t)$ has direct physical interpretation of a step in time delay. To illustrate the connection between a differential equation and the associated difference equation consider the simple case of the first-order equation

$$
\frac{dy(t)}{dt} + a_0\, y(t) = f(t). \tag{6.69}
$$

With the forward difference approximation we get

$$
\frac{y(t + \Delta t) - y(t)}{\Delta t} + a_0\, y(t) \approx f(t). \tag{6.70}
$$

If we are interested in $y(t)$ only at discrete time intervals we can set $t = n\Delta t$ so that $y(t + \Delta t) = y[\Delta t(n+1)] \equiv y[n+1]$, $y(t) = y(n\Delta t) \equiv y[n]$ and $\Delta t f(t) = \Delta t f(n\Delta t) \equiv f[n]$. Making these changes transforms (6.70) into the difference equation

$$y[n+1] + (a_0\Delta t - 1) y[n] = f[n].\tag{6.71}$$

With $\mathcal{Z}\{y[n]\} = Y(z)$, $\mathcal{Z}\{f[n]\} = F(z)$ we get

$$Y(z) = \frac{F(z) + z\, y[0]}{z + a_0\Delta t - 1} = \frac{F(z)}{z + a_0\Delta t - 1} + \frac{z\, y[0]}{z + a_0\Delta t - 1}$$

and upon inversion the solution of (6.71):

$$y[n] = \mathcal{Z}^{-1}\{\frac{F(z)}{z + a_0\Delta t - 1}\} + (a_0\Delta t - 1)^n\, y[0].\tag{6.72}$$

As far as the difference equation (6.71) is concerned (6.72) is its exact solution. The actual solution to the differential equation (the analogue problem) has to be gotten via a limiting process. For simplicity we do this when $F(z) = 0$ and look for the limit of

$$y[n] = (1 - a_0\Delta t)^n\, y[0]\tag{6.73}$$

as $\Delta t \to 0$. This is easily done by noting that discretization tells us that t may be replaced by $n\Delta t$. If in addition we replace $a_0\Delta t$ by $-\delta$, (6.73) becomes

$$y[n] = y[0]\,(1+\delta)^{-(1/\delta)a_0 t}.\tag{6.74}$$

Recalling the definition of e we get

$$\lim_{\delta \to 0} y[n] = (1+\delta)^{-(1/\delta)} \to e^{-1}$$

so that (6.73) approaches $y[0]\, e^{-a_0 t}$ which is the solution of (6.69). On the other hand, the physical problem of interest may have been formulated ab initio as a difference equation (6.71) where a finite increment Δt has a direct physical significance. In that case the limiting form would constitute a wrong answer.

The application of the forward difference operation to an N-th order differential equation with constant coefficients (3.145) in 3.3 leads, after some unpleasant algebra, to the N-th order difference equation

$$\sum_{k=0}^{N} a_k\, y[n-k] = f[n].\tag{6.75}$$

Taking the Z-Transform yields

$$\sum_{k=0}^{N} a_k z^k Y(z) = F(z) + \sum_{k=0}^{N} a_k \sum_{\ell=0}^{k-1} f[\ell]\, z^{k-\ell},\tag{6.76}$$

where the excitation on the right includes, in addition to the forced excitation $F(z)$, the contribution from the initial conditions as given by (6.29a). Solving for the transform of the output

$$Y(z) = \frac{F(z)}{\sum_{k=0}^{N} a_k z^k} + \frac{\sum_{k=0}^{N} a_k \sum_{\ell=0}^{k-1} y[\ell] z^{k-\ell}}{\sum_{k=0}^{N} a_k z^k} \tag{6.77}$$

we identify the quantity

$$H_+(z) = \frac{1}{\sum_{k=0}^{N} a_k z^k} \tag{6.78}$$

as the system transfer function.[2] If all poles are within the unit circle then, in accordance with the results in the preceding section, the application of the inversion integral to (6.78) along the unit circle yields a causal and stable sequence $h[n]$. Evaluation of (6.78) on the unit circle gives the FT

$$H(\theta) = \frac{1}{\sum_{k=0}^{N} a_k e^{ik\theta}} = A(\theta) e^{i\psi(\theta)}, \tag{6.79}$$

where $A(\theta)$ is the amplitude and $\psi(\theta)$ the phase. The amplitude is an even and the phase an odd function of frequency just like for continuous signals. An important class of transfer functions is characterized by the absence of zeros outside the unit circle. They are called minimum phase-shift functions. For functions of this type the phase can be determined from the amplitude (see 6.4.2).

Similar to the filter structures in Fig. 3.16 and 3.17 that were derived from differential equations, difference equations lead to topologically similar representations. They play an important role in the design of DSP algorithms. Here we will switch from the representation based on forward differencing we used to derive (6.78) to backward differencing which is more common in DSP applications. Figure 6.9 shows a feedback-type structure similar to that in Fig. 3.16.

The integrators have been replaced by unit delay elements (denoted by -1 within the upper circles). For each delay element an input $y[n]$ gives an output $y[n-1]$ consistent with backward differencing (6.68). Referring to the figure, if we subtract the sum of the outputs from the difference operators from the input $f[n]$ and multiply the result by $1/a_0$ we get

$$\sum_{k=0}^{N} a_k y[n-k] = f[n]. \tag{6.80}$$

Assuming $y[n] = 0$ for $n < 0$ we have for the Z-transform

$$Y(z) \sum_{k=0}^{N} a_k z^{-k} = F(z) \tag{6.81}$$

---

[2]The subscript $(+)$ identifies that it is based on forward differencing.

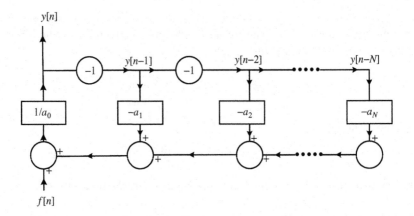

Figure 6.9: Infinite impulse response filter (IIR)

and for the transfer function

$$H_-(z) = \frac{1}{\sum_{k=0}^{N} a_k z^{-k}}. \tag{6.82}$$

Since (6.82) is based on backward differencing it does not agree with (6.78). We can see the relationship between the two by changing the summation index from k to $m = N - k$ with the result

$$H_-(z) = \frac{z^N}{\sum_{m=0}^{N} a_{N-m} z^m}. \tag{6.83}$$

The denominator has the form of (6.78) but the polynomial coefficients have been interchanged so that the pole positions are different.

The transfer function (6.82) forms the basis for the design of infinite impulse response (IIR) filters. The name derives from the property that with poles within the unit circle (the usual case) the impulse response is of infinite duration. This is also true for all pole analogue filters. In fact design procedures for digital IIR filters are essentially the same as for analogue filters. A second class of filters is finite impulse response (FIR) filters. A representative structure of an FIR filter is shown in Fig. 6.10.

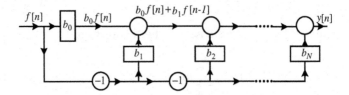

Figure 6.10: Finite impulse response filter (FIR)

Adding the tapped and in unit steps delayed input to the input that has been delayed and multiplied by the last tap we get

$$b_0 f[n] + b_1 f[n] + b_2 f[n-2] + \ldots b_{N-1} f[n-(N-1)] + b_N f[n-N] = y[n]$$

or, in compact, notation

$$y[n] = \sum_{k=0}^{N} b_k \, f[n-k]. \tag{6.84}$$

Here the impulse response is the finite length sequence

$$h[n] = b_n \; ; n = 0, 1, 2 \ldots N \tag{6.85}$$

with the transfer function

$$H_{FIR}(z) = \sum_{k=0}^{N} b_k \, z^{-k} \tag{6.86}$$

and an FT[3]

$$H_{FIR}(\theta) = \sum_{k=0}^{N} b_k \, e^{-ik\theta}. \tag{6.87}$$

Since this is also an FS we can easily compute the coefficients for a prescribed $H(\theta)$ and hence the filter parameters in Fig. 6.10. The practical problem here is that finite length sequences cannot be contained within a finite bandwidth. Generally the biggest offenders here are steep changes in the frequency spectrum as would be the case e.g., for band-pass filters with steep skirts. These problems can be in part alleviated by tapering the sequence (i.e., in the time domain). However, generally for this and other reasons FIR filters require many taps.

A general filter structure in DSP applications combines IIR and FIR transfer functions into the form

$$H(z) = \frac{\sum_{k=0}^{N} b_k \, z^{-k}}{\sum_{k=0}^{M} a_k z^{-k}}. \tag{6.88}$$

Such transfer functions can be realized either by using FIR and IIR filters in tandem or combining them into a single structure similar to that in Fig. 3.17.

6.4 Amplitude and Phase Relations Using the Discrete Hilbert Transform

6.4.1 Explicit Relationship Between Real and Imaginary Parts of the FT of a Causal Sequence

We recall from 2.2.6 that the real and imaginary parts of the FT of a real causal analogue signal are related by the Hilbert transform. In 2.4.2 this relationship

---

[3] Here and in the entire discussion of difference equations we have increased the sequence length from that used with the DFT in Chap. 5 from N to $N+1$. Consistency is easily restored by setting the Nth term to zero.

is reestablished and is shown to be a direct consequence of the analytic properties of the FT of causal signals; its extension to amplitude and phase of transfer
functions is discussed in 2.4.3. In the following we show that a similar set of relationships holds also for causal discrete signals, or, equivalently, for bandlimited
functions whose Nyquist samples are identically zero for negative indices.

As in the analogue case, in (2.168) in 2.2 we start directly from the definition
of a causal signal, in this case a real sequence $f[n]$ which we decompose into its
even and odd parts as follows:

$$f[n] = f_e[n] + f_o[n],\tag{6.89}$$

where

$$f_e[n] = \frac{f[n] + f[-n]}{2},\tag{6.90a}$$

$$f_o[n] = \frac{f[n] - f[-n]}{2}.\tag{6.90b}$$

With

$$f[n] \overset{\mathcal{F}}{\Longleftrightarrow} F(\theta) = R(\theta) + iX(\theta),\tag{6.91}$$

wherein $R(\theta)$ and $X(\theta)$ are real and imaginary parts of $F(\theta)$ it is not hard to
show that

$$f_e[n] \overset{\mathcal{F}}{\Longleftrightarrow} R(\theta),\tag{6.92a}$$

$$f_o[n] \overset{\mathcal{F}}{\Longleftrightarrow} iX(\theta),\tag{6.92b}$$

i.e., just as in the case of analogue signals, the odd and even parts are defined,
respectively, by the real and the imaginary parts of the FT of the sequence. In
order to carry out the transformations analogous to in (2.172) 2.2 we need the
FT of the discrete sign function $sign[n]$ defined as $+1$ for all positive integers,
-1 for negative integers and zero for $n = 0$. We can find this transform from
(6.45) by first expressing the unit step as follows:

$$u[n] = \frac{1}{2}(sign[n] + \delta[n] + 1),\tag{6.93}$$

where $\delta[n] = 1$ for $n = 0$ and 0 otherwise. With the aid of (6.45) we then obtain

$$sign[n] \overset{\mathcal{F}}{\Longleftrightarrow} -i\cot(\theta/2).\tag{6.94}$$

Suppose the sequence $f[n]$ is causal. Then in view of (6.89) and (6.90) we have

$$f[n] = 2f_e[n]u[n]\tag{6.95}$$

and also

$$f_o[n] = sign[n]f_e[n]\tag{6.96a}$$

$$f_e[n] = f[0]\delta[n] + sign[n]f_o[n].\tag{6.96b}$$

Taking account of (6.92) and (6.94) and applying the frequency convolution theorem to (6.96) yield

$$X(\theta) \;=\; -\frac{1}{2\pi}CPV\int_{-\pi}^{\pi} R(\theta')\cot\left[(\theta-\theta')/2\right]d\theta', \qquad (6.97a)$$

$$R(\theta) \;=\; f[0]+\frac{1}{2\pi}CPV\int_{-\pi}^{\pi} X(\theta')\cot\left[(\theta-\theta')/2\right]d\theta'. \qquad (6.97b)$$

These relations are usually referred to as the discrete Hilbert transforms (DHT).

6.4.2   Relationship Between Amplitude and Phase of a Transfer Function

We now suppose that $A(\theta)$ is the amplitude of the transfer function of a causal digital filter with a real unit sample response. This means that there exists a phase function $\psi(\theta)$ such that the FT $A(\theta)e^{i\psi(\theta)}=H(\theta)$ has a causal inverse. The problem before us is to find $\psi(\theta)$ given $A(\theta)$. As we shall demonstrate in the sequel, a $\psi(\theta)$ that results in a causal transfer function can always be found provided $\ln A(\theta)$ can be expanded in a convergent Fourier series in $(-\pi,\pi)$. As in the corresponding analogue case the solution is not unique for we can always multiply the resulting transfer function by an all pass factor of the form $e^{i\psi_0(\theta)}$ which introduces an additional time delay (and hence does not affect causality) but leaves the amplitude response unaltered.

To find $\psi(\theta)$ we proceed as follows. Assuming that the FS expansion

$$\ln A(\theta) = \sum_{n=-\infty}^{\infty} w[n]e^{-in\theta} \qquad (6.98)$$

exists we form the function

$$Q(\theta) = \ln H(\theta) = \ln A(\theta) + i\psi(\theta). \qquad (6.99)$$

Since the unit sample response is real, $A(\theta)$ is an even function of θ so that we may regard $\ln A(\theta)$ as the real part of the FT of a causal sequence $q[n]$ with $\psi(\theta)$ the corresponding imaginary part. In that case $\psi(\theta)$ can be expressed in terms of $\ln A(\theta)$ by the DHT (6.97a) so that the solution to our problem appears in the form

$$\psi(\theta) = -\frac{1}{2\pi}CPV\int_{-\pi}^{\pi} \ln A(\theta')\cot\left[(\theta-\theta')/2\right]d\theta'. \qquad (6.100)$$

The evaluation of this integral is numerically much less efficient than the following alternative approach. It is based on the observation that for the causal sequence $q[n]$ the FT of its even part is just the given log amplitude $\ln A(\theta)$, i.e.,

$$w[n] = q_e[n] = \frac{q[n]+q[-n]}{2} \;\overset{\mathcal{F}}{\Longleftrightarrow}\; \ln A(\theta). \qquad (6.101)$$

But in view of (6.95) we can write

$$q[n] = 2q_e[n] u[n] \overset{\mathcal{F}}{\longleftrightarrow} \ln A(\theta) + i\psi(\theta). \tag{6.102}$$

Thus to find $\psi(\theta)$ from $A(\theta)$ we first take the inverse FT of $\ln A(\theta)$, truncate the sequence to positive n, multiply the result by 2, and compute the FT of the new sequence. The desired phase function is then given by the imaginary part of this FT. Expressed in symbols the procedure reads:

$$\psi(\theta) = \mathrm{Im}\, \mathcal{F}\left\{2\mathcal{F}^{-1}\{\ln A(\theta)\}u[n]\right\}. \tag{6.103}$$

The complete transfer function is then

$$H(\theta) = A(\theta)\, e^{i\,\mathrm{Im}\,\mathcal{F}\{2\mathcal{F}^{-1}\{2\ln A(\theta)\}u[n]\}}. \tag{6.104}$$

To prove that this $H(\theta)$ has a causal inverse we note that, by construction,

$$\ln H\left(e^{i\theta}\right) = \sum_{n=0}^{\infty} q[n]\, e^{-in\theta} \tag{6.105}$$

so that $\sum_{n=0}^{\infty} q[n] z^{-n}$ is analytic in $|z| > 1$. But then $\ln H(z)$ must also be analytic outside the unit circle. In particular, this means that $H(z)$ cannot have any poles or zeros outside the unit circle. Since the exponential is an analytic function

$$H(z) = e^{\sum_{n=0}^{\infty} q[n] z^{-n}} \tag{6.106}$$

is necessarily analytic in $|z| > 1$ and hence has a causal inverse.

Analytic functions devoid of zeros outside the unit circle are referred to as minimum phase-shift functions, a terminology shared with Laplace transforms of analogue signals that are analytic in the right half of the s plane and free of zeros there. Despite this common terminology minimum phase-shift functions of sampled causal signals are endowed with an important feature not shared by transfer functions of causal analogue signals, viz., the absence of zeros as well as analyticity outside the unit circle also ensures the analyticity of $1/H(z)$ outside the unit circle. Hence the reciprocal of minimum phase-shift function also possesses a causal inverse, a feature of great importance in the design of feedback control systems.

6.4.3   Application to Design of FIR Filters

The preceding procedure can be used to synthesize a minimum phase FIR filter having a prescribed amplitude response. With $A[m]$, $m = 0, 1, \ldots N-1$ the prescribed amplitude function we compute the transfer function via (6.104)

$$H[m] = A[m]\, e^{i\,\mathrm{Im}\,\mathcal{F}\{2\mathcal{F}^{-1}\{2\ln A(k)\}u[n]\}}. \tag{6.107}$$

The unit sample response $h[n]$ is then given by

$$h[n] = \frac{1}{N} \sum_{m=0}^{N-1} H[m]\, e^{i\frac{2\pi n m}{N}} \; ; \quad n = 0, 1, \ldots N-1. \tag{6.108}$$

Equations (6.108) and (6.109) are exact but unfortunately the required number of filter coefficients (taps) equals the number of samples, which would normally be impractically large. It turns out, however, that for many practically useful filter functions the $h[n]$ decay rapidly with increasing n so that the number of taps can be chosen much less than N. Thus instead of the exact form of $H[m]$ we shall use the first M values of $h[n]$ in (6.108) and define

$$H^{(M)}[m] = \sum_{n=0}^{M-1} h[n] e^{i\frac{2\pi nm}{N}} \quad ; m = 0, 1.2 \ldots N-1. \tag{6.109}$$

The degree of permissible truncation depends on the specified performance level (e.g., stopband attenuation, relative bandwidth) and is strongly dependent on the functional form of the chosen amplitude function. The recommended procedure is to start with (6.108) and progressively reduce the number of taps until the observed amplitude spectrum begins to show significant deviations from the prescribe amplitude response.

Problems

1. The Z-transform

$$F(z) = \frac{z}{(z^2 + 1/2)(z + 16)}$$

   can represent several sequences.

   (a) Find all the sequences.

   (b) One of the sequences represents the coefficients of the Fourier series expansion of the Fourier transform of a band-limited function. Identify the sequence and find the Fourier transform of the corresponding bandlimited function.

2. The Fourier transform of a bandlimited function $f(t)$ is given by

$$F(\omega) = \begin{cases} \frac{1}{5/4 - \cos(\pi\omega/\Omega)}; |\omega| \le \Omega \\ 0; |\omega| > \Omega \end{cases}$$

   (a) The function is sampled at intervals $\Delta t = \pi/\Omega$. Find the Z-transform of the sequence $f[n] = f(n\,\Delta t)\Delta t$.

   (b) Find the n-th sampled value $f(n\,\Delta t)$ of $f(t)$.

   (c) Suppose the function is sampled at intervals $\Delta t' = 2\pi/\Omega$ and its Fourier transform $\hat{F}(\omega)$ is reconstructed using the formula.

$$\hat{F}(\omega) = \Delta t' \sum_{n=-\infty}^{\infty} f(n\Delta t')e^{-i\omega n\Delta t'}$$

   Compute $\hat{F}(\omega)$ and sketch $\left|\hat{F}(\omega)\right|$ in the interval $|\omega| \le \Omega$.

3. Using Z-transforms solve the following difference equation:

$$y[n+2] + 2y[n+1] + \frac{1}{4}y[n] = (1/2)^n, \ n \geq 0, y[0] = 1, \ y[1] = -1.$$

4. The Fourier transform $F(\omega)$ of a bandlimited signal $f(t)$ is given by

$$F(\omega) = \begin{cases} \sin^4\left(\frac{\pi\omega}{\Omega}\right) \ ; |\omega| \leq \Omega, \\ 0 \ ; |\omega| > \Omega. \end{cases}$$

(a) The signal is sampled at intervals $\Delta t = \pi/\Omega$. Find the samples.

(b) Find the Z-transform of the sampled sequence $f[n] = f(n\Delta t)\Delta t$.

(c) Find $f(t)$.

(d) Suppose the signal is sampled at intervals $\Delta t' = 2\pi/\Omega$ and its Fourier transform is approximated by

$$\hat{F}(\omega) = \Delta t' \sum_{n=-\infty}^{\infty} f(n\Delta t') e^{-i\omega n\Delta t'}.$$

Compute and sketch $\hat{F}(\omega)$ within the band $|\omega| \leq \Omega$.

5. The sequence $y[n]$ satisfies the following difference equation:

$$y[n+2] + \frac{1}{6}y[n+1] - \frac{1}{6}y[n] = u[n] \ ,$$

where

$$u[n] = \begin{cases} 1 \ ; \ n \geq 0, \\ 0 \ ; \ n < 0. \end{cases}$$

(a) Assuming $y[0] = 0$ and $y[1] = 0$ find the causal solution of the difference equation using Z transforms.

(b) Suppose the $y(n\Delta t) = y[n]/\Delta t$ represent samples of a bandlimited signal with a bandwidth of 10 Hz. Find the Fourier transform of the signal.

6. Using Z-transforms solve the following difference equation:

$$y[n+2] + y[n+1] + \frac{1}{4}y[n] = (1/3)^n, \ n \geq 0, y[0] = 1, \ y[1] = 2.$$

Appendix A

Introduction to Functions of a Complex Variable

A.1 Complex Numbers and Complex Variables

A.1.1 Complex Numbers

A complex number z is defined as a combination of two real numbers x and y and the imaginary unit $i = \sqrt{-1}$ as follows: $z = x + iy$. The x and y are then referred to, respectively, as the real and imaginary parts of z. In symbols, we write $x = \Re e\,(z)$ and $y = \Im m\,(z)$. When $z = iy$ the complex number is said to be imaginary (sometimes pure imaginary) and when $z = x$ it is said to be purely real. It is customary to represent a complex number graphically as a point in a rectangular coordinate system wherein the real part is placed along the abscissa and the imaginary part along the ordinate. This representation is sometimes referred to as the Argand[1] representation and the plane as the Argand plane. We shall refer to it simply as the complex plane, which is the more common designation. For example, the three points in Fig. A.1 represent the complex numbers $3 + i4$ and $2 - i3$ and $4 - i2$. Of course, we could just as well have identified the three points by the coordinate pairs $(3, 4)$, $(2, -3)$, $(4, -2)$ so that this geometrical representation appears indistinguishable from that of a of two-dimensional real Cartesian vector with x-component $x = \Re e\,(z)$ and y-component $y = \Im m\,(z)$. The magnitude (length) of such a vector is the nonnegative quantity $\sqrt{x^2 + y^2}$ represented by an arrow drawn for the coordinate origin while its direction may be specified by the angle $\arctan(y/x)$. By analogy with real two-dimensional vectors we associate with the complex number $z = x + iy$ the *magnitude* $r \equiv \sqrt{x^2 + y^2} \equiv |z|$ and an angle $\theta = \arctan(y/x)$

---

[1] after Jean Robert Argand (1768–1822)

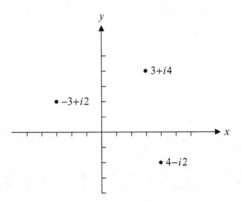

Figure A.1: Complex number representation in the Argand plane

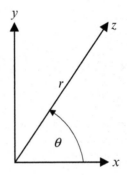

Figure A.2: Polar form of a complex number

with the latter referred to either as the *argument* of z (and written $\theta = \arg(z)$) or as the *phase angle* of z. Taking account of these definitions we have

$$z = x + iy = r\cos\theta + ir\sin\theta = re^{i\theta},\tag{A.1}$$

where we have used the Euler formula[2]

$$e^{i\theta} = \cos\theta + i\sin\theta.\tag{A.2}$$

The last term in (A.1) is referred to as the polar form of a complex number which can be represented as in Fig. A.2. The addition of complex numbers is carried out by adding their real and imaginary parts and appending the imaginary unit i to the imaginary part. Thus with $z_1 = x_1 + iy_1$ and $z_2 = x_2 + iy_2$ the sum is simply $z_3 = z_1 + z_2 = x_1 + x_2 + i(y_1 + y_2)$ so that $\Re(z_3) = x_1 + x_2 = x_3$ and

---

[2]The usual proof of this remarkable formula proceeds by showing equality of the Taylor series expansions of $e^{i\theta}$ and $\cos\theta + i\sin\theta$. A more elementary proof is the following. Define a function $f(\theta) = (\cos\theta + i\sin\theta)e^{-i\theta}$. Differentiating the right side with respect to θ yields $(-\sin\theta + i\cos\theta)e^{-i\theta} - i(\cos\theta + i\sin\theta)e^{-i\theta} = 0$ for all θ. Hence $f(\theta)$ is a constant, independent of θ. Since $f(0) = 1$, this constant equals unity and Euler's formula follows.

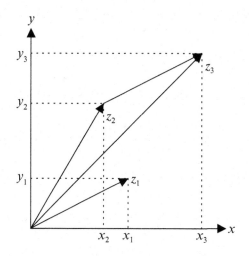

Figure A.3: Parallelogram law of addition

$\Im m\,(z_3) = y_1 + y_2 = y_3$. This leads to the parallelogram construction illustrated in Fig. A.3 just as for real two-dimensional vectors. Despite this similarity complex numbers are not vectors since they do not share all properties of vectors. Nevertheless the term vector had been applied to complex numbers particularly in the older Engineering literature dealing with electric circuit analysis. In accordance with current terminology in electric circuit analysis a representation of parallelogram addition of complex numbers as in Fig. A.3 would be referred to as a phasor diagram.

Multiplication of complex numbers follows from the defining relation of the imaginary unit i. Thus $i^2 = -1$, $(-i)^2 = -1$, $(-i)\,(i) = 1$. Also, since in accordance with (A.2) $i = e^{i\pi/2}$, we obtain $i^n = e^{i\pi n/2} = \cos n\pi/2 + i\sin n\pi/2$ and $i^{-n} = e^{-i\pi n/2} = \cos n\pi/2 - i\sin n\pi/2$. For $n > 0$ these correspond, respectively, to counterclockwise and clockwise rotations by $\pi n/2$ along the unit circle in the complex plane. Thus multiplication of two arbitrary complex numbers $z_1 = x_1 + iy_1$, $z_1 = x_1 + iy_1$ results in

$$z_1 z_2 = (x_1 + iy_1)\,(x_2 + iy_2) = x_1 x_2 - y_1 y_2 + i\,(y_1 x_2 + y_2 x_1)\,. \qquad (A.3)$$

A simple geometrical interpretation can be had by writing z_1 and z_2 in polar form. Thus with $z_1 = |z_1|\,e^{i\theta_1}$ and $z_2 = |z_2|\,e^{i\theta_2}$ we get

$$z_1 z_2 = |z_1|\,|z_2|\,e^{i(\theta_1 + \theta_2)}$$

so that magnitudes get multiplied and the angles (arguments) are added, as shown in Fig. A.4. A similar geometrical interpretation obtains for division where the magnitudes are divided and the angles subtracted. Thus

$$\frac{z_1}{z_2} = \frac{|z_1|}{|z_2|}e^{i(\theta_1 - \theta_2)}\,.$$

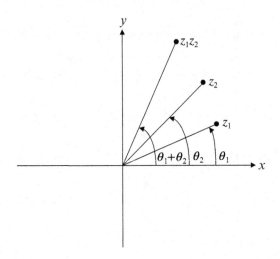

Figure A.4: Multiplication of complex numbers

The *conjugate* of $z = x + iy$ is obtained by changing i to $-i$. For this purpose we use the symbol $*$. Thus $z^* = x - iy$. Obviously $(z^*)^* = z$. Multiplication of a complex number by its conjugate results in

$$zz^* = (x + iy)(x - iy) = x^2 + y^2 = r^2 = |z|^2 .$$

The operation conjugation is a useful tool when division is done in terms of the (Cartesian) component rather than the polar form and a separation in real and imaginary parts is desired. This is facilitated by a multiplication of numerator and denominator by the conjugate of the denominator. Thus

$$\frac{z_1}{z_2} = \frac{x_1 + iy_1}{x_2 + iy_2} = \frac{(x_1 + iy_1)(x_2 - iy_2)}{(x_2 + iy_2)(x_2 - iy_2)} = \frac{x_1 x_2 + y_1 y_2 + i(y_1 x_2 - y_2 x_1)}{x_2^2 + y_2^2}$$

so that

$$\Re(z_1/z_2) = (x_1 x_2 + y_1 y_2) / (x_2^2 + y_2^2)$$

and

$$\Im(z_1/z_2) = (y_1 x_2 - y_2 x_1) / (x_2^2 + y_2^2) .$$

Exponentiation is best performed by using the polar form of the complex number. Thus

$$z^w = |z|^w e^{iw\theta} = e^{w \ln|z|} e^{iw\theta}$$

with $w = u + iv$ an arbitrary complex number. Of particular interest is the inverse of the above problem with $w = N$ a positive integer, i.e., that of finding z when the right side is known. Stated differently, we want to solve

$$z^N = q$$

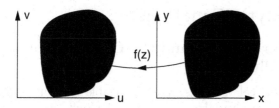

Figure A.5: Representation of $f(z)$ as a mapping from the z to the w-plane

for z for a given (possibly complex) $q = |q| e^{i\theta}$. Since this is an N-th order equation, by the fundamental theorem of algebra it must have N roots. We find these roots by noting that

$$q = qe^{i2\pi k} \; ; \; k = 0, \pm 1, \pm 2, \ldots$$

so that

$$z \equiv z_k = q^{1/N} e^{i2\pi k/N} = |q|^{1/N} e^{i\theta/N} e^{i2\pi k/N}; k = 0, 1, 2, \ldots N - 1.$$

Also, since the coefficient of the $N-1$ power in the algebraic equation $z^N - q = 0$ is zero the sum of the roots is likewise zero, i.e.,

$$\sum_{k=0}^{N-1} e^{i2\pi k/N} = 0. \tag{A.4}$$

A.1.2   Function of a Complex Variable

In general, a function of a complex variable z may be defined as a mapping of a set of complex numbers $z = x + iy$ in a region the complex z-plane (the domain of the function) into another set complex of numbers $w = u + iv$ falling into a region of the complex w-plane (the range of the function) in accordance with a rule denoted symbolically by $f(z)$. We represent this mapping analytically by writing

$$w = f(z) = u(x, y) + iv(x, y) \tag{A.5}$$

and may be thought of geometrically as shown in Fig. A.5. We shall be primarily interested in functions whose domain is the entire z-plane as, e.g.,

$$w = u + iv = f(z) = z^2 = (x + iy)^2 = x^2 - y^2 + i2xy$$

so that $u = x^2 - y^2$ and $v = 2xy$.

A.2   Analytic Functions

A.2.1   Differentiation and the Cauchy–Riemann Conditions

By analogy with real variables, it is reasonable to define the derivative of a function of a complex variable at $z = z_0$ as the limit

$$\lim_{\Delta z \longrightarrow 0} \frac{f(z_0 + \Delta z) - f(z_0)}{\Delta z}. \tag{A.6}$$

However, unlike for a real variable such a limit is not unique for it may depend on the direction (i.e., argument) assumed by Δz as its magnitude tends to zero. Functions

This potential lack of uniqueness can be seen from the phasor diagram in Fig. A.6 where Δz may approach zero from any initial position on the rim of in the circular region with radius $|\Delta z|$ and for each such position the limits may differ. Evidently for a derivative of $f(z)$ to possess a unique value at $z = z_0$ requires that the limit (A.6) be independent of the direction (argument) of the increment Δz. Functions with this property are referred to analytic functions.

We can readily establish the necessary conditions for the existence of a derivative at a point $z_0 = x_0 + iy_0$ by first rewriting the limit (A.6) in terms of real and imaginary parts of $f(z)$. Thus with

$$f(x + iy) = u(x, y) + iv(x, y)$$

Eq. (A.6) reads

$$\lim_{\Delta z \to 0} \frac{f(z_0 + \Delta z) - f(z_0)}{\Delta z}$$
$$= \lim_{\Delta z \to 0} \frac{u(x_0 + \Delta x, y_0 + \Delta y) + iv(x_0 + \Delta x, y_0 + \Delta y) - u(x_0, y_0) - iv(x_0, y_0)}{\Delta x + i\Delta y}. \tag{A.7}$$

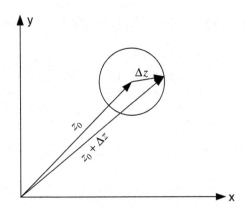

Figure A.6: Directions though which the increment Δz can approach zero

A necessary condition for the existence of the derivative is that the approach to the limit in (A.7), when evaluated for any two distinct directions in Fig. A.6, remains the same. For example, suppose we evaluate the limits for $\Delta z = \Delta x$ and for $\Delta z = i\Delta y$. In the first instance we obtain

$$
\lim_{\Delta x \to 0} \frac{f(z_0 + \Delta x) - f(z_0)}{\Delta x}
$$

$$
= \lim_{\Delta x \to 0} \frac{u(x_0 + \Delta x, y_0) - u(x_0, y_0) + iv(x_0 + \Delta x, y_0) - iv(x_0, y_0)}{\Delta x}
$$

$$
= \lim_{\Delta x \to 0} \frac{\Delta u + i\Delta v}{\Delta x} = \frac{\partial u}{\partial x} + i\frac{\partial v}{\partial x}. \tag{A.8a}
$$

Next, taking the increment along y, we get

$$
\lim_{\Delta v \to 0} \frac{f(z_0 + i\Delta y) - f(z_0)}{i\Delta y}
$$

$$
= \lim_{\Delta y \to 0} \frac{u(x_0, y_0 + \Delta y) - u(x_0, y_0) + iv(x_0, y_0 + \Delta y) - iv(x_0, y_0)}{i\Delta y}
$$

$$
= \lim_{\Delta y \to 0} \frac{\Delta u + i\Delta v}{i\Delta y} = -i\frac{\partial u}{\partial y} + \frac{\partial v}{\partial y}. \tag{A.8b}
$$

Thus, equating (A.8a) to (A.8b) provides us with the necessary set of conditions for the existence of a derivative at $z = z_0$:

$$
\frac{\partial u}{\partial x} = \frac{\partial v}{\partial y}, \tag{A.9a}
$$

$$
\frac{\partial u}{\partial y} = -\frac{\partial v}{\partial x}. \tag{A.9b}
$$

Equations (A.9a) and (A.9b) are known as the Cauchy–Riemann (CR) conditions. It turns out that they are also sufficient conditions for the existence of a derivative. To prove sufficiency we take an arbitrary increment $\Delta f(z)$ and write it in terms of the partial derivatives of u and v as follows:

$$
\begin{aligned}
\Delta f(z) &= \Delta u(x, y) + i\Delta v(x, y) \\
&= \frac{\partial u}{\partial x}\Delta x + \frac{\partial u}{\partial y}\Delta y + i\left[\frac{\partial v}{\partial x}\Delta x + \frac{\partial v}{\partial y}\Delta y\right].
\end{aligned}
$$

Next we substitute the CR conditions from (A.9) to obtain

$$
\begin{aligned}
\Delta f(z) &= \frac{\partial u}{\partial x}\Delta x - \frac{\partial v}{\partial x}\Delta y + i\left[\frac{\partial v}{\partial x}\Delta x + \frac{\partial u}{\partial x}\Delta y\right] \\
&= \frac{\partial u}{\partial x}(\Delta x + i\Delta y) + \frac{\partial v}{\partial x}(-\Delta y + i\Delta x) \\
&= \left(\frac{\partial u}{\partial x} + i\frac{\partial v}{\partial x}\right)(\Delta x + i\Delta y) = \left(\frac{\partial u}{\partial x} + i\frac{\partial v}{\partial x}\right)\Delta z.
\end{aligned}
$$

From the last relation we note that the increment Δz is arbitrary. Hence we are entitled to the assertion that the direction along which Δz approaches 0 is immaterial, and write

$$\lim_{\Delta z \to 0} \frac{\Delta f(z)}{\Delta z} = \frac{\partial u}{\partial x} + i \frac{\partial v}{\partial x} = \frac{\partial v}{\partial y} - i \frac{\partial u}{\partial y} = \frac{df(z)}{dz} = f'(z) \qquad (A.10)$$

A.2.2 Properties of Analytic Functions

A function satisfying the CR conditions at a point is said to be analytic at that point. A function which possesses a derivative at all points within a region $\Re$ of the complex plane is said to be analytic in $\Re$. It is easy to see that a sum of analytic functions is analytic. Likewise the product of two analytic functions is analytic. To show this, consider two functions $f_1(z)$ and $f_2(z)$ both analytic in $\Re$. We decompose these into their respective real and imaginary parts

$$\begin{align}
f_1(z) &= u_1 + iv_1 & \text{(A.11a)}\\
f_2(z) &= u_2 + iv_2 & \text{(A.11b)}
\end{align}$$

and, as a notational convenience, denote the partial derivatives of the real and imaginary parts by subscripts. The CR conditions for each function then assume the form

$$\begin{align}
u_{1x} &= v_{1y}, \; u_{1y} = -v_{1x}, & \text{(A.12a)}\\
u_{2x} &= v_{21y}, \; u_{2y} = -v_{2x}. & \text{(A.12b)}
\end{align}$$

With

$$g(z) = f_1(z) f_2(z) = U + iV$$

we then obtain via the substitution of (A.11)

$$\begin{align}
U &= u_1 u_2 - v_1 v_2, & \text{(A.13a)}\\
V &= v_1 u_2 + u_1 v_2. & \text{(A.13b)}
\end{align}$$

Analyticity of $g(z)$ requires

$$\begin{align}
U_x &= V_y, & \text{(A.14a)}\\
U_y &= -V_x. & \text{(A.14b)}
\end{align}$$

By direct calculation we obtain

$$\begin{align}
U_x &= u_1 u_{2x} + u_{1x} u_2 - v_1 v_{2x} - v_{1x} v_2, & \text{(A.15a)}\\
V_y &= v_1 u_{2y} + v_{1y} u_2 + u_1 v_{2y} + u_{1y} v_2, & \text{(A.15b)}\\
U_y &= u_1 u_{2y} + u_{1y} u_2 - v_1 v_{2y} - v_{1y} v_2, & \text{(A.15c)}\\
V_x &= v_1 u_{2x} + v_{1x} u_2 + u_1 v_{2x} + u_{1x} v_2. & \text{(A.15d)}
\end{align}$$

Using (A.12) in (A.15b) we set $u_{2y} = -v_{2x}$, $v_{1y} = u_{1x}$, $v_{2y} = u_{2x}$, $u_{1y} = -v_{1x}$ and 14a follows. Similarly we can verify (A.14b).

The simplest analytic function is a constant. On the next level of complexity is $f(z) = z$ which clearly is analytic for all finite values of z. This is also the case for $f(z) = z^2$ as can be seen by writing

$$z^2 = x^2 - y^2 + i2xy = u + iv$$

and showing that the CR conditions are satisfied. Thus

$$\frac{\partial u}{\partial x} = 2x, \frac{\partial v}{\partial y} = 2x = \frac{\partial u}{\partial x}, \frac{\partial u}{\partial y} = -2y, \frac{\partial v}{\partial x} = 2y = -\frac{\partial u}{\partial y}.$$

From this and the fact that a product of two analytic functions is analytic follows that z^n is analytic for any nonnegative integer. Thus a polynomial, being a sum of powers of z, is an analytic function for all $|z| < \infty$.

Examples of analytic functions that are not polynomials are the trigonometric functions $\sin z$ and $\cos z$. By decomposing them into real and imaginary parts with the aid of the trigonometric addition formulas one can show that the CR conditions are satisfied in the entire finite complex plane and that $d(\sin z)/dz = \cos z$ and $d(\cos z)/dz = -\sin z$. Thus not only are the sinusoids analytic for all $|z| < \infty$, but all the derivatives are analytic as well.[3]

A.2.3   Integration

The formal definition of an integral of a function of a complex variable is quite similar to that of a line integral in a plane for real variables. The integration is carried out over a curve C connecting two points in the complex plane, say, $z_1 = x_1 + iy_1$ and $z_2 = x_2 + iy_2$, as shown in Fig. A.7, and the integral defined by the usual Riemann sums applied to the real and imaginary constituents:

$$\begin{aligned} I &= \int_{z_1}^{z_2} f(z)\, dz = \int_{(x_1,y_1)}^{(x_2,y_2)} [u(x,y) + iv(x,y)]\, (dx + idy) \\ &= \int_C \{u[x, y(x)]\, dx - v[x, y(x)]\, dy\} + i \int_C \{u[x, y(x)]\, dx + v[x, y(x)]\, dy\} \end{aligned}$$

$$\text{(A.16)}$$

In general, the value of the integral I will depend on the choice of the curve connecting z_1 and z_2. However, if the path of integration is confined entirely to the region of analyticity of $f(z)$, then I will be independent of the chosen curve C. The proof of this assertion requires an auxiliary result, known as Green's theorem.

---

[3] In fact we shall show in the sequel that this is a general property of analytic functions, i.e., an analytic function possesses derivatives of all orders.

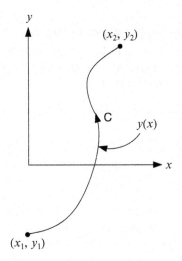

Figure A.7: Integration path for line integral in the complex plane

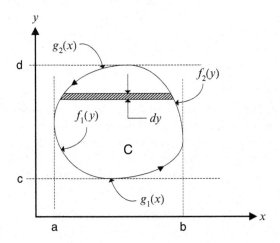

Figure A.8: Closed integration contour

Green's Theorem.   Any two real functions $P(x,y)$ and $Q(x,y)$ with continuous partial derivatives satisfy

$$\oint_C (Pdx + Qdy) = \iint_A \left(\frac{\partial Q}{\partial x} - \frac{\partial P}{\partial y} \right) dxdy, \qquad (A.17)$$

where the line integral is taken in a counterclockwise direction along a closed planar curve C and double integral over the area A enclosed by the curve, as shown in Fig. A.8.

We note from the figure that the closed contour C can be traced using the two functions $f_2(y)$ and $f_1(y)$ and that the elementary contribution to the area A is

$$dA = [f_2(y) - f_1(y)] \, dy.$$

Hence the integral of $\partial Q / \partial x$ over the area enclosed by the closed contour is

$$\iint_A \frac{\partial Q}{\partial x} dx dy = \int_c^d dy \int_{f_1(y)}^{f_2(y)} \frac{\partial Q}{\partial x} dx = \int_c^d dy \left\{ Q\left[f_2(y), y\right] - Q\left[f_1(y), y\right] \right\}.$$

(A.18)

Evidently the last integral can be interpreted as a line integral taken in the counterclockwise direction along the close contour C encompassing the area A. Accordingly, we rewrite it in the notation used in (A.17):

$$\oint_C Q \, dy = \iint_A \frac{\partial Q}{\partial x} dx dy.$$

(A.19)

Alternatively, we can trace out the closed contour using the functions $g_2(x)$ and $g_1(x)$ so that the elementary area contribution becomes

$$dA = [g_2(x) - g_1(x)] \, dx.$$

Hence

$$\iint_A \frac{\partial P}{\partial y} dx dy = \int_a^b dx \int_{g_1(x)}^{g_2(x)} \frac{\partial P}{\partial y} dy = \int_a^b dx \left\{ P\left[x, g_2(x)\right] - P\left[x, g_1(x)\right] \right\}$$

$$= - \oint_C P \, dx$$

(A.20)

so that

$$\oint_C P \, dx = - \iint_A \frac{\partial P}{\partial y} dx dy.$$

(A.21)

Adding (A.19) and (A.21) we obtain (A.17).

Q.E.D.

We the aid of Green's theorem we can now establish a fundamental property of analytic functions.

Theorem 1 *Let $f(z)$ be an analytic function in $\Re$. Then*

$$\oint_C f(z) \, dz = 0$$

(A.22)

for any closed curve lying entirely in $\Re$.

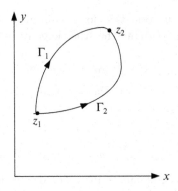

Figure A.9: Integrals along two different paths

Proof.

$$\oint_C f\,(z)\,dz = \oint_C (u+iv)\,(dx+idy) = \oint_C (udx-vdy) + i \oint_C (vdx+udy)\,. \quad \text{(A.23)}$$

Using Green's theorem

$$\oint_C (udx - vdy) = \iint_A \left(-\frac{\partial v}{\partial x} - \frac{\partial u}{\partial y} \right) dxdy, \quad \text{(A.24a)}$$

$$\oint_C (vdx + udy) = \iint_A \left(\frac{\partial u}{\partial x} - \frac{\partial v}{\partial y} \right) dxdy. \quad \text{(A.24b)}$$

In accordance with the CR conditions the right side of (A.24a) and (A.24b) is identically zero validating (A.22). □

One corollary of the preceding theorem is that an integral of an analytic function taken along a path connecting two fixed points is dependent on the path as long as the curve connecting the points lies entirely within the region of analyticity of the function. Thus assuming that the closed curve in Fig. A.9 lies within the region of analyticity of $f\,(z)$ we have

$$\oint f(z)dz = 0. \quad \text{(A.25)}$$

Alternatively,[4] we can write this as a sum of two integrals over paths Γ_1 and Γ_2

$$\oint f(z)dz = \int_{z_1}^{z_2} f\,(z)\,dz - \int_{z_1}^{z_2} f\,(z)\,dz = 0 \quad \text{(A.26)}$$
$$\quad\;\; (\Gamma_2) \qquad\qquad (\Gamma_1)$$

---

[4]Here and in the following the symbol $\oint$ denotes an integral carried out over a closed path in the *counterclockwise* direction.

so that

$$\int_{z_1 \atop (\Gamma_1)}^{z_2} f(z)\, dz = \int_{z_1 \atop (\Gamma_2)}^{z_2} f(z)\, dz \qquad (A.27)$$

over two arbitrary curves Γ_1 and Γ_2.

A.3   Taylor and Laurent Series

A.3.1   The Cauchy Integral Theorem

We have shown that the product of two analytic functions is analytic. Thus for any analytic function $f(z)$, the function $g(z)$ defined by the product

$$g(z) = \frac{f(z)}{z - z_0} \qquad (A.28)$$

will be analytic in the same region as $f(z)$ with a possible exception of the point $z = z_0$. Therefore an integral of $g(z)$ carried out along any closed path lying entirely within the region of analyticity must yield zero, and, in particular, along the closed path shown in Fig. A.10. Tracing out this path in the direction indicated by the arrows, we first traverse the circular segment C_0 in the clockwise direction and then, moving along the linear segment L_1, reach the outer contour C. Continuing in the counterclockwise direction along C we connect to path segment L_2 which brings us back to C_0. The corresponding contributions to the closed path integral are

$$- \oint_{C_0} g(z)dz + \int_{L_1} g(z)dz + \oint_{C} g(z)dz + \int_{L_2} g(z)dz = 0. \qquad (A.29)$$

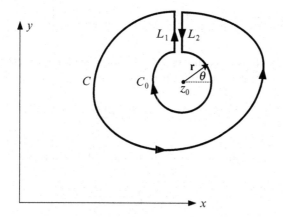

Figure A.10: Integration path

In the limit as the two line segments approach coincidence the contributions from L_1 and L_2 cancel so that (A.29) reduces to

$$\oint_C \frac{f(z)}{z - z_0} dz = \oint_{C_0} \frac{f(z)}{z - z_0} dz. \tag{A.30}$$

On the circular contour C_0 we set $z - z_0 = re^{i\theta}$ and obtain

$$-\oint_{C_0} \frac{f(z)}{z - z_0} dz = -\int_0^{-2\pi} \frac{ire^{i\theta} f(z_0 + re^{i\theta})}{re^{i\theta}} d\theta = i \int_0^{2\pi} f(z_0 + re^{-i\theta}) d\theta. \tag{A.31}$$

Using this in (A.30) we have[5]

$$\oint_C \frac{f(z)}{z - z_0} dz = i \int_0^{2\pi} f(z_0 + re^{-i\theta}) d\theta. \tag{A.32}$$

After setting $r = 0$ we solve for $f(z_0)$ and obtain the Cauchy integral formula

$$f(z_0) = \frac{1}{2\pi i} \oint_C \frac{f(z)}{z - z_0} dz. \tag{A.33}$$

As indicated in Fig. A.10, the integration is to be carried out in the counter-clockwise direction along a contour enclosing the point z_0 but can otherwise be arbitrary. If the contour does not enclose z_0, we are integrating an analytic function and the integral vanishes. This result can be rephrased more formally as follows. Let $f(z)$ be an analytic function in a region $\Re$ and C a closed contour within $\Re$. Let $\Re_C$ represent a region within $\Re$ entirely enclosed by C. Then

$$\frac{1}{2\pi i} \oint_C \frac{f(z)}{z - z_0} dz = \begin{cases} 0, & \text{when } z_0 \notin \Re_C, \\ f(z_0), & \text{when } z_0 \in \Re_C. \end{cases} \tag{A.34}$$

Since for $z_0 \neq z$ the integrand in (A.33) is an analytic function of z_0 we are entitled to differentiation with respect to z_0 under the integral sign. We then obtain

$$f^{(1)}(z_0) = \frac{1}{2\pi i} \oint_C \frac{f(z)}{(z - z_0)^2} dz. \tag{A.35}$$

The same argument applies also to an n-fold differentiation. This leads to the formula

$$\frac{f^{(n)}(z_0)}{n!} = \frac{1}{2\pi i} \oint_C \frac{f(z)}{(z - z_0)^{n+1}} dz, \tag{A.36}$$

---

[5]Note that according to (A.31) the radius of the circle C_0 along which we are integrating can be arbitrarily small.

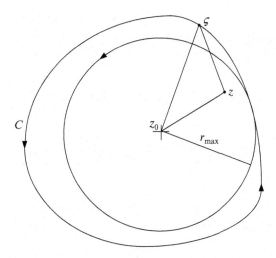

Figure A.11: Integration paths in the derivation of the Taylor series

where the integral again vanishes when the contour fails to enclose z_0. Formula (A.36) embodies an extremely important property of analytic functions. It states that an analytic function possesses derivatives of all orders and that all of these derivatives are also analytic functions. This result forms the basis for the Taylor as well as the Laurent expansions to be discussed in the sequel. Another important consequence of the Cauchy integral formula is that a function analytic and bounded for all z must be a constant, a result referred to a Liouville's Theorem. Its proof is quite straightforward. Thus if $f(z)$ is analytic for any z_0 and bounded i.e., $|f(z_0)| < M$, then with $|z - z_0| = r$

$$\left| f^{(1)}(z_0) \right| = \left| \frac{1}{2\pi i} \oint_{|z-z_0|} \frac{f(z)}{(z-z_0)^2} dz \right| \le \frac{1}{2\pi} \left| \oint_{|z-z_0|} \frac{f(z)}{(z-z_0)^2} dz \right| \le \frac{M}{r}.$$

Since M is independent of r, we can set $r = \infty$. Then $\left| f^{(1)}(z_0) \right| = 0$ for all z_0 so that $f(z_0)$ is a constant. Functions that are analytic for all finite z are called entire (or integral) functions. Examples are polynomials, $e^z, \sin z, \cos z, J_n(z)$ (with n an integer).

A.3.2   The Taylor Series

Let $f(z)$ be an analytic function within the region enclosed by C as shown in Fig. A.11. We apply the Cauchy integral theorem (A.33) to the closed contour C where we denote the variable of integration by ς. The function $f(z)$ is then given by

$$f(z) = \frac{1}{2\pi i} \oint_C \frac{f(\varsigma)}{\varsigma - z} d\varsigma. \tag{A.37}$$

Recall that z, which in the notation of (A.34) now plays the role of z_0, must lie within the contour C. Presently the point z_0 shown in the figure is just an arbitrary point within C. We use it to rewrite the denominator of (A.37) by setting

$$\varsigma - z = \varsigma - z_0 + z_0 - z = (\varsigma - z_0)\left[1 - \frac{z - z_0}{\varsigma - z_0}\right] \qquad (A.38)$$

$$\frac{1}{\varsigma - z} = \frac{1}{\varsigma - z_0}\left[\frac{1}{1 - \frac{z - z_0}{\varsigma - z_0}}\right]. \qquad (A.39)$$

In the region

$$\left|\frac{z - z_0}{\varsigma - z_0}\right| < 1 \qquad (A.40)$$

(A.39) can be represented by the convergent geometric series

$$\frac{1}{\varsigma - z} = \frac{1}{\varsigma - z_0}\sum_{n=0}^{\infty}\left(\frac{z - z_0}{\varsigma - z_0}\right)^n, \qquad (A.41)$$

which we substitute into the Cauchy integral formula (A.37) and obtain

$$f(z) = \sum_{n=0}^{\infty}\left[\frac{1}{2\pi i}\oint_{r_{max}^+}\frac{f(\varsigma)}{(\varsigma - z_0)^{n+1}}d\varsigma\right](z - z_0)^n, \qquad (A.42)$$

where in view of (A.40) we have shifted the integration contour from C to r_{max}. Comparing the terms in braces with (A.36) we see that (A.42) is equivalent to

$$f(z) = \sum_{n=0}^{\infty}\frac{f^{(n)}(z_0)}{n!}(z - z_0)^n, \qquad (A.43)$$

which is the Taylor series expansion of $f(z)$ about the point $z = z_0$. Its region of convergence is defined by (A.40) and represented geometrically in Fig. A.11 by the circle with center at z_0 and radius r_{max}. From the geometrical construction in the figure it is evident that r_{max} is the shortest distance from z_0 to the integration contour C. The radius of convergence can be increased by moving z_0 further away from the integration contour provided the circle remains entirely within region of analyticity of $f(z)$ as prescribed by (A.40). Thus for $f(z)$ analytic in the region $\Re$ shown in Fig. A.12 every point can serve as an expansion point of a Taylor series. These expansions are alternative representations of the same function where each representation has its own circle of convergence whose maximum radius is governed by the proximity of the expansion point to the boundary of the region of analyticity. For example, if we were to choose the three points z_1, z_2, and z_3 as shown in the figure as Taylor series expansion points we would obtain convergence only within the circles with radii r_1, r_2, and r_3, respectively. When the convergence regions corresponding to two or more expansion points overlap one has the choice of several Taylor representations

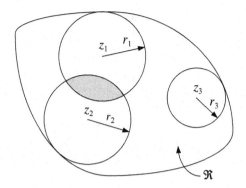

Figure A.12: Convergence regions for Taylor series

all converging within the same region. For example, the common region of convergence for expansions about z_1 and z_2 corresponds to the shaded area in Fig. A.12. In this case $f(z)$ can be represented by two different series both of which converge within the shaded region. When the circles of convergence are disjoint, as, for example, z_1 and z_3, the two respective Taylor series still represent the same function. This identification is referred to as analytic continuation.

In Fig. A.12 the boundary between the region where a function is analytic and where it fails to be analytic is represented by a continuous curve. More frequently a function will fail to be analytic only at a set of discrete points as, e.g., the function in (A.28). As a special case let us set $f(z) = 1$ and determine Taylor expansions of

$$g(z) = \frac{1}{z - 2}. \tag{A.44}$$

Evidently this function is analytic for all finite z with the exception of the point $z = 2$. Hence we should be able to find a convergent Taylor series about any point $z \neq 2$. For example, let us pick $z = 0$. Dividing the numerator and the denominator in (A.44) by 2 and using the geometric series expansion in $z/2$, we get

$$g(z) = \sum_{n=0}^{\infty} \left(-\frac{1}{2^{n+1}} \right) z^n \tag{A.45}$$

a Taylor series that converges for $|z - 1| < 2$. Suppose we expand $g(z)$ about the point $z = 1$. We again use the geometric series approach and obtain[6]

$$g(z) = \frac{1}{z - 2} = -\frac{1}{1 - (z - 1)} = \sum_{n=0}^{\infty} -(z - 1)^n, \tag{A.46}$$

which converges for $|z - 1| < 1$. The circles of convergence of (A.45) and (A.46) are shown in Fig. A.13.

---

[6]We could have also used (A.36) but in this particular case the indirect approach is simpler.

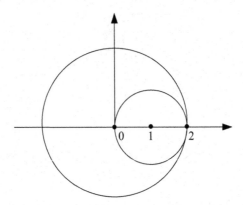

Figure A.13: Circles of convergence for series (A.45) and (A.46)

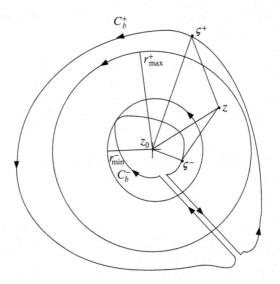

Figure A.14: Convergence regions for the Laurent series

A.3.3   Laurent Series

Just like the Taylor series the Laurent series it is based on the Cauchy integral formula. We start by positioning the expansion point z_0 within the region bounded by the closed curve C_b^- shown in Fig. A.14 where $f(z)$ is not necessarily analytic. The region of analyticity currently of interest to us lies between the two closed boundaries: the inner boundary C_b^- and the outer boundary C_b^+. The evaluation of the Cauchy integral proceeds along the path in the direction indicated by the arrows. To close the path we introduce, in addition to C_b^+ and C_b^-, two straight closely spaced line segments. The integrations along these segments are carried out in opposite directions so that their contributions cancel.

Thus the curves C_b^+ and C_b^- alone enclose the region of analyticity of $f(z)$ so that

$$\oint_{C_b^+} f(z)dz - \oint_{C_b^-} f(z)dz = 0. \tag{A.47}$$

Applying the Cauchy integral formula to the same closed path we have

$$f(z) = \frac{1}{2\pi i} \oint_{C_b^+} \frac{f(\varsigma^+)d\varsigma^+}{\varsigma^+ - z} - \frac{1}{2\pi i} \oint_{C_b^-} \frac{f(\varsigma^-)d\varsigma^-}{\varsigma^- - z}. \tag{A.48}$$

As in the Taylor series we set

$$\frac{1}{\varsigma^+ - z} = \frac{1}{\varsigma^+ - z_0 - (z - z_0)} = \frac{1}{(\varsigma^+ - z_0)\left[1 - \frac{z - z_0}{\varsigma^+ - z_0}\right]}$$

$$= \frac{1}{\varsigma^+ - z_0} \sum_{n=0}^{\infty} \left(\frac{z - z_0}{\varsigma^+ - z_0}\right)^n, \tag{A.49}$$

which converges for $|z - z_0| < |\varsigma^+ - z_0|$. In Fig. A.14 this corresponds to the interior of the circle with radius r_{max}^+. Thus we can represent the first of the two integrals in (A.48) by a series that converges within the same circle provided we shift the path of integration in each term of the series from the original contour C_b^+ to r_{max}^+. We then get

$$\frac{1}{2\pi i} \oint_{C_b^+} \frac{f(\varsigma^+)d\varsigma^+}{\varsigma^+ - z} = \sum_{n=0}^{\infty} a_n (z - z_0)^n, \tag{A.50}$$

where

$$a_n = \frac{1}{2\pi i} \oint_{r_{\mathrm{max}}^+} \frac{f(\varsigma^+)d\varsigma^+}{(\varsigma^+ - z_0)^{n+1}}. \tag{A.51}$$

Eq. (A.50) converges within the circle $|z - z_0| < r_{\mathrm{max}}^+$. We note that (A.50) appears to be identical with the Taylor series expansion in (A.36). This identity is, however, purely formal because despite the fact that (A.50) converges within the same region as (A.36) it does not converge to $f(z)$ but only to a function represented by the first of the two integrals in (A.48). This function is referred to as the regular part of $f(z)$.

To develop a series representation for the second integral in (A.48) we again utilize the geometric series. Thus

$$\frac{1}{\varsigma^- - z} = -\frac{1}{(z - z_0)\left[1 - \frac{\varsigma^- - z_0}{z - z_0}\right]} = -\frac{1}{z - z_0} \sum_{n=0}^{\infty} \left(\frac{\varsigma^- - z_0}{z - z_0}\right)^n, \tag{A.52}$$

which converges for $|\varsigma^- - z_0| < |z - z_0|$. In Fig. A.14 this corresponds to points outside of the circle with radius r_{min}^-. With the aid of (A.52) we now construct a series representation of the second integral in (A.48):

$$\frac{1}{2\pi i} \oint_{C_b^-} \frac{f(\varsigma^-)d\varsigma^-}{\varsigma^- - z_0} = \sum_{n=-\infty}^{-1} a_n(z - z_0)^n, \tag{A.53}$$

where

$$a_n = \frac{1}{2\pi i} \oint_{r_{min}^-} (\varsigma^- - z_0)^{-n-1} f(\varsigma^-)d\varsigma^-. \tag{A.54}$$

The function represented by (A.53) is called the principal part of $f(z)$. The coefficient a_{-1} plays a special role in complex variable theory and is referred to as the residue of $f(z)$. The sum of (A.50) and (A.53) is the Laurent series

$$f(z) = \sum_{n=-\infty}^{\infty} a_n(z - z_0)^n, \tag{A.55}$$

which converges within the annular region $r_{min}^- < |z| < r_{max}^+$ to the analytic function $f(z)$. In virtue of the analyticity of $f(z)$ the expansion coefficients a_n represented by the two separate expressions (A.51) and (A.54) can be combined into the single formula

$$a_n = \frac{1}{2\pi i} \oint (\varsigma - z_0)^{-n-1} f(\varsigma)d\varsigma, \tag{A.56}$$

where the integral may be evaluated over any closed path within the annular region of convergence of (A.56). In particular the residue of $f(z)$ is

$$a_{-1} = \frac{1}{2\pi i} \oint f(\varsigma)d\varsigma. \tag{A.57}$$

A.4   Singularities of Functions and the Calculus of Residues

A.4.1   Classification of Singularities

Isolated Singularities

Definitions. In regions where a function fails to be analytic it is termed singular. A function can be singular on a discrete set of points, or over a continuum (i.e., on an uncountable set of points). Singularities at discrete points will be referred to as isolated those over a continuum extended. Isolated singularities can be classified as poles or as essential singularities. A function $f(z)$ has pole of order N at $z = z_0$ if it

$$f(z)(z - z_0)^n$$

is analytic for $n \geq N$ but fails to be analytic for $n < N$. For example,

$$f(z) = \frac{z^2}{z - 2} \tag{A.58}$$

has a pole of order 1 (also called a simple pole) at $z = 2$ and

$$f(z) = \frac{z}{(z - 1)^2} \tag{A.59}$$

has a pole of order 2 at $z = 1$. One can generalize this definition to encompass poles at infinity. In that case z, z^2, z^3 represent, respectively, poles of order one, two, and three. In accordance with this definition (A.58) has, in addition to the pole at $z = 2$, a simple pole at infinity. We note that an N-th order pole at $z = z_0$ can be removed by multiplying the function by $(z - z_0)^N$. We shall call a singularity that can be removed by multiplying the function by a polynomial, a removable singularity. For poles at infinity we will widen this definition by including multiplications by z^{-n}. A function can heave multiple removable singularities. For a ratio of polynomials these are all the zeros of the denominator. For example, for $f(z) = z^5/(z - 1)(z - 2)^3$, we have a simple poles at $z = 1$ and a pole of order 3 at $z = 2$ and a simple pole at infinity. A function with removable singularities need not be a polynomial as, for example, $\sin z/z^2$ which has a removable simple pole at $z = 0$. An isolated singularity that cannot be removed by multiplying it by a polynomial is referred to as an essential singularity. For example, the function $e^{1/z}$ fails to be analytic only at $z = 0$. This singularity cannot be removed by multiplying it by z^n, as is apparent from the Laurent expansion

$$e^{1/z} = 1 + \frac{1}{z} + \frac{1}{2\,z^2} + \frac{1}{3!\,z^3} + \frac{1}{4!\,z^4} + \cdots \tag{A.60}$$

Replacing $1/z$ by z transforms (A.60) into a Taylor series for e^z which has an essential singularity at infinity but is otherwise analytic. Functions that have no singularities other than essential singularities at infinity are called entire functions. Many elementary functions fall into this category. Examples are: $\sin z, \cos z$, the Bessel function $J_n(z)$.

One necessary characteristic of entire functions is that they can be represented by infinite series of ascending powers in z, i.e., a Taylor expansions about $z = 0$. However, this is not a sufficient property, for such expansions can also be gotten for ratios of polynomials where the singularities can be removed. This is not possible for essential singularities in general, and for entire functions in particular.

A function can have both removable and essential singularities. For example, $\cos z/z$ has a simple pole at $z = 0$ and an essential singularity at infinity.

Laurent Series for Functions with Isolated Singularities. In Sect. A.2 the type of singularities outside the regions of convergence for the Taylor and Laurent series was left unspecified. In the following we assume

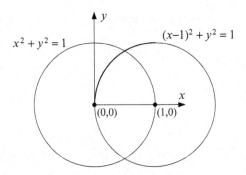

Figure A.15: Convergence regions for Laurent expansions of $1/z(z-1)$

that they are poles. In this case the Taylor and Laurent series coefficients can generally be found without the explicit use of (A.56). For example, consider the function

$$f(z) = \frac{1}{(z-1)z} \qquad (A.61)$$

with its two simple poles shown in Fig. A.15

There are four possible Laurent series. The two for the pole at $z = 0$ converge in the annular region[7] defined by $0 < z < 1$, and $1 < z < \infty$, corresponding to the circle at the origin in Fig. A.15. For the pole at $z = 1$ the boundary between the convergence regions $0 < |1 - z| < 1$ and $1 < |1 - z| < \infty$ is the circle with its origin is at $z = 0$.

For $z = 0$ we obtain the first of the two Laurent series by expanding $1/(1-z)$ in a geometric series and dividing each term by $-z$. This yields

$$f(z) = -\frac{1}{z} - 1 - z - z^2 - z^3 \dots, \qquad (A.62)$$

which converges within the unit circle $|z| < 1$. We note that the residue, as defined by (A.56), equals -1. To obtain the second series we divide the numerator and denominator of (A.61) by z so that

$$f(z) = \frac{1}{z^2} \frac{1}{1 - \frac{1}{z}} \qquad (A.63)$$

and expand $1/(1 - z^{-1})$ in a geometric series. The result is the Laurent expansion

$$f(z) = \frac{1}{z^2} + \frac{1}{z^3} + \frac{1}{z^4} \dots, \qquad (A.64)$$

which converges for $|z| > 1$.

---

[7]In all four cases the inner circle of the annulus encloses the pole with a vanishingly small radius.

For the pole at $z = 1$ we can again rearrange $f(z)$ and use the geometric series. The series converging for $0 < |1 - z| < 1$ is

$$
f(z) = \frac{1}{(z - 1 + 1)(z - 1)} = \frac{1}{(z - 1)}[1 - (z - 1) + (z - 1)^2 - (z - 1)^3 \dots]
$$

$$
= \frac{1}{(z - 1)} - 1 + (z - 1) - (z - 1)^2 + (z - 1)^3 \dots \tag{A.65}
$$

and for $1 < |1 - z| < 1$

$$
f(z) = \frac{1}{(z - 1 + 1)(z - 1)} = \frac{1}{(z - 1)^2} \frac{1}{\left[1 + \frac{1}{z-1}\right]}
$$

$$
= \frac{1}{(z - 1)^2} - \frac{1}{(z - 1)^3} + \frac{1}{(z - 1)^4} - \frac{1}{(z - 1)^5} \dots \tag{A.66}
$$

Multivalued Functions and Extended Singularities. Functions such as $f^*(z)$ and $|z|$ that are not analytic anywhere on the complex plane are only of peripheral interest in analytic function theory. Here we will be dealing with functions that fail to be analytic only along a line or curve in the complex plane. They generally result from the imposition of constraints on multivalued functions. As an example consider the function

$$
f(z) = \sqrt{z}. \tag{A.67}
$$

Using the polar form $z = re^{i\theta}$

$$
f(z) = r^{1/2}e^{i\theta/2}, \tag{A.68}
$$

where $r^{1/2}$ will be defined as nonnegative. Along the positive segment of the x-axis $\theta = 0$ and $x = r$ so that $f(z) = r^{1/2} > 0$. When θ is incremented by 2π we return to the same point $x = re^{i2\pi} = r$ but not to the same value of $f(z)$. Instead, it equals $-r^{1/2}$. To return to $r^{1/2}$ requires an additional rotation of 2π. Thus at any point of the complex plane $\sqrt{z}$ equals either $r^{1/2}e^{i\theta/2}$ or $-r^{1/2}e^{i\theta/2}$. To return to the same value of the function smaller fractional powers require more than two rotations. For example, $z^{1/3}$ requires 2 rotations. As a result at the point $z_1 = r_1 e^{i\theta_1} = r_1 e^{i\theta_1} e^{i2\pi k}$, $k = 0, 1, 2, \dots$ $z^{1/3}$ can assume one 3 possible values, i.e.,

$$
z_1^{1/3} = \begin{cases} r_1^{1/3}e^{i\theta_1/3}, \\ r_1^{1/3}e^{i\theta_1/3}e^{i2\pi/3} \\ r_1^{1/3}e^{i\theta_1/3}e^{i4\pi/3}. \end{cases} \tag{A.69}
$$

Each of these values can be plotted on a separate complex plane, referred to as a Riemann sheet. For example, for $\sqrt{z}$ we can identify two Riemann sheets for $z^{1/3}$ three and for $z^{1/n}$ n. With this construction, each branch of the function is single valued on the corresponding Riemann sheet. Points at which all branches of a function have the same value are referred to as branch points. For $z^{1/n}$ the

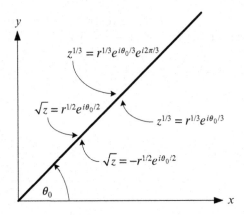

Figure A.16: Values of $z^{-1/2}$ and $z^{-1/3}$ on both sides of the branch cut on the top Riemann sheet

branch point is at the coordinate origin. We still have to define the boundaries among the different Riemann sheets, i.e., the domains of the branches of the function. This is done by introducing a branch cut, which in the present examples can be a straight line running from the branch point to infinity. Upon crossing this branch cut one winds up on the next Riemann sheet. Continuing, one returns to the top sheet. Any direction of the line is permitted but the specification of a particular direction defines a specific set of Riemann sheets. For example, a branch cut at $\theta = \theta_0$ restricts the angle on the top Riemann sheet to $\theta_0 \leq \theta < 2\pi + \theta_0$ on the second Riemann sheet to $2\pi + \theta_0 \leq \theta < 4\pi + \theta_0$ and on the n-th Riemann sheet to $2\pi(n-1) + \theta_0 \leq \theta < 2\pi n + \theta_0$. For $z^{1/3}$ and $\theta_1 > \theta_0$ the three values in (A.69) lie, respectively, on the first, second, and third Riemann sheet. Crossing a branch cut provides a smooth transition to the next Riemann sheet. This is not the case if the domain of the function is restricted to only one of the Riemann sheets since this forces the function to assume different values on the two sides of the branch cut. In contrast to point singularities discussed in the previous subsection, here we have a step discontinuity across the branch cut and continuity along it. We shall call this type of singularity an extended singularity to distinguish it from point singularities discussed in the preceding subsection. Figure A.16 shows the branch cut at $\theta = \theta_0$ as well as values on the top Riemann sheet on both sides of the branch cut for $z^{1/3}$ and $z^{1/2}$. Across the branch cut the discontinuities for $z^{1/2}$ and $z^{1/3}$ are, respectively, $2r^{1/2}e^{i\theta_0/2}$ and $r^{1/3}\sqrt{3}e^{i\theta_0/3}$. Excluding points directly on the branch cut these functions are analytic on the entire top Riemann sheet. They are also analytic at points arbitrarily close to the branch cut as can be demonstrated by a direct application of the limiting forms of the CR conditions.

One novelty not encountered with single-valued functions is that path deformations and, in particular, the construction of closed paths, may require integration around branch cuts. As an example, consider the function $z^{-1/2}$,

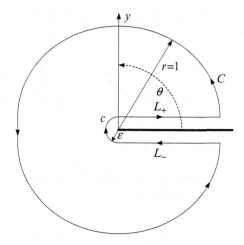

Figure A.17: Integration around a branch cut

again with its branch point at the origin. We use branch cut in Fig A.16 with $\theta_0 = 0$ and evaluate the integral

$$I = \oint \frac{dz}{\sqrt{z}} \tag{A.70}$$

on the top Riemann sheet over the closed path shown in Fig. A.17.

Integrating in the counterclockwise direction, the contribution from the two straight line segments L_- and L_+ for sufficiently small ε is

$$\int_1^0 -z^{-1/2}dz + \int_0^1 z^{-1/2}dz = 4 \tag{A.71}$$

and around the circle C we get

$$\int_0^{2\pi} ie^{i\theta}e^{-i\theta/2} = -4. \tag{A.72}$$

Finally the contribution of the semicircle c with center at the branch point and radius ε is

$$\int_{3\pi/2}^{\pi/2} i\varepsilon\, e^{i\theta}\varepsilon^{-1/2}e^{-i\theta/2}, \tag{A.73}$$

which vanishes as ε approaches 0. Adding the three contributions we get $I = 0$, as expected of an analytic function.

A.4.2 Calculus of Residues

The Residue Theorem

Integration of a function along a contour enclosing isolated singularities plays an important role in the application of complex variables. In the following we present the underlying theory.

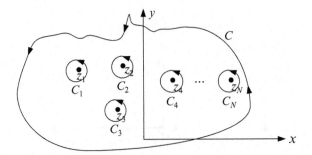

Figure A.18: Equivalence between the integration along the contour C and the circles enclosing the isolated singularities

Let $f(z)$ be an analytic function within the closed curve C except at N isolated singularities at z_ℓ, $\ell = 1, 2, \ldots N$ as shown in Fig. A.18. We enclose each of the N singularities within C by circles C_ℓ. Since an integral of an analytic function along a closed path yields is zero we can deform curve C so that the integral along C equals the sum of the integrals along the N circles within C. This is expressed by

$$\oint f(z)\, dz = \sum_{\ell=1}^{N} \oint_{C_\ell} f_\ell(z)\, dz, \tag{A.74}$$

where $f_\ell(z)$ is the local value of $f(z)$ surrounding the ℓ-th singularity and where all the integrations run in the counterclockwise direction. Because the singularities are isolated there is a finite circular region where $f_\ell(z)$ can be represented by the Laurent series

$$f_\ell(z) = \sum_{n=-\infty}^{\infty} a_{n\ell}(z - z_\ell)^n, \tag{A.75}$$

where $a_{n\ell}$ is the n-th coefficient of the Laurent series within the ℓ-th circle with $\ell = 1, 2, \ldots N$. Integrating over C_ℓ as indicated in (A.74)

$$\oint_{C_\ell} f_\ell(z)\, dz = \sum_{n=-\infty}^{\infty} a_{n\ell} \oint_{C_\ell} (z - z_\ell)^n\, dz. \tag{A.76}$$

The values of the integrals within the sum are independent of the paths enclosing the singularities. Since we have chosen a circle the integrations result in

$$\oint_{C_\ell} (z - z_\ell)^n\, dz = \int_0^{2\pi} r_\ell e^{in\theta_\ell} i r_\ell e^{i\theta_\ell}\, d\theta_\ell = \begin{cases} 2\pi i \; ; n = -1\,, \\ 0 \; ; n \neq -1. \end{cases} \tag{A.77}$$

Substituting into (A.76) and then into (A.74) we obtain

$$\frac{1}{2\pi i}\oint f\left(z\right)dz = \sum_{\ell=1}^{N} a_{-1\ell}. \tag{A.78}$$

In the special case in (A.57) the a_{-1} coefficient was referred to as the residue of $f\left(z\right)$. Formula (A.77) states that a function that has only isolated singularities when integrated along a closed path is equal to the sum of the residues multiplied by $2\pi i$. It should be noted that the preceding derivation does not assume that the singularities are poles. For example, for the function $e^{1/z}$

$$\oint e^{1/z}dz = 2\pi i,$$

which is agreement with the first term of the Laurent series in (A.60).

Computation of Residues

Suppose the function $f(z)$ has an M-th order pole at $z = z_\ell$. Since this is a removable singularity the function $g\left(z\right) = \left(z - z_\ell\right)^M f\left(z\right)$ is analytic at $z = z_\ell$ and has the Taylor expansion

$$g\left(z\right) = \left(z - z_\ell\right)^M f\left(z\right) = \sum_{k=0}^{\infty} \frac{g^{(k)}\left(z_\ell\right)}{k!}\left(z - z_\ell\right)^k. \tag{A.79}$$

On the other hand, the Laurent expansion for $f\left(z\right)$ is

$$f\left(z\right) = \frac{a_{-M}}{\left(z - z_\ell\right)^M} + \frac{a_{-M+1}}{\left(z - z_\ell\right)^{M-1}} + \cdots + \frac{a_{-1}}{\left(z - z_\ell\right)} + a_0 + a_1\left(z - z_\ell\right) + \cdots \tag{A.80}$$

Multiplying (A.80) by $\left(z - z_\ell\right)^M$ we get

$$g\left(z\right) = a_{-M} + a_{-M+1}\left(z - z_\ell\right) + \cdots + a_{-1}\left(z - z_\ell\right)^{M-1} + a_0 + a_1\left(z - z_\ell\right)^M + \cdots$$

we identify the residue term

$$a_{-1} = \frac{1}{\left(M - 1\right)!}\frac{d^{M-1}}{dz^{M-1}}\left[\left(z - z_\ell\right)^M f(z)\right]\Bigg|_{z=z_\ell}. \tag{A.81}$$

For a simple pole this formula reduces to

$$a_{-1} = \lim_{z \longrightarrow z_\ell}\left(z - z_\ell\right)\frac{g\left(z\right)}{f\left(z\right)}\Bigg|_{z=z_\ell}. \tag{A.82}$$

An alternative expression to (A.82) can be found by taking explicit account of the generic form of $f\left(z\right)$ i.e.,

$$f\left(z\right) = \frac{g\left(z\right)}{h\left(z\right)}. \tag{A.83}$$

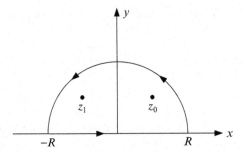

Figure A.19: Integration path for evaluation of I

Since $h(z)$ has a simple zero at $z = z_\ell$ its Taylor expansion is

$$h(z) = (z - z_\ell) h_\ell^{(1)}(z_\ell) + (z - z_\ell)^2 h_\ell^{(2)}(z_\ell)/2 + \cdots$$

With the Taylor expansions for $g(z)$ we form the ratio

$$f(z) = \frac{g(z_\ell) + (z - z_\ell) g_\ell^{(1)}(z_\ell) + (z - z_\ell)^2 g_\ell^{(2)}(z_\ell)/2 + \cdots}{(z - z_\ell) h^{(1)}(z_\ell) + (z - z_\ell)^2 h^{(2)}(z_\ell)/2 + \cdots}, \qquad (A.84)$$

where $g(z_\ell) \neq 0$. Long division gives the Laurent expansion

$$f(z) = \frac{g(z_\ell)}{h_\ell^{(1)}(z_\ell)} \frac{1}{z - z_\ell} + Taylor\ Series$$

so that

$$a_{-1} = \frac{g(z_\ell)}{h^{(1)}(z_\ell)}. \qquad (A.85)$$

This formula avoids the limiting process in (A.82) and is generally preferred.

Evaluation of Integrals.   One of the applications of Residue Calculus is the evaluation of improper integrals (Integrals with infinite limits). For example, consider the integral

$$I = \int_0^\infty \frac{dx}{1 + x^4} = \frac{1}{2} \int_{-\infty}^\infty \frac{dx}{1 + x^4} = \frac{1}{2} \lim_{R \to \infty} \int_{-R}^R \frac{dz}{1 + z^4}. \qquad (A.86)$$

The denominator is a product of four linear factors with zeros at $z_k = e^{i\pi(2k+1)/4}$; $k = 0, 1, 2, 3$ that correspond to the four simple poles of $1/(1 + z^4)$. We enclose the poles in the upper half plane, i.e., z_0 and z_1 by the path consisting of the segment $-R < x < R$ of the real axis and a semicircle of radius R in the upper half plane, as shown in Fig. A.19.

Integrating $1/(1 + z^4)$ in the counterclockwise direction along the semicircle

$$I_R = \int_0^\pi \frac{ie^{i\theta} R d\theta}{1 + R^4 e^{i4\theta}}$$

we get the bound

$$|I_R| \leq \left| \int_0^\pi \frac{ie^{i\theta} R d\theta}{1 + R^4 e^{i4\theta}} \right| < \int_0^\pi \left| \frac{ie^{i\theta} R}{1 + R^4 e^{i4\theta}} \right| d\theta < \frac{\pi}{R^3} \tag{A.87}$$

so that

$$\lim_{R \to \infty} I_R = 0.$$

Thus for sufficiently large R the integration along the x-axis in (A.86) is equivalent to the integration of $1/(1 + z^4)$ along the closed path in Fig. A.19. In accordance with (A.78) this equals $2\pi i$ times the sum of the residues at the two enclosed poles. Since the poles are simple we are entitled to using (A.85). A simple calculation yields

$$2\pi i \left\{ 1/4z^3 \big|_{z = e^{i\pi/4}} + 1/4z^3 \big|_{z = e^{i\pi 3/4}} \right\} = \pi/\sqrt{2}$$

and taking account of the factor of two in (A.86) we get the final result $I = \pi/2\sqrt{2}$. This evaluation procedure can be generalized to integrals of the form

$$I_N = \int_{-\infty}^\infty \frac{P_{N-2}(x)}{Q_N(x)} dx, \tag{A.88}$$

where $P_{N-2}(x)$ and $Q_N(x)$ are polynomials of order $N - 2$ and N. The residue evaluation then gives

$$I_N = 2\pi i \sum_{\ell=1}^N a_{-1\ell}. \tag{A.89}$$

A different situation arises for integrals of the form

$$I = \int_{-\infty}^\infty f(x) e^{ixt} dx, \tag{A.90}$$

where t is a real parameter. In this case the vanishing of the contribution from the integration along the semicircle (as in Fig. A.19) as its radius tends to infinity depends on $f(x)$ and the sign of t. Conditions under which this contribution vanishes are given by the Jordan Lemma which for our purpose may be stated as follows:

Referring to Fig. A.20, let $f(z)$ be an analytic function in the region $R > R_0^-$ or R_0^+ that includes the semi-circular contours C_- or C_+. If on C_- (C_+)

$$\lim_{R \to \infty} f(R e^{i\theta}) \longrightarrow 0 \tag{A.91}$$

then for $t < 0$ $(t > 0)$

$$\lim_{R \to \infty} \int_{C_- (C_+)} f(z) e^{izt} dz \longrightarrow 0. \tag{A.92}$$

Proof. We first deal with C_- and denote the corresponding integral by I_R.

$$I_R = \int_{C_-} f(z) e^{izt} dz = \int_0^{-\pi} f(R e^{i\theta}) e^{itR e^{i\theta}} iR d\theta$$

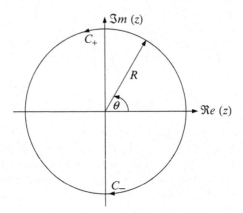

Figure A.20: Coordinates in the proof of the Jordan Lebesgue lemma

$$= \int_0^{-\pi} e^{iRt\cos\theta} f\left(R\, e^{i\theta}\right) e^{-Rt\sin\theta} iR d\theta$$

$$= \int_0^{\pi} e^{iRt\cos\theta} f\left(R\, e^{-i\theta}\right) e^{Rt\sin\theta} iR d\theta. \qquad (A.93)$$

We bound it as follows:

$$|I_R| \leq \int_0^{\pi} \left|f\left(R\, e^{-i\theta}\right)\right| e^{Rt\sin\theta} R d\theta \leq 2\left|f\left(R\, e^{-i\theta}\right)\right| \int_0^{\pi/2} e^{Rt\sin\theta} R d\theta. \qquad (A.94)$$

For $0 < \theta < \pi/2$, $\sin\theta > 2\theta/\pi$. Since the argument of the exponential in (A.94) is negative we can increase the integral by replacing $\sin\theta$ with $2\theta/\pi$. Making this replacement we get

$$\int_0^{\pi/2} e^{-R|t|\sin\theta} R d\theta < \int_0^{\pi/2} e^{-2|t|\theta R/\pi} R d\theta = \frac{\pi}{2\,|t|}\left(1 - e^{-R|t|}\right) \leq \frac{\pi}{2\,|t|}. \qquad (A.95)$$

Substituting (A.95) in (A.94) we obtain the final result:

$$|I_R| \leq \frac{\pi}{|t|}\left|f\left(R\, e^{-i\theta}\right)\right|. \qquad (A.96)$$

In view of (A.91), $|I_R|$ approaches zero which proves the lemma for $t < 0$. The steps in the proof for $t > 0$ are identical. What distinguishes the two cases is that implicit in the proof for $t < 0$ is the requirement of analyticity of $f(z)$ in the lower half plane, whereas for $t > 0$ it is the upper half plane. $\qquad \square$

In general, to evaluate an integral by Residue Calculus one must be able to identify in the complex plane an integration path that can be closed by adding to it the integration range of the integral to be evaluated without changing its value. In the case of (A.86), the path was closed by the addition of a semicircle along which the integration was reduced to zero. This reduction was achieved entirely by the algebraic decay of the integrand which, when expressed in polar coordinates, reduces to $1/R^3$. As indicated in (A.88), for integrands that are ratios of polynomials the required minimum ratio of the denominator to numerator degrees is 2. Multiplication by the polar coordinate differential on the circle introduces an additional factor of R so that the net decay rate is only $1/R$.

Consider now the integral

$$I(t) = \int_{-\infty}^{\infty} \frac{-ix}{1+x^2} e^{ixt} dx. \tag{A.97}$$

Here as in (A.88) we can form a closed path by supplementing the integration range with a semicircle and let its radius approach infinity. The decay rate of the integrand is now exponential and in accordance with the Jordan Lemma the decay rate of the multiplier of the exponential is immaterial as long as it approaches zero at infinity. The integrand in (A.97) satisfies this criterion. The corresponding closed path can be constructed using either C_+ or C_- depending on the sign of t. Continuing the integrand into the complex domain and closing the path with one of the semicircles we write

$$I(t) = \begin{cases} \oint_{C_+} \frac{-iz}{1+z^2} e^{ixt} dz; & t > 0, \\ \oint_{C_-} \frac{-iz}{1+z^2} e^{ixt} dz; & t < 0, \end{cases} \tag{A.98}$$

where, in accordance with Fig. A.20, the integral on C_+ runs in the counter clockwise direction (positive residue summation) and on C_- in the clockwise direction (negative residue summation). The integrand has two simple poles one at $z = i$ and the other at $z = -i$. For $t > 0$ the exponent decays in the upper half plane. In accordance with the Jordan Lemma the contribution from C_+ vanishes so that we can equate (A.97) to the first integral in (A.98) and hence to $2\pi i$ times the enclosed residue. We get $I(t) = \pi e^{-t}$. For $t < 0$ the integrand decays on C_-. Closing the path of integration in the lower half plane we pick up the residue at $z = -i$ and obtain $I(t) = \pi e^{t}$. Both results can be included in the compact form

$$I(t) = \pi e^{-|t|}. \tag{A.99}$$

One important application of the Jordan Lemma is in the inversion of Fourier and Laplace Transforms. Because the direct Laplace transform is defined in terms of the real parameter s, the lower half plane and its association with the negative time range as used herein correspond to the right half plane in the Laplace Transform. Thus to rephrase the statement of the Jordan Lemma in terms of the Laplace Transform variable one should replace z by $-is$.

Bibliography

[1] Abramovitz M, Stegun IA (eds) (1965) Handbook of mathematical function: with formulas, graphs, and mathematical tables. Dover Publications, Inc., New York.

[2] Chen CT(1984) Linear systems theory and design. Oxford University Press, New York.

[3] Chen CT(1994) System and signal analysis. Saunders College Publishing, Fort Worth.

[4] Cohen L (1995) Time-frequency analysis. Prentice Hall, Englewood Cliffs.

[5] Courant, Hilbert (1953) Methods of mathematical physics. Interscience Publishers, New York.

[6] Gabor D (1946) Theory of communications. J Inst Electr Eng 93:249–457.

[7] Giordano AA, HsuFM (1985) Least square estimation with applications to digital signal processing. Wiley, New York.

[8] Golub GH, Van Loan CF (1996). Matrix computations, 3rd edn. Johns Hopkins Studies in Mathematical Sciences, Baltimore.

[9] Golub GH, Van Loan CF (1980). An analysis of the total least square problem. SIAM J Numer Anal 17:883–893.

[10] Haykin S, Van Veen B (1999). Signals and systems. Wiley New York.

[11] Lebensbaum MC(1963) Further study of the black box problem. In: Proceedings of the IEEE (Correspondence), vol 51, p. 864.

[12] Mandal M, Asif A (2007). Continuous and discrete time signals and systems. Cambridge University Press, Cambridge.

[13] Mitra SK (1998) Digital signal processing: a computer-based approach. The McGraw-Hill Companies, Inc., New York.

[14] Oppenheim AV, Willsky AS, Hamid Nawab S (1997) Signals & systems, 2nd edn. Prentice Hall, Upper Saddle River.

W. Wasylkiwskyj, *Signals and Transforms in Linear Systems Analysis*, 369
DOI 10.1007/978-1-4614-3287-6, © Springer Science+Business Media, LLC 2013

[15] Paley REAC, Wiener N (1934) Fourier transforms in the complex domain. American Society Collegium Publication, New York, vol 19.

[16] Papoulis A (1962) The fourier integral and its applications. McGraw-Hill Book Company, Inc., New York.

[17] Phillips CL, Parr JM, Riskin EA (2008) Signals, systems, and transforms. Pearson Prentice Hall, Upper Saddle River.

[18] Poor VH (1994) An introduction to signal detection and estimation, 2nd edn. Springer New York.

[19] Poularikas AD, Samuel Seely S (1991). Signals and systems, 2nd edn. PWS KENT Publishing Company, Boston.

[20] Poularikas AD (ed) (1996) The transforms and applications handbook. CRC Press. CRC Handbook Published in Cooperation with IEEE Press, Boca Raton.

[21] Riesz F, SZ.-Nagy B (1990) Functional analysis (translated from 2nd French edition by Boron LF). Dover Publications, Inc., New York.

[22] Slepian J (1949) Letters to the editor. Electr Eng 63:377.

[23] Slepian J, Pollard HO (1961). Polate spherical functions: fourier analysis and uncertainty. Bell Syst Tech J 40:43–64.

[24] Tikhonov AN (1963) Solution of incorrectly forumalted problems and the regularization method. Soviet Math Doklady 4:1035–1038 (English translation of doklad nauk SSSR 151:501–504).

[25] Van Huffel S, Vandewalle J (1991) The total least square problems: computational aspects and analysis. Society for Industrial and Applied Mathematics, Philadelphia.

[26] Ziemer RE, Tranter WH, Fannin RD (1993). Signals and systems: continuous and discrete 3rd edn. Macmillan Publishing Company, New York.

Index

W. Wasylkiwskyj, *Signals and Transforms in Linear Systems Analysis*,
DOI 10.1007/978-1-4614-3287-6, © Springer Science+Business Media, LLC 2013